Inventing Temperature

Oxford Studies in Philosophy of Science
General Editor: Paul Humphreys, University of Virginia

The Book of Evidence
Peter Achinstein

Science, Truth, and Democracy
Philip Kitcher

The Devil in the Details: Asymptotic Reasoning in Explanation, Reduction, and Emergence
Robert W. Batterman

Science and Partial Truth: A Unitary Approach to Models and Scientific Reasoning
Newton C. A. da Costa and Steven French

Inventing Temperature: Measurement and Scientific Progress
Hasok Chang

Inventing Temperature

Measurement and Scientific Progress

Hasok Chang

2004

OXFORD
UNIVERSITY PRESS

Oxford   New York
Auckland   Bangkok   Buenos Aires   Cape Town   Chennai
Dar es Salaam   Delhi   Hong Kong   Istanbul   Karachi   Kolkata
Kuala Lumpur   Madrid   Melbourne   Mexico City   Mumbai   Nairobi
São Paulo   Shanghai   Taipei   Tokyo   Toronto

Copyright © 2004 by Oxford University Press, Inc.

Published by Oxford University Press, Inc.
198 Madison Avenue, New York, New York 10016
www.oup.com

Oxford is a registered trademark of Oxford University Press

All rights reserved. No part of this publication may be reproduced,
stored in a retrieval system, or transmitted, in any form or by any means,
electronic, mechanical, photocopying, recording, or otherwise,
without the prior permission of Oxford University Press.

Library of Congress Cataloging-in-Publication Data
Chang, Hasok.
Inventing temperature : measurement and scientific progress / Hasok Chang.
p. cm.—(Oxford studies in philosophy of science)
Includes bibliographical references and index.
ISBN 0-19-517127-6
1. Temperature measurements—History. 2. Thermometers—History.
3. Interdisciplinary approach to knowledge. 4. Science—Philosophy.
I. Title. II. Series.
QC271.6.C46 2004
536'.5'0287—dc22   2003058489

2 4 6 8 9 7 5 3 1
Printed in the United States of America
on acid-free paper

To my parents

Acknowledgments

As this is my first book, I need to thank not only those who helped me directly with it but also those who helped me become the scholar and person who could write such a book.

First and foremost I thank my parents, who raised me not only with the utmost love and intellectual and material support but also with the basic values that I have proudly made my own. I would also like to thank them for the faith and patience with which they supported even those of my life decisions that did not fit their own visions and hopes of the best possible life for me.

While I was studying abroad, there were many generous people who took care of me as if they were my own parents, particularly my aunts and uncles Dr. and Mrs. Young Sik Jang of Plattsburgh, N.Y., and Mr. and Mrs. Chul Hwan Chang of Los Angeles. Similarly I thank my mother-in-law, Mrs. Elva Siglar.

My brother and sister have not only been loving siblings but emotional and intellectual guiding lights throughout my life. They also had the good sense to marry people just as wonderful as themselves, who have helped me in so many ways. My loving nieces and nephews are also essential parts of this family without whom I would be nothing. In the best Korean tradition, my extended family has also been important, including a remarkable community of intellectual cousins.

The long list of teachers who deserve the most sincere thanks begins with Mr. Jong-Hwa Lee, my first-grade teacher, who first awakened my love of science. I also thank all the other teachers I had at Hong-Ik Elementary School in Seoul. I would like to record the most special thanks to all of my wonderful teachers at Northfield Mount Hermon School, who taught me to be my own whole person as well as a scholar. To be thanked most directly for their influences on my eventual intellectual path are Glenn Vandervliet and Hughes Pack. Others that I cannot go without mentioning include Jim Antal, Fred Taylor, Yvonne Jones, Vaughn Ausman, Dick and Joy Unsworth, Mary and Bill Compton, Juli and Glenn Dulmage, Bill Hillenbrand, Meg Donnelly, James Block, and the late Young Il Shin. There is something I once promised to say, and I will say it now in case I never achieve anything better than this book in my life: "Northfield Mount Hermon has made all the difference."

As an undergraduate at Caltech, I was very grateful to be nurtured by the excellent Division of the Humanities and Social Sciences. I would not have become a scholar in the humanities, or even survived college at all, without the tutelage and kindness of the humanists, especially Bruce Cain, Robert Rosenstone, Dan Kevles, Randy Curren, Nicholas Dirks, and Jim Woodward (whom I have the honor of following in the Oxford Studies series). The SURF program at Caltech was also very important. My year away at Hampshire College changed my intellectual life so significantly, and I would like to thank Herbert Bernstein and Jay Garfield particularly.

I went to Stanford specifically to study with Nancy Cartwright and Peter Galison, and ever since then they have been so much more than Ph.D. advisors to me. They have opened so many doors into the intellectual and social world of academia that I have completely lost count by now. What I did not know to expect when I went to Stanford was that John Dupré would leave such a permanent mark on my thinking. I would also like to thank many other mentors at Stanford including Tim Lenoir, Pat Suppes, Marleen Rozemond, and Stuart Hampshire, as well as my fellow graduate students and the expert administrators who made the Philosophy Department such a perfect place for graduate work.

Gerald Holton, the most gracious sponsor of my postdoctoral fellowship at Harvard in 1993–94, has taught me more than can be measured during and since that time. My association with him and Nina Holton has been a true privilege. I also thank Joan Laws for all her kindness and helpful expertise during my time at Harvard.

Many other mentors taught and supported me although they never had any formal obligation to do so. My intellectual life would have been so much poorer had it not been for their generosity. The kind interest expressed by the late Thomas Kuhn was a very special source of strength for the young undergraduate struggling to find his direction. Evelyn Fox Keller showed me how to question science while loving it. Jed Buchwald helped me enormously in my post-Ph.D. education in the history of science and gave me confidence that I could do first-rate history. Alan Chalmers first taught me by his wonderful textbook and later occasioned the first articulation of the intellectual direction embodied in this book. Jeremy Butterfield has helped me at every step of my intellectual and professional development since I arrived in England a decade ago. Sam Schweber has given me the same gentle and generous guidance with which he has blessed so many other young scholars. In similar ways, I also thank Olivier Darrigol, Kostas Gavroglu, Simon Schaffer, Michael Redhead, Simon Saunders, Nick Maxwell, and Marcello Pera.

To all of my colleagues at the Department of Science and Technology Studies (formerly History and Philosophy of Science) at University College London, I owe sincere thanks for a supportive and stimulating interdisciplinary environment. In an approximate order of seniority within the department, the permanent members are: Piyo Rattansi, Arthur Miller, Steve Miller, Jon Turney, Brian Balmer, Joe Cain, Andrew Gregory, and Jane Gregory. I also want to thank our dedicated administrators who have put so much of their lives into the department, especially Marina

Ingham and Beck Hurst. I would also like to thank Alan Lord and Jim Parkin for their kindness and guidance.

My academic life in London has also been enriched by numerous associations outside my own department. A great source of support and intellectual stimulation has been the Centre for the Philosophy of the Natural and Social Sciences at the London School of Economics. I would like to thank especially my co-organizers of the measurement project: Nancy Cartwright, Mary Morgan, and Carl Hoefer. This project also allowed me to work with collaborators and assistants who helped with various parts of this book. I would also like to thank my colleagues in the London Centre for the History of Science, Medicine and Technology, who helped me complete my education as a historian, particularly Joe Cain, Andy Warwick, David Edgerton, Janet Browne, and Lara Marks.

Many other friends and colleagues helped me nurture this brain-child of mine as would good aunts and uncles. Among those I would like to note special thanks to Sang Wook Yi, Nick Rasmussen, Felicia McCarren, Katherine Brading, Amy Slaton, Brian Balmer, Marcel Boumans, Eleonora Montuschi, and Teresa Numerico.

There are so many other friends who have helped enormously with my general intellectual development, although they did not have such a direct influence on the writing of this book. Among those I must especially mention: the late Seung-Joon Ahn and his wonderful family, Sung Ho Kim, Amy Klatzkin, Deborah and Phil McKean, Susannah and Paul Nicklin, Wendy Lynch and Bill Bravman, Jordi Cat, Elizabeth Paris, Dong-Won Kim, Sungook Hong, Alexi Assmus, Mauricio Suárez, Betty Smocovitis, David Stump, Jessica Riskin, Sonja Amadae, Myeong Seong Kim, Ben Harris, Johnson Chung, Conevery Bolton, Celia White, Emily Jernberg, and the late Sander Thoenes.

I must also give my hearty thanks to all of my students who taught me by allowing me to teach them, especially those whom I have come to regard as dear friends and intellectual equals rather than mere former students. Among that large number are, in the order in which I had the good fortune to meet them and excluding those who are still studying with me: Graham Lyons, Guy Hussey, Jason Rucker, Grant Fisher, Andy Hammond, Thomas Dixon, Clint Chaloner, Jesse Schust, Helen Wickham, Alexis de Greiff, Karl Galle, Marie von Mirbach-Harff, and Sabina Leonelli. They have helped me maintain my faith that teaching is the ultimate purpose of my career.

Over the years I received gratefully occasional and more than occasional help on various aspects of this project from numerous other people. I cannot possibly mention them all, but they include (in alphabetical order): Rachel Ankeny, Theodore Arabatzis, Diana Barkan, Matthias Dörries, Robert Fox, Allan Franklin, Jan Golinski, Graeme Gooday, Roger Hahn, Rom Harré, John Heilbron, Larry Holmes, Keith Hutchison, Frank James, Catherine Kendig, Mi-Gyung Kim, David Knight, Chris Lawrence, Cynthia Ma, Michela Massimi, Everett Mendelsohn, Marc-Georges Nowicki, Anthony O'Hear, John Powers, Stathis Psillos, Paddy Ricard, George Smith, Barbara Stepansky, Roger Stuewer, George Taylor, Thomas Uebel, Rob Warren, Friedel Weinert, Jane Wess, Emily Winterburn, and Nick Wyatt.

Various institutions have also been crucial in supporting this work. I could have not completed the research and writing without a research fellowship from the

Leverhulme Trust, whom I thank most sincerely. I would like to thank the helpful librarians, archivists, and curators at many places including the following, as well as the institutions themselves: the British Library, the London Science Museum and its Library, University College London, Harvard University, Yale University, the Royal Society, the University of Cambridge, and the National Maritime Museum.

This book would not have come into being without crucial and timely interventions from various people. The project originated in the course of my postdoctoral work with Gerald Holton. Peter Galison initially recommended the manuscript for the Oxford Studies in the Philosophy of Science. Mary Jo Nye made a crucial structural suggestion. Three referees for Oxford University Press provided very helpful comments that reoriented the book substantially and productively, as well as helped me refine various details. Carl Hoefer and Jeremy Butterfield provided much-needed last-minute advice. Paul Humphreys, the series editor, encouraged me along for several years and guided the improvement of the manuscript with great patience and wisdom. Peter Ohlin and Bob Milks directed the process of reviewing, manuscript preparation, and production with kind and expert attention. Lynn Childress copyedited the manuscript with meticulous and principled care.

Finally, I would like to record my deep thanks to Gretchen Siglar for her steadfast love and genuine interest in my work, with which she saw me through all the years of labor on this project as well as my life in general.

Contents

Note on Translation   xv

Chronology   xvii

Introduction   3

1. Keeping the Fixed Points Fixed   8
   Narrative: What to Do When Water Refuses to Boil at the Boiling Point   8
       Blood, Butter, and Deep Cellars: The Necessity and Scarcity of Fixed Points   8
       The Vexatious Variations of the Boiling Point   11
       Superheating and the Mirage of True Ebullition   17
       Escape from Superheating   23
       The Understanding of Boiling   28
       A Dusty Epilogue   35
   Analysis: The Meaning and Achievement of Fixity   39
       The Validation of Standards: Justificatory Descent   40
       The Iterative Improvement of Standards: Constructive Ascent   44
       The Defense of Fixity: Plausible Denial and Serendipitous Robustness   48
       The Case of the Freezing Point   53

2. Spirit, Air, and Quicksilver   57
   Narrative: The Search for the "Real" Scale of Temperature   57
       The Problem of Nomic Measurement   57
       De Luc and the Method of Mixtures   60
       Caloric Theories against the Method of Mixtures   64
       The Calorist Mirage of Gaseous Linearity   69

Regnault: Austerity and Comparability   74
The Verdict: Air over Mercury   79
Analysis: Measurement and Theory in the Context of Empiricism   84
The Achievement of Observability, by Stages   84
Comparability and the Ontological Principle of Single Value   89
Minimalism against Duhemian Holism   92
Regnault and Post-Laplacian Empiricism   96

3. To Go Beyond   103
Narrative: Measuring Temperature When Thermometers Melt and Freeze   103
Can Mercury Be Frozen?   104
Can Mercury Tell Us Its Own Freezing Point?   107
Consolidating the Freezing Point of Mercury   113
Adventures of a Scientific Potter   118
It Is Temperature, but Not As We Know It?   123
Ganging Up on Wedgwood   128
Analysis: The Extension of Concepts beyond Their Birth Domains   141
Travel Advisory from Percy Bridgman   142
Beyond Bridgman: Meaning, Definition, and Validity   148
Strategies for Metrological Extension   152
Mutual Grounding as a Growth Strategy   155

4. Theory, Measurement, and Absolute Temperature   159
Narrative: The Quest for the Theoretical Meaning of Temperature   159
Temperature, Heat, and Cold   160
Theoretical Temperature before Thermodynamics   168
William Thomson's Move to the Abstract   173
Thomson's Second Absolute Temperature   182
Semi-Concrete Models of the Carnot Cycle   186
Using Gas Thermometers to Approximate Absolute Temperature   192
Analysis: Operationalization—Making Contact between Thinking and Doing   197
The Hidden Difficulties of Reduction   197
Dealing with Abstractions   202
Operationalization and Its Validity   205
Accuracy through Iteration   212
Theoretical Temperature without Thermodynamics?   217

5. Measurement, Justification, and Scientific Progress   220
Measurement, Circularity, and Coherentism   221
Making Coherentism Progressive: Epistemic Iteration   224
Fruits of Iteration: Enrichment and Self-Correction   228
Tradition, Progress, and Pluralism   231
The Abstract and the Concrete   233

6. Complementary Science—History and Philosophy of Science as a Continuation of Science by Other Means   235
   The Complementary Function of History and Philosophy of Science   236
   Philosophy, History, and Their Interaction in Complementary Science   238
   The Character of Knowledge Generated by Complementary Science   240
   Relations to Other Modes of Historical and Philosophical Study of Science   247
   A Continuation of Science by Other Means   249

Glossary of Scientific, Historical, and Philosophical Terms   251

Bibliography   259

Index   275

Note on Translation

Where there are existing English translations of non-English texts, I have relied on them in quotations except as indicated. In other cases, translations are my own.

Chronology

c. 1600	Galileo, Sanctorio, Drebbel, etc.: first recorded use of thermometers
c. 1690	Eschinardi, Renaldini, etc.: first use of the boiling and melting points as fixed points of thermometry
1710s	Fahrenheit: mercury thermometer
1733	First Russian expedition across Siberia begins, led by Gmelin
c. 1740	Celsius: centigrade thermometer
1751–	Diderot et al.: *L'Encyclopédie*
1760	Accession of George III in England.
1764–	Black: measurements of latent and specific heats
	Watt: improvements on the steam engine
1770s	Irvine: theory of heat capacity
1772	De Luc: *Recherches sur les modifications de l'atmosphère*
1776	Declaration of American Independence
1777	Report of the Royal Society committee on thermometry
1782–83	Compound nature of water argued; spread of Lavoisier's ideas
1782	Wedgwood: clay pyrometer
1783	Cavendish/Hutchins: confirmation of the freezing point of mercury
1789	Lavoisier: *Traité élémentaire de chimie*
	Onset of the French Revolution.
1793	Execution of Louis XVI
	Beginning of the "Terror" in France and war with Great Britain
1794	Execution of Lavoisier; death of Robespierre, end of the Terror
	Establishment of the École Polytechnique in Paris
1798	Laplace: first volume of *Traité de mécanique céleste*
1799	Rise of Napoleon as First Consul
1800	Rumford: founding of the Royal Institution
	Herschel: observation of infrared heating effects
	Volta: invention of the "pile" (battery)
	Nicholson and Carlisle: electrolysis of water

1801	Berthollet/Proust: beginning of controversy on chemical proportions
1802	Dalton; Gay-Lussac: works on the thermal expansion of gases
1807	Davy: isolation of potassium and sodium
1808	Dalton: first part of *A New System of Chemical Philosophy*
1815	Fall of Napoleon
c. 1820	Fresnel: establishment of the wave theory of light
1820	Oersted: discovery of electromagnetic action
1824	Carnot: *Réflexions sur la puissance motrice du feu*
1827	Death of Laplace
1831	Faraday: discovery of electromagnetic induction
1837	Pouillet: reliable low-temperature measurements down to −80°C
1840s	Joule, Mayer, Helmholtz, etc.: conservation of energy
1847	Regnault: first extensive set of thermal measurements published
1848	William Thomson (Lord Kelvin): first definition of absolute temperature
1854	Joule and Thomson: operationalization of Thomson's second absolute temperature, by means of the porous-plug experiment
1871	End of Franco-Prussian War; destruction of Regnault's laboratory

Inventing Temperature

Introduction

This book aspires to be a showcase of what I call "complementary science," which contributes to scientific knowledge through historical and philosophical investigations. Complementary science asks scientific questions that are excluded from current specialist science. It begins by re-examining the obvious, by asking why we accept the basic truths of science that have become educated common sense. Because many things are protected from questioning and criticism in specialist science, its demonstrated effectiveness is also unavoidably accompanied by a degree of dogmatism and a narrowness of focus that can actually result in a loss of knowledge. History and philosophy of science in its "complementary" mode can ameliorate this situation, as I hope the following chapters will illustrate in concrete detail.

Today even the most severe critics of science actually take a lot of scientific knowledge for granted. Many results of science that we readily believe are in fact quite extraordinary claims. Take a moment to reflect on how unbelievable the following propositions would have appeared to a keen and intelligent observer of nature from 500 years ago. The earth is very old, well over 4 billion years of age; it exists in a near-vacuum and revolves around the sun, which is about 150 million kilometers away; in the sun a great deal of energy is produced by nuclear fusion, the same kind of process as the explosion of a hydrogen bomb; all material objects are made up of invisible molecules and atoms, which are in turn made up of elementary particles, all far too small ever to be seen or felt directly; in each cell of a living creature there is a hypercomplex molecule called DNA, which largely determines the shape and functioning of the organism; and so on. Most members of today's educated public subscribing to the "Western" civilization would assent to most of these propositions without hesitation, teach them confidently to their children, and become indignant when some ignorant people question these truths. However, if they were asked to say why they believe these items of scientific common sense, most would be unable to produce any convincing arguments. It may even be that the more basic and firm the belief is, the more stumped we tend to feel in trying to justify it. Such a correlation would indicate that unquestioning belief has served as a substitute for genuine understanding.

Nowhere is this situation more striking than in our scientific knowledge of heat, which is why it is an appropriate subject matter of this study. Instead of revisiting debates about the metaphysical nature of heat, which are very well known to historians of science, I will investigate some basic difficulties in an area that is usually considered much less problematic, and at the same time fundamental to all empirical studies of heat. That area of study is *thermometry*, the measurement of temperature. How do we know that our thermometers tell us the temperature correctly, especially when they disagree with each other? How can we test whether the fluid in our thermometer expands regularly with increasing temperature, without a circular reliance on the temperature readings provided by the thermometer itself? How did people without thermometers learn that water boiled or ice melted always at the same temperature, so that those phenomena could be used as "fixed points" for calibrating thermometers? In the extremes of hot and cold where all known thermometers broke down materially, how were new standards of temperature established and verified? And were there any reliable theories to support the thermometric practices, and if so, how was it possible to test those theories empirically, in the absence of thermometry that was already well established?

These questions form the topics of the first four chapters of this book, where they will be addressed in full detail, both historically and philosophically. I concentrate on developments in the eighteenth and nineteenth centuries, when scientists established the forms of thermometry familiar today in everyday life, basic experimental science, and standard technological applications. Therefore I will be discussing quite simple instruments throughout, but simple epistemic questions about these simple instruments quickly lead us to some extremely complex issues. I will show how a whole host of eminent past scientists grappled with these issues and critically examine the solutions they produced.

I aim to show that many simple items of knowledge that we take for granted are in fact spectacular achievements, obtained only after a great deal of innovative thinking, painstaking experiments, bold conjectures, and serious controversies, which may in fact never have been resolved quite satisfactorily. I will point out deep philosophical questions and serious technical challenges lurking behind very elementary results. I will bring back to life the loving labors of the great minds who created and debated these results. I will attempt to communicate my humble appreciation for these achievements, while sweeping away the blind faith in them that is merely a result of schoolroom and media indoctrination.

It is neither desirable nor any longer effective to try bullying people into accepting the authority of science. Instead, all members of the educated public can be invited to participate in science, in order to experience the true nature and value of scientific inquiry. This does not mean listening to professional scientists tell condescending stories about how they have discovered wonderful things, which you should believe for reasons that are too difficult for you to understand in real depth and detail. Doing science ought to mean asking your own questions, making your own investigations, and drawing your own conclusions for your own reasons. Of course it will not be feasible to advance the "cutting edge" or "frontier" of modern science without first acquiring years of specialist training. However, the cutting edge is not all there is to science, nor is it necessarily the most valuable part of science. Questions that have been answered are still worth asking again, so you can understand for yourself how to arrive at the standard answers, and possibly discover new answers or recover forgotten answers that are valuable.

In a way, I am calling for a revival of an old style of science, the kind of "natural philosophy" that was practiced by the European "gentlemen" of the eighteenth and nineteenth centuries with such seriousness and delight. But the situation in our time is indeed different. On the encouraging side, today a much larger number of women and men can afford to engage in activities that are not strictly necessary for their immediate survival. On the other hand, science has become so much more advanced, professionalized, and specialized in the last two centuries that it is no longer very plausible for the amateurs to interact with the professionals on an equal footing and contribute in an immediate sense to the advancement of specialist knowledge.

In this modern circumstance, science for the non-specialist and by the non-specialist should be historical and philosophical. It is best practiced as "complementary science" (or the complementary mode of history and philosophy of science), as I explain in detail in chapter 6. The studies contained in the first four chapters are presented as illustrations. They are offered as exemplars that may be followed in pursuing other studies in complementary science. I hope that they will convince you that complementary science can improve our knowledge of nature. Most of the scientific material presented there is historical, so I am not claiming to have produced much that is strictly new. However, I believe that the rehabilitation of discarded or forgotten knowledge does constitute a form of knowledge creation. Knowing the historical circumstances will also set us free to agree or disagree with the best judgments reached by the past masters, which form the basis of our modern consensus.

Each of the first four chapters takes an item of scientific knowledge regarding temperature that is taken for granted now. Closer study, however, reveals a deep puzzle that makes it appear that it would actually be quite impossible to obtain and secure the item of knowledge that seemed so straightforward at first glance. A historical look reveals an actual scientific controversy that took place, whose vicissitudes are followed in some detail. The conclusion of each episode takes the form of a judgment regarding the cogency of the answers proposed and debated by the past scientists, a judgment reached by my own independent reflections—sometimes in agreement with the verdict of modern science, sometimes not quite.

Each of those chapters consists of two parts. The narrative part states the philosophical puzzle and gives a problem-centered narrative about the historical attempts to solve that puzzle. The analysis part contains various in-depth analyses of certain scientific, historical, and philosophical aspects of the story that would have distracted the flow of the main narrative given in the first part. The analysis part of each chapter will tend to contain more philosophical analyses and arguments than the narrative, but I must stress that the division is not meant to be a separation of history and philosophy. It is not the case that philosophical ideas and arguments cannot be embodied in a narrative, and it is also not the case that history should always be presented in a narrative form.

The last parts of the book are more abstract and methodological. Chapter 5 presents in a more systematic and explicit manner a set of abstract epistemological ideas that were embedded in the concrete studies in the first four chapters. In that discussion I identify measurement as a locus where the problems of foundationalism are revealed with stark clarity. The alternative I propose is a brand of coherentism buttressed by the method of "epistemic iteration." In epistemic iteration we start by adopting an existing system of knowledge, with some respect for it but without any firm assurance that it is correct; on the basis of that initially affirmed system we launch inquiries that result in the refinement and even correction of the original system. It is this self-correcting progress that justifies (retrospectively) successful courses of development in science, not any assurance by reference to some indubitable foundation. Finally, in chapter 6, I close with a manifesto that articulates in explicit methodological terms what it is that I am trying to achieve with the kind of studies that are included in this book. The notion of complementary science, which I have sketched only very briefly for now, will be developed more fully and systematically there.

As this book incorporates diverse elements, it could be read selectively. The main themes can be gathered by reading the narrative parts of the first four chapters; in that case, various sections in the analysis parts of those chapters can be sampled according to your particular interests. If you have little patience for historical details, it may work to read just the analysis parts of chapters 1 to 4 (skipping the obviously historical sections), then chapter 5. If you are simply too busy and also prefer to take philosophy in the more abstract vein, then chapter 5 could be read by itself; however, the arguments there will be much less vivid and convincing unless you have seen at least some of the details in earlier chapters. Chapter 6 is intended mainly for professional scholars and advanced students in the history and philosophy of science. However, for anyone particularly excited, puzzled, or disturbed by the work contained in the first five chapters, it will be helpful to read chapter 6 to get my own explanation of what I am trying to do. In general, the chapters could be read independently of each other and in any order. However, they are arranged in roughly chronological order and both the historical and the philosophical discussions contained in them do accumulate in a real sense, so if you have the time and intention to read all of the chapters, you would do well to read them in the order presented.

As indicated by its inclusion in the Oxford Studies in the Philosophy of Science, this book is intended to be a work of philosophy. However, the studies presented here are works of philosophy, science, and history simultaneously. I am aware that they may cross some boundaries and offend the sensibilities of particular academic disciplines. And if I go into explanations of various elementary points well known to specialists, that is not a sign of condescension or ignorance, but only an allowance for the variety of intended readership. I fear that professional philosophy today is at risk of becoming an ailing academic discipline shunned by large numbers of students and seemingly out of touch with other human concerns. It should not be that way, and this book humbly offers one model of how philosophy might engage more productively with endeavors that are perceived to be more practically significant, such as empirical scientific research. I hope that this book will serve as a reminder that interesting and useful philosophical insights can emerge from a critical study of concrete scientific practices.

The intended audience closest to my own professional heart is that small band of scholars and students who are still trying to practice and promote history-and-philosophy of science as an integrated discipline. More broadly, discussions of epistemology and scientific methodology included in this book will interest philosophers of science, and perhaps philosophers in general. Discussions of physics and chemistry in the eighteenth and nineteenth centuries will be of interest to historians of science. Much of the historical material in the first four chapters is not to be found in the secondary literature and is intended as an original contribution to the history of science. I also hope that the stories of how we came to believe what we believe, or how we discovered what we know, will interest many practicing scientists, science students, and non-professional lovers of science. But, in the end, professional labels are not so relevant to my main aspirations. If you can glimpse through my words any of the fascination that has forced me to write them, then this book is for you.

1

Keeping the Fixed Points Fixed

Narrative: What to Do When Water Refuses to Boil at the Boiling Point

> The *excess* of the heat of water *above the boiling point* is influenced by a great variety of circumstances.
>
> Henry Cavendish, "Theory of Boiling," c. 1780

The scientific study of heat started with the invention of the thermometer. That is a well-worn cliché, but it contains enough truth to serve as the starting point of our inquiry. And the construction of the thermometer had to start with the establishment of "fixed points." Today we tend to be oblivious to the great challenges that the early scientists faced in establishing the familiar fixed points of thermometry, such as the boiling and freezing points of water. This chapter is an attempt to become reacquainted with those old challenges, which are no less real for being forgotten. The narrative of the chapter gives a historical account of the surprising difficulties encountered and overcome in establishing one particular fixed point, the boiling point of water. The analysis in the second half of the chapter touches on broader philosophical and historical issues and provides in-depth discussions that would have interrupted the flow of the narrative.

Blood, Butter, and Deep Cellars: The Necessity and Scarcity of Fixed Points

Galileo and his contemporaries were already using thermometers around 1600. By the late seventeenth century, thermometers were very fashionable but still notoriously unstandardized. Witness the complaint made about the existing thermometers in 1693 by Edmond Halley (1656–1742), astronomer of the comet fame and secretary of the Royal Society of London at the time:

> I cannot learn that any of them...were ever made or adjusted, so as it might be concluded, what the Degrees or Divisions...did mean; neither were they ever otherwise graduated, but by Standards kept by each particular Workman, without any agreement or reference to one another. (Halley 1693, 655)

Most fundamentally, there were no standard "fixed points," namely phenomena that could be used as thermometric benchmarks because they were known to take place always at the same temperature. Without credible fixed points it was impossible to create any meaningful temperature scale, and without shared fixed points used by all makers of thermometers there was little hope of making a standardized scale.

Halley himself recommended using the boiling point of alcohol ("spirit of wine") as a fixed point, having seen how the alcohol in his thermometer always came up to the same level when it started to boil. But he was also quick to add a cautionary note: "Only it must be observed, that the Spirit of Wine used to this purpose be highly rectified or dephlegmed, for otherwise the differing Goodness of the Spirit will occasion it to boil sooner or later, and thereby pervert the designed Exactness" (1693, 654). As for the lower fixed point, he repudiated Robert Hooke's and Robert Boyle's practice of using the freezing points of water and aniseed oil, either of which he thought was "not so justly determinable, but with a considerable latitude." In general Halley thought that "the just beginning of the Scales of Heat and Cold should not be from such a Point as freezes any thing," but instead recommended using the temperature of deep places underground, such as "the Grottoes under the Observatory at Paris," which a "certain Experiment of the curious Mr. Mariotte" had shown to be constant in all seasons (656).[1]

Halley's contribution clearly revealed a basic problem that was to plague thermometry for a long time to come: in order to ensure the stability and usefulness of thermometers, we must be quite certain that the presumed fixed points are actually fixed sharply, instead of having "a considerable latitude." There are two parts to this problem, one epistemic and the other material. The epistemic problem is to know how to judge whether a proposed fixed point is actually fixed: how can that judgment be made in the absence of an already-trusted thermometer? This problem will not feature prominently on the surface of the narrative about the history of fixed points to be given now; however, in the analysis part of this chapter, it will be discussed as a matter of priority (see especially "The Validation of Standards" section). Assuming that we know how to judge fixedness, we can face the material problem of finding or creating some actual points that are fixed.

Throughout the seventeenth century and the early parts of the eighteenth century, there was a profusion of proposed fixed points, with no clear consensus as to which ones were the best. Table 1.1 gives a summary of some of the fixed

---

[1] He did not name Hooke and Boyle explicitly. See Birch [1756–57] 1968, 1:364–365, for Hooke's suggestion to the Royal Society in 1663 to use the freezing point of water; see Barnett 1956, 290, for Boyle's use of aniseed oil.

TABLE 1.1. Summary of fixed points used by various scientists

Person	Year	Fixed points ("and" indicates a two-point system)	Source of information
Sanctorius	c. 1600	candle flame *and* snow	Bolton 1900, 22
Accademia del Cimento	c. 1640?	most severe winter cold *and* greatest summer heat	Boyer 1942, 176
Otto Von Guericke	c. 1660?	first night frost	Barnett 1956, 294
Robert Hooke	1663	freezing distilled water	Bolton 1900, 44–45; Birch [1756] 1968, 1:364–365
Robert Boyle	1665?	congealing oil of aniseed *or* freezing distilled water	Bolton 1900, 43
Christiaan Huygens	1665	boiling water *or* freezing water	Bolton 1900, 46; Barnett 1956, 293
Honoré Fabri	1669	snow *and* highest summer heat	Barnett 1956, 295
Francesco Eschinardi	1680	melting ice *and* boiling water	Middleton 1966, 55
Joachim Dalencé	1688	freezing water *and* melting butter *or* ice *and* deep cellars	Bolton 1900, 51
Edmond Halley	1693	deep caves *and* boiling spirit	Halley 1693, 655–656
Carlo Renaldini	1694	melting ice *and* boiling water	Middleton 1966, 55
Isaac Newton	1701	melting snow *and* blood heat	Newton [1701] 1935, 125, 127
Guillaume Amontons	1702	boiling water	Bolton 1900, 61
Ole Rømer	1702	ice/salt mixture *and* boiling water	Boyer 1942, 176
Philippe de la Hire	1708	freezing water *and* Paris Observatory cellars	Middleton 1966, 56
Daniel Gabriel Fahrenheit	c. 1720	ice/water/salt mixture *and* ice/water mixture *and* healthy body temperature	Bolton 1900, 70
John Fowler	c. 1727	freezing water *and* water hottest to be endured by a hand held still	Bolton 1900, 79–80
R. A. F. de Réaumur	c. 1730	freezing water	Bolton 1900, 82
Joseph-Nicolas De l'Isle	1733	boiling water	Middleton 1966, 87–89
Anders Celsius	by 1741	melting ice *and* boiling water	Beckman 1998
J. B. Micheli du Crest	1741	Paris Observatory cellars *and* boiling water	Du Crest 1741, 8
Encyclopaedia Britannica	1771	freezing water *and* congealing wax	*Encyclopaedia Britannica*, 1st ed., 3:487

points used by the most respectable scientists up to the late eighteenth century. One of the most amusing to our modern eyes is a temperature scale proposed by Joachim Dalencé (1640–1707?), which used the melting point of butter as its upper fixed point. But even that was an improvement over previous proposals like the "greatest summer heat" used in the thermometers of the Accademia del Cimento, a group of experimental philosophers in Florence led by Grand Duke Ferdinand II and his brother Leopold Medici. Even the great Isaac Newton (1642–1727) seems to have made an unwise choice in using what was often called "blood heat," namely human body temperature, as a fixed point in his 1701 scale of temperatures.[2]

By the middle of the eighteenth century, a consensus was emerging about using the boiling and freezing of water as the preferred fixed points of thermometry, thanks to the work of the Swedish astronomer Anders Celsius (1701–1744), among others.[3] However, the consensus was neither complete nor unproblematic. In 1772 Jean-André De Luc (1727–1817), whose work I shall be examining in great detail shortly, published these words of caution:

> Today people believe that they are in secure possession of these [fixed] points, and pay little attention to the uncertainties that even the most famous men had regarding this matter, nor to the kind of anarchy that resulted from such uncertainties, from which we still have not emerged at all. (De Luc 1772, 1:331, §427[4])

To appreciate the "anarchy" that De Luc was talking about, it may be sufficient to witness the following recommendation for the upper fixed point given as late as 1771, in the first edition of the *Encyclopaedia Britannica*: "water just hot enough to let wax, that swims upon it, begin to coagulate" (3:487).[5] Or there is the more exotic case of Charles Piazzi Smith (1819–1900), astronomer royal for Scotland, who proposed as the upper fixed point the mean temperature of the King's Chamber at the center of the Great Pyramid of Giza.[6]

The Vexatious Variations of the Boiling Point

In 1776 the Royal Society of London appointed an illustrious seven-member committee to make definite recommendations about the fixed points of thermometers.[7] The chair of this committee was Henry Cavendish (1731–1810), the reclusive

---

[2] See Newton [1701] 1935, 125, 127. Further discussion can be found in Bolton 1900, 58, and Middleton 1966, 57. Blood heat may actually not have been such a poor choice in relative terms, as I will discuss further in "The Validation of Standards" in the analysis part of this chapter. Middleton, rashly in my view, berates Newton's work in thermometry as "scarcely worthy of him." According to modern estimates, the temperatures of healthy human bodies vary by about 1 degree centigrade.

[3] On Celsius's contributions, see Beckman 1998. According to the consensus emerging in the late eighteenth century, both of these points were used together to define a scale. However, it should be noted that it is equally cogent to use only one fixed point, as emphasized in Boyer 1942. In the one-point method, temperature is measured by noting the volume of the thermometric fluid in relation to its volume at the one fixed point.

[4] In citing from this work, I will give both the paragraph number and the page number from the two-volume edition (quarto) that I am using, since there was also a four-volume edition (octavo) with different pagination.

[5] Newton ([1701] 1935, 125) had assigned the temperature of 20 and 2/11 degrees on his scale to this point. It was not till the 3d edition of 1797 that *Britannica* caught on to the dominant trend and noted: "The fixed points which are now universally chosen...are the boiling and freezing water points." See "Thermometer," *Encyclopaedia Britannica*, 3d ed., 18:492–500, on pp. 494–495.

[6] The information about Piazzi Smith is from the display in the Royal Scottish Museum, Edinburgh.

[7] This committee was appointed at the meeting of 12 December 1776 and consisted of Aubert, Cavendish, Heberden, Horsley, De Luc, Maskelyne, and Smeaton. See the Journal Book of the Royal Society, vol. 28 (1774–1777), 533–534, in the archives of the Royal Society of London.

aristocrat and devoted scientist who was once described as "the wisest of the rich and the richest of the wise."[8] The Royal Society committee did take it for granted that the two water points should be used, but addressed the widespread doubts that existed about their true fixity, particularly regarding the boiling point. The committee's published report started by noting that the existing thermometers, even those made by the "best artists," differed among themselves in their specifications of the boiling point. The differences easily amounted to 2–3 degrees Fahrenheit. Two causes of variation were clearly identified and successfully dealt with.[9] First, the boiling temperature was by then widely known to vary with the atmospheric pressure,[10] and the committee specified a standard pressure of 29.8 English inches (roughly 757 mm) of mercury, under which the boiling point should be taken. Drawing on De Luc's previous work, the committee also gave a formula for adjusting the boiling point according to pressure, in case it was not convenient to wait for the atmosphere to assume the standard pressure. The second major cause of variation was that the mercury in the stem of the thermometer was not necessarily at the same temperature as the mercury in the thermometer bulb. This was also dealt with in a straightforward manner, by means of a setup in which the entire mercury column was submerged in boiling water (or in steam coming off the boiling water). Thus, the Royal Society committee identified two main problems and solved both of them satisfactorily.

However, the committee's report also mentioned other, much less tractable questions. One such question is represented emblematically in a thermometric scale from the 1750s that is preserved in the Science Museum in London. That scale (shown in fig. 1.1), by George Adams the Elder (?–1773), has two boiling points: at 204° Fahrenheit "water begins to boyle," and at 212°F "water boyles vehemently." In other words, Adams recognized a temperature interval as wide as 8°F in which various stages of boiling took place. This was not an aberrant quirk of an incompetent craftsman. Adams was one of Britain's premier instrument-makers, the official "Mathematical Instrument Maker" to George III, starting from 1756 while the latter was the Prince of Wales.[11] Cavendish himself had addressed the question of whether there was a temperature difference between "fast" and "slow" boiling ([1766] 1921, 351). The notion that there are different temperatures associated with different "degree of boiling" can be traced back to Newton ([1701] 1935, 125), who recorded that water began to boil at 33° of his scale and boiled vehemently at 34° to 34.5°, indicating a range of about 5–8°F. Similar observations were made by

---

[8]This description was by Jean-Baptiste Biot, quoted in Jungnickel and McCormmach 1999, 1. Cavendish was a grandson of William Cavendish, the Second Duke of Devonshire, and Rachel Russell; his mother was Anne de Grey, daughter of Henry de Grey, Duke of Kent. See Jungnickel and McCormmach 1999, 736–737, for the Cavendish and Grey family trees.

[9]For further details, see Cavendish et al. 1777, esp. 816–818, 853–855.

[10]Robert Boyle had already noted this in the seventeenth century, and Daniel Gabriel Fahrenheit knew the quantitative relations well enough to make a barometer that inferred the atmospheric pressure from the boiling point of water. See Barnett 1956, 298.

[11]The description of Adams's scale is from Chaldecott 1955, 7 (no. 20). For information about his status and work, see Morton and Wess 1993, 470, and passim.

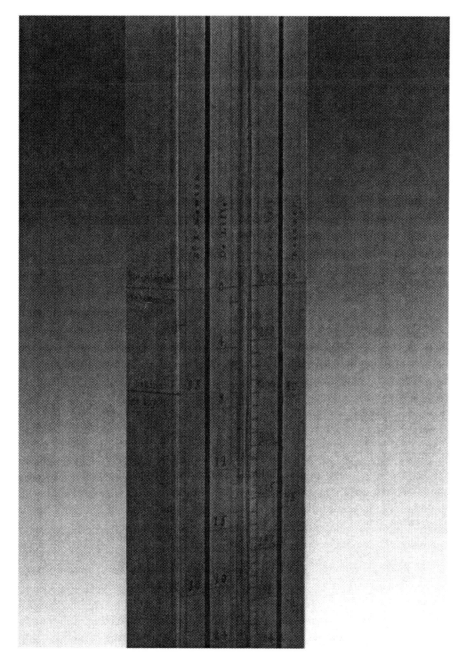

FIGURE 1.1. George Adams's thermometric scale, showing two boiling points (inventory no. 1927-1745). Science Museum/Science & Society Picture Library.

De Luc, who was a key member of the Royal Society committee and perhaps the leading European authority in thermometry in the late eighteenth century.

As Jean-André De Luc (fig. 1.2) is a little-known figure today even among historians of science, a brief account of his life and work is in order.[12] In his own day De Luc had a formidable reputation as a geologist, meteorologist, and physicist. He received his early education from his father, François De Luc, a clockmaker, radical politician, and author of pious religious tracts, who was once described by Jean-Jacques Rousseau as "an excellent friend, the most honest and boring of men" (Tunbridge 1971, 15). The younger De Luc maintained equally active interests in science, commerce, politics, and religion. To his credit were some very popular natural-theological explanations of geological findings, strenuous arguments against Lavoisier's new chemistry, and a controversial theory of rain postulating the transmutation of air into water.[13] One of the early "scientific mountaineers," De Luc made pioneering excursions into the Alps (with his younger brother Guillaume-Antoine), which stimulated and integrated his scientific interests in natural history, geology, and meteorology. His decisive improvement of the method of measuring the heights of mountains by barometric pressure was a feat that some considered sufficient to qualify him as one of the most important physicists in all of Europe.[14] More generally he was famous for his inventions and improvements of meteorological instruments and for the keen observations he made with them. Despite his willingness to theorize, his empiricist leanings were clearly encapsulated in statements such as the following: "[T]he progress made towards perfecting [measuring instruments] are the most effectual steps which have been made towards the knowledge of Nature; for it is they that have given us a disgust to the jargon of systems...spreading fast into metaphysics" (De Luc 1779, 69). In 1772 De Luc's business in Geneva collapsed, at which point he retired from commercial life and devoted himself entirely to scientific work. Soon thereafter he settled in England, where he was welcomed as a Fellow of the Royal Society (initially invited to the Society by Cavendish) and also given the prestigious position of "Reader" to Queen Charlotte. De Luc became an important member of George III's court and based himself in Windsor to his dying day, though he did much traveling and kept up his scientific connections particularly with the Lunar Society of Birmingham and a number of German scholars, especially in Göttingen.

De Luc's first major scientific work, the two-volume *Inquiries on the Modifications of the Atmosphere*, published in 1772, had been eagerly awaited for the promised discussion of the barometric measurement of heights. When it was finally published after a delay of ten years, it also contained a detailed discourse on the

---

[12]The most convenient brief source on De Luc's life and work is the entry in the *Dictionary of National Biography*, 5:778–779. For more detail, see De Montet 1877–78, 2:79–82, and Tunbridge 1971. The entry in the *Dictionary of Scientific Biography*, 4:27–29, is also informative, though distracted by the contributor's own amazement at De Luc's seemingly unjustified renown. Similarly, W. E. K. Middleton's works contain a great deal of information about De Luc, but suffer from a strong bias against him.

[13]On the controversy surrounding De Luc's theory of rain, which was also the cornerstone of his objections to Lavoisier's new chemistry, see Middleton 1964a and Middleton 1965.

[14]For this appraisal, see *Journal des Sçavans*, 1773, 478.

FIGURE 1.2. Jean-André De Luc. Geneva, Bibliothèque publique et universitaire, Collections iconographiques.

construction and employment of thermometers, with an explanation that De Luc had originally become interested in thermometers because of the necessity to correct barometer readings for variations in temperature.[15] I will have occasion to discuss other aspects of De Luc's work in thermometry in chapter 2, but for now let us return to the subject of possible variations in the boiling temperature according to the "degree of boiling." Initially De Luc asserted:

> When water begins to boil, it does not yet have the highest degree of heat it can attain. For that, the entire mass of the water needs to be in movement; that is to say, that the boiling should start at the bottom of the vessel, and spread all over the surface of the water, with the greatest impetuosity possible. From the commencement of ebullition to its most intense phase, the water experiences an increase in heat of more than a degree. (De Luc 1772, 1:351–352, §439)

In further experiments, De Luc showed that there was an interval of 76 to 80 degrees on his thermometer (95–100°C, or 203–212°F) corresponding to the spectrum of ebullition ranging from "hissing" to full boil, which is quite consistent with the range of 204–212°F indicated in Adams's thermometer discussed earlier. The weakest degree of genuine boiling started at 78.75° on De Luc's thermometer, in which the full-boiling point was set at 80°, so there was a range of 1.25° (over 1.5°C) from the commencement of boiling to the highest boiling temperature.[16]

The Royal Society committee investigated this issue carefully, which is not surprising given that its two leading members, Cavendish and De Luc, had been concerned by it previously. The committee's findings were somewhat reassuring for the stability of the boiling point:

> For the most part there was very little difference whether the water boiled fast or very gently; and what difference there was, was not always the same way, as the thermometer sometimes stood higher when the water boiled fast, and sometimes lower. The difference, however, seldom amounted to more than 1/10th of a degree. (Cavendish et al. 1777, 819–820)

Still, some doubts remained. The trials were made in metallic pots, and it seemed to matter whether the pots were heated only from the bottom or from the sides as well:

> In some trials which we made with the short thermometers in the short pot, with near four inches of the side of the vessel exposed to the fire, they constantly stood lower when the water boiled fast than when slow, and the height was in general greater than when only the bottom of the pot was exposed to the fire. (820)

Not only was that result in disagreement with the other trials made by the committee but also it was the direct opposite of the observations by Adams and De Luc,

---

[15]See De Luc 1772, 1:219–221, §408.
[16]See De Luc 1772, 2:358, §983. De Luc's own thermometer employed what came to be known as the "Réaumur" scale, which had 80 points between the freezing and the boiling points. R. A. F. Réaumur had used an 80-point scale, but his original design was considerably modified by De Luc.

according to which water boiling vigorously had a higher temperature than water boiling gently.

There were other factors to worry about as well. One was the depth of the boiling water: "[I]f the ball be immersed deep in the water, it will be surrounded by water which will be compressed by more than the weight of the atmosphere, and on that account will be rather hotter than it ought to be" (817–818). Experiments did vindicate this worry, revealing a variation of about 0.06° per inch in the depth of the water above the ball of the thermometer. However, the committee was reluctant to advance that observation as a general rule. For one thing, though this effect clearly seemed to be caused by the changes of pressure, it was only half as large as the effect caused by changes in the atmospheric pressure. Even more baffling was the fact that "the boiling point was in some measure increased by having a great depth of water *below* the ball...[T]his last effect, however, did not always take place" (821–822; emphasis added). Although the committee made fairly definite recommendations on how to fix the boiling point in the end, its report also revealed a lingering sense of uncertainty:

> Yet there was a very sensible difference between the trials made on different days, even when reduced to the same height of the barometer, though the observations were always made either with rain or distilled water.... We do not at all know what this difference could be owing to.... (826–827)

Superheating and the Mirage of True Ebullition

The work of the Royal Society committee on the boiling point is a lively testimony to the shakiness of the cutting-edge knowledge of the phenomenon of boiling in the late eighteenth century. No one was more clearly aware of the difficulties than De Luc, who had started worrying about them well before the Royal Society commission. Just as his book was going to the press in 1772, De Luc added a fifteen-chapter supplement to his discussion of thermometers, entitled "inquiries on the variations of the heat of boiling water." The logical starting point of this research was to give a precise definition of boiling, before disputing whether its temperature was fixed. What, then, is boiling? De Luc (1772, 2:369, §1008) conceived "true ebullition" ("la vraie ébullition") as the phenomenon in which the "first layer" of water in contact with the heat source became saturated with the maximum possible amount of heat ("fire" in his terminology), thereby turning into vapor and rising up through the water in the form of bubbles. He wanted to determine the temperature acquired by this first layer. That was a tall order experimentally, since the first layer was bound to be so thin that no thermometer could be immersed in it. Initial experiments revealed that there must indeed be a substantial difference between the temperature of the first layer and the rest of the water under normal conditions. For example, when De Luc heated water in a metallic vessel put into an oil bath, the thermometer in the middle of the water reached 100°C only when the oil temperature was 150°C or above. One could only surmise that the first layer of water must have been brought to a temperature somewhere between 100°C and 150°C. De Luc's best estimate, from an experiment in which small drops of water introduced into hot oil exploded into vapor when the oil was hot enough, was that the first layer of water had to be at about 112°C, for true ebullition to occur.[17]

Was it really the case that water could be heated to 112°C before boiling? Perhaps incredulous about his own results, De Luc devised a different experiment (1772, 2:362–364, §§994–995). Thinking that the small drops of water suspended in oil may have been too much of an unusual circumstance, in the new experiment he sought to bring all of a sizeable body of water up to the temperature of the first layer. To curtail heat loss at the open surface of the water, he put the water in a glass flask with a long narrow neck (only about 1 cm wide) and heated it slowly in an oil bath as before. The water boiled in an unusual way, by producing very large occasional bubbles of vapor, sometimes explosive enough to throw off some of the liquid water out of the flask. While this strange boiling was going on, the temperature of the water fluctuated between 100°C and over 103°C. After some time, the water filled only part of the flask and settled into a steadier boil, at the temperature of 101.9°C. De Luc had observed what later came to be called "superheating," namely the heating of a liquid beyond its normal boiling point.[18] It now seemed certain to De Luc that the temperature necessary for true ebullition was higher than the normally recognized boiling point of water. But how much higher?

There was one major problem in answering that question. The presence of dissolved air in water induced an ebullition-like phenomenon before the temperature of true ebullition was reached. De Luc knew that ordinarily water contained a good deal of dissolved air, some of which was forced out by heating and formed small bubbles (often seen sticking to the inner surface of vessels), before the boiling point was reached. He was also well aware that evaporation from the surface of water happened at a good rate at temperatures well below boiling. Putting the two points together, De Luc concluded that significant evaporation must happen at the inner surfaces of the small air bubbles at temperatures much lower than that of true ebullition. Then the air bubbles would swell up with vapor, rise, and escape, releasing a mixture of air and water vapor. Does that count as boiling? It surely has the appearance of boiling, but it is not true ebullition as De Luc defined it. He identified this action of dissolved air as "the greatest obstacle" that he had to overcome in his research: "that is, the production of internal vapors, which is occasioned by this emergence of air, before there is true ebullition."[19]

De Luc was determined to study true ebullition, and that meant obtaining water that was completely purged of dissolved air. He tried everything. Luckily,

---

[17] For further details on these experiments, see De Luc 1772, 2:356–362, §§980–993.

[18] The term "superheating" was first used by John Aitken in the 1870s, as far as I can ascertain; see Aitken 1878, 282. The French term *surchauffer* was in use quite a bit earlier.

[19] For the discussion of the role of air in boiling, see De Luc 1772, 2:364–368, §§996–1005; the quoted passage is from p. 364.

sustained boiling actually tended to get much of the air out of the water.[20] And then he filled a glass tube with hot boiled water and sealed the tube; upon cooling, the contraction of the water created a vacuum within the sealed tube, and further air escaped into that vacuum.[21] This process could be repeated as often as desired. De Luc also found that shaking the tube (in the manner of rinsing a bottle, as he put it) facilitated the release of air; this is a familiar fact known to anyone who has made the mistake of shaking a can of carbonated beverage. After these operations, De Luc obtained water that entered a steady boil only in an oil bath as hot as 140°C.[22] But as before, he could not be sure that the water had really taken the temperature of the oil bath, though this time the water was in a thin tube. Sticking a thermometer into the water in order to verify its temperature had the maddening side effect of introducing some fresh air into the carefully purified water. There was no alternative except to go through the purging process with the thermometer already enclosed in the water, which made the already delicate purging operation incredibly frustrating and painful. He reported:

> This operation lasted four weeks, during which I hardly ever put down my flask, except to sleep, to do business in town, and to do things that required both hands. I ate, I read, I wrote, I saw my friends, I took my walks, all the while shaking my water.... (De Luc 1772, 2:387, §§1046–1049)

Four mad weeks of shaking had its rewards. The precious airless water he obtained could stand the heat of 97.5°C even in a vacuum, and under normal atmospheric pressure it reached 112.2°C before boiling off explosively (2:396–397, §§1071–1072). The superheating of pure water was now confirmed beyond any reasonable doubt, and the temperature reached in this experiment was very much in agreement with De Luc's initial estimate of the temperature reached by the "first layer" of water in ebullition.

Superheating was an experimental triumph for De Luc. However, it placed him into a theoretical dilemma, if not outright confusion. Ordinary water was full of air

---

[20] Compare this observation with the much later account by John Aitken (1923, 10), whose work will be discussed in some detail in "A Dusty Epilogue": "After the water has been boiling some time there is less and less gas in it. A higher temperature is therefore necessary before the gas will separate from the water. Accordingly, we find that the water rises in temperature after boiling some time. The boiling-point depends in fact not on the temperature at which steam is formed, but on the temperature at which a free surface is formed."

[21] De Luc had used this technique earlier in preparing alcohol for use in thermometers. Alcohol boils at a lower temperature than water (the exact temperature depending on its concentration), so there was an obvious problem in graduating alcohol thermometers at the boiling point of water. De Luc (1772, 1:314–318, §423) found that purging the alcohol of dissolved air made it capable of withstanding the temperature of boiling water. If W. E. Knowles Middleton (1966, 126) had read De Luc's discussion of superheating, he would have thought twice about issuing the following harsh judgment: "If there was any more in this than self-deception, Deluc must have removed nearly all the alcohol by this process. Nevertheless, this idea gained currency on the authority of Deluc."

[22] For a description of the purging process, see De Luc 1772, 2:372–380, §§1016–1031. The boiling experiment made with the airless water (the "sixth experiment") is described in 2:382–384, §§1037–1041.

and not capable of attaining true ebullition, but his pure airless water was not capable of normal boiling at all, only explosive puffing with an unsteady temperature. To complicate matters further, the latter type of boiling also happened in a narrow-necked flask even when the water had not been purged of air. De Luc had started his inquiry on boiling by wanting to know the temperature of true boiling; by the time he was done, he no longer knew what true boiling was. At least he deserves credit for realizing that boiling was not a simple, homogeneous phenomenon. The following is the phenomenology of what can happen to water near its boiling point, which I have gathered from various parts of De Luc's 1772 treatise. It is not a very neat classification, despite my best efforts to impose some order.

1. Common boiling: numerous bubbles of vapor (probably mixed with air) rise up through the surface at a steady rate. This kind of boiling can happen at different rates or "degrees" of vigorousness, depending on the power of the heat source. The temperature is reasonably stable, though possibly somewhat variable according to the rate of boiling.
2. Hissing (*sifflement* in De Luc's French): numerous bubbles of vapor rise partway through the body of water, but they are condensed back into the liquid state before they reach the surface. This happens when the middle or upper layers of the water are cooler than the bottom layers. The resulting noise just before full boiling sets in is a familiar one to serious tea-drinkers, once known as the "singing" of the kettle.
3. Bumping (*soubresaut* in French; both later terminology): large isolated bubbles of vapor rise occasionally; the bubbles may come only one at a time or severally in an irregular pattern. The temperature is unstable, dropping when the large bubbles are produced and rising again while no bubbles form. There is often a loud noise.
4. Explosion: a large portion of the body of water suddenly erupts into vapor with a bang, throwing off any remaining liquid violently. This may be regarded as an extreme case of bumping.
5. Fast evaporation only: no bubbles are formed, but a good deal of vapor and heat escape steadily through the open surface of the water. The temperature may be stable or unstable depending on the particular circumstance. This phenomenon happens routinely below the normal boiling point, but it also happens in superheated water; in the latter case, it may be a stage within the process of bumpy or explosive boiling.
6. Bubbling (*bouillonement* in De Luc's French): although this has the appearance of boiling, it is only the escape of dissolved air (or other gases), in the manner of the bubbling of fizzy drinks. It is especially liable to happen when there is a sudden release of pressure.[23]

---

[23] For a discussion of bubbling, see De Luc 1772, 2:380–381, §1033.

Now which of these is "true" boiling? None of the options is palatable, and none can be ruled out completely, either. Bubbling would not seem to be boiling at all, but as we will see in "The Understanding of Boiling" section, a popular later theory of boiling regarded boiling as the release of water vapor (gas) dissolved in liquid water. Hissing and fast evaporation can probably be ruled out easily enough as "boiling" as we know it, since in those cases no bubbles of vapor come from within the body of water through to its surface; however, we will see in "A Dusty Epilogue" that there was a credible theoretical viewpoint in which evaporation at the surface was regarded as the essence of "boiling." Probably closest to De Luc's original conception of "true ebullition" is bumping (and explosion as a special case of it), in which there is little or no interference by dissolved air and the "first layer" of water is probably allowed to reach something like saturation by heat. But defining bumping as true boiling would have created a good deal of discomfort with the previously accepted notions of the boiling point, since the temperature of bumping is not only clearly higher than the temperature of common boiling but also unstable in itself. The only remaining option was to take common boiling as true boiling, which would have implied that the boiling point was the boiling temperature of impure water, mixed in with air. In the end, De Luc seems to have failed to reach any satisfactory conclusions in his investigation of boiling, and there is no evidence that his results were widely adopted or even well known at the time, although there was to be a powerful revival of his ideas many decades later as we will see shortly.

In the course of the nineteenth century, further study revealed boiling to be an even more complex and unruly phenomenon than De Luc had glimpsed. A key contribution was made in the 1810s by the French physicist-chemist Joseph-Louis Gay-Lussac (1778–1850). His intervention was significant, since he was regarded as one of the most capable and reliable experimenters in all of Europe at the time, and his early fame had been made in thermal physics. Gay-Lussac (1812) reported (with dubious precision) that water boiled at 101.232°C in a glass vessel, while it boiled at 100.000°C in a metallic vessel. However, throwing in some finely powdered glass into the glass vessel brought the temperature of the boiling water down to 100.329°C, and throwing in iron filings brought it to 100.000°C exactly. Gay-Lussac's findings were reported in the authoritative physics textbook by his colleague Jean-Baptiste Biot (1774–1862), who stressed the extreme importance of ascertaining whether the fixed points of thermometry were "perfectly constant." Biot (1816, 1:41–43) admitted that Gay-Lussac's phenomena could not be explained by the thermal physics of his day, but thought that they contributed to a more precise definition of the boiling point by leading to the specification that the boiling needed to be done in a metallic vessel. If Gay-Lussac and Biot were right, the members of the Royal Society committee had got reasonably fixed results for the boiling point only because they happened to use metallic vessels. The reasons for that choice were not explained in their reports, but De Luc may have advised the rest of the committee that his troublesome superheating experiments had been carried out in glass vessels.

Gay-Lussac's results, unlike De Luc's, were widely reported and accepted despite some isolated criticism. However, it took another thirty years for the next significant step to be taken, this time in Geneva again, by the professor of physics

François Marcet (1803–1883)—son of Alexandre, the émigré physician in London, and Jane, the well-known author of popular science. Marcet (1842) produced superheating beyond 105°C in ordinary water, by using glass vessels that had contained strong sulphuric acid; clearly, somehow, the acid had modified the surface in such a way as to make boiling more difficult. Superheating became a clearly recognized object of study after Marcet's work, stimulating a string of virtuoso experimental performances vying for record temperatures. François Marie Louis Donny (1822–?), chemist at the University of Ghent, combined this insight on adhesion with a revival of De Luc's ideas about the role of air and produced a stunning 137°C using airless water in his own special instrument. Donny declared: "The faculty to produce ordinary ebullition cannot in reality be considered as an inherent property of liquids, because they show it only when they contain a gaseous substance in solution, which is to say only when they are not in a state of purity" (1846, 187–188).

In 1861 the work of Louis Dufour (1832–1892), professor of physics at the Academy of Lausanne, added yet another major factor for consideration. Dufour (1861, esp. 255) argued that contact with a solid surface was the crucial factor in the production of ebullition and demonstrated the soundness of his idea by bringing drops of water floating in other liquids up to 178°C, without even purging the air out of the water. Even Dufour was outdone, when Georg Krebs (1833–1907) in 1869 achieved an estimated 200°C with an improvement of Donny's technique.[24]

The superheating race must have been good fun to watch, but it also presented a great theoretical challenge.[25] All investigators now agreed that the raising of temperature to the "normal" boiling point was not a sufficient condition to produce boiling. What they could not agree on was what the additional conditions needed for the production of boiling were. And if these additional conditions were not met, it was not clear how far superheating could go. Donny in 1846 had already expressed bemused uncertainty on this point: "[O]ne cannot predict what would happen if one could bring the liquid to a state of perfect purity." Krebs, in the work cited earlier, opined that water completely purged of air could not boil at all. In the more careful view of Marcel Émile Verdet (1824–1866), renowned professor of physics in Paris credited with introducing thermodynamics to France, there was probably a limit to the degree of superheating, namely that point at which there is enough heat to vaporize the whole mass of water instantly. Verdet, however, admitted that there was only one experiment in support of that view, namely the now-classic work of Charles Cagniard de la Tour (1777?–1859) on the critical point, a temperature above which a gas cannot be liquefied regardless of pressure.[26] There was sufficient uncertainty on this question even toward the end

---

[24]Krebs's work is reported in Gernez 1875, 354.

[25]An intriguing parallel might be drawn between this superheating race and the modern-day race to reach higher and higher temperatures in superconductivity.

[26]See the review of the works on superheating given in Gernez 1875. Donny's statement is quoted on p. 347, and the report of Verdet's view can be found on p. 353.

of the nineteenth century. In 1878 the 9th edition of the *Encyclopaedia Britannica* reported: "It has been stated that the boiling of *pure* water has not yet been observed."[27]

Escape from Superheating

The discussion in the last section leaves a puzzle: if there were such unmanageable and ill-understood variations in the temperatures required for the boiling of water, how could the boiling point have served as a fixed point of thermometry at all? It seems that superheating would have threatened the very notion of a definite "boiling point," but all the thermometers being used for the investigation of superheating were graduated with sharp boiling points that agreed increasingly well with each other. The philosopher can only conjecture that there must have been an identifiable class of boiling phenomena with sufficiently stable and uniform temperatures, which allowed the calibration of thermometers with which scientists could go on to study the more exotic instances. Fortunately, a closer look at the history bears out that philosophical conjecture. There were three main factors that allowed the boiling point to be used as a fixed point despite its vagaries.

First of all, an immediate relief comes in realizing the difference between the temperature that water can withstand *without boiling*, and the temperature that water maintains *while boiling*. All observers of superheating from De Luc onward had noted that the temperature of superheated water went down as soon as steady boiling was induced (or each time a large bubble was released, in the case of bumping). Extreme temperatures were reached only before boiling set in, so the shocking results obtained by Donny, Dufour, and Krebs could be disregarded for the purpose of fixing the boiling point. De Luc got as far as 112°C without boiling, but the highest temperature he recorded while the water was boiling was 103°C. Still, the latter is 3°C higher than the "normal" boiling temperature, and there was also Gay-Lussac's observation that the temperature of boiling water was 101.232°C in a glass vessel. Marcet (1842, 397 and 404) investigated this question with more care than anyone else. In ordinary glass vessels, he observed the temperature of boiling water to range from 100.4 to 101.25°C. In glass treated with sulphuric acid, the temperature while boiling went easily up to 103 or 104°C and was very unsteady in each case due to bumping.

That is where the second factor tending to stabilize the boiling point enters. In fact, this "second factor" is a whole set of miscellaneous factors, which might cause embarrassment to misguided purists. The spirit of dealing with them was to do whatever happened to prevent superheating. I have already mentioned that the Royal Society committee avoided superheating by using metallic vessels instead of glass. Gay-Lussac had shown how to prevent superheating in glass vessels by

---

[27] "Evaporation," *Encyclopaedia Britannica*, 9th ed., vol. 8 (1878), 727–733, on p. 728; emphasis original. This entry was by William Garnett.

throwing in metal chippings or filings (or even powdered glass). Other investigators found other methods, such as the insertion of solid objects (especially porous things like charcoal and chalk), sudden localized heating, and mechanical shocks. But in most practical situations the prevention of superheating simply came down to not bothering too much. If one left naturally occurring water in its usual state full of dissolved air (rather than taking the trouble to purge air out of it), and if one left the container vessels just slightly dirty or rough (instead of cleaning and smoothing it off with something like concentrated sulphuric acid), and if one did not do anything else strange like isolating the water from solid surfaces, then common boiling did take place. Serious theoretical arguments about the factors that facilitate ebullition continued into the twentieth century as we will see in the next section, but all investigators agreed sufficiently on how to break superheating and prevent bumping in practice. Verdet observed that under "ordinary conditions," there would be dissolved air in the water and the water would be in contact with solid walls, and hence boiling would "normally" set in at the normal boiling point (see Gernez 1875, 351). It was a great blessing for early thermometry that the temperature of boiling was quite fixed under the sort of circumstances in which water tended to be boiled by humans living in ordinary European-civilization conditions near the surface of the earth without overly advanced purification technologies.

However, happy-go-lucky sloppiness is not the most robust strategy of building scientific knowledge in the end, as the Royal Society committee realized quite well. The committee's lasting contribution, the last of our three factors contributing to the fixity of the boiling point, was to find one clear method of reducing the variations of the boiling point due to miscellaneous causes. The following was the committee's chief recommendation: "The most accurate way of adjusting the boiling point is, not to dip the thermometer into the water, but to expose it only to the steam, in a vessel closed up in the manner represented," shown in figure 1.3 (Cavendish et al. 1777, 845). Somehow, using the boiled-off steam rather than the boiling water itself seemed to eliminate many of the most intractable variations in the temperature:

> The heat of the steam therefore appears to be not sensibly different in different parts of the same pot; neither does there appear to be any sensible difference in its heat, whether the water boil fast or slow; whether there be a greater or less depth of water in the pot; or whether there be a greater or less distance between the surface of the water and the top of the pot; so that the height of a thermometer tried in steam, in vessels properly closed, seems to be scarce sensibly affected by the different manner of trying the experiment. (824)

The recommendation to use steam came most strongly from Cavendish (1776, 380), who had already made the same proposal in his review of the instruments used at the Royal Society. The committee report only noted that using steam did in fact produce more stable results, but Cavendish went further and gave theoretical reasons for preferring steam, in an unpublished paper that followed but modified

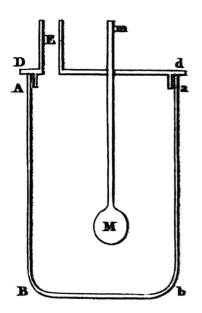 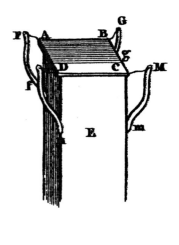

FIGURE 1.3. A scheme of the metallic pots used for fixing the boiling point by the Royal Society committee on thermometry. In the figure *mM* is the thermometer, and *E* is the "chimney" for the escape of steam. *ABCD* in the separate figure is a loose-fitting tin plate to cover the chimney in just the right way. These are fig. 4 and fig. 3 of Cavendish et al. 1777, from the plate opposite p. 856. The full description of the vessel and its proper employment can be found on 845–850. Courtesy of the Royal Society.

De Luc's theoretical ideas.[28] Cavendish stated as the first two of his four "principles of boiling":

> Water as soon as it is heated ever so little above that degree of heat which is acquired by the steam of water boiling in vessels closed as in the experiments tried at the Royal Society, is immediately turned into steam, provided that it is in contact either with steam or air; this degree I shall call the boiling heat, or boiling point. It is evidently different according to the pressure of the atmosphere, or more properly to the pressure acting on the water. But 2ndly, if the water is not in contact with steam or air, it will bear a much greater heat without being changed into steam, namely that which Mr. De Luc calls the heat of ebullition. (Cavendish [n.d.] 1921, 354)

Cavendish believed that the temperature of boiling water was variable, probably always hotter than the temperature of the steam, but to different degrees depending on the circumstances. The boiling water itself was not fixed in its temperature, and

---

[28]The essay, entitled "Theory of Boiling," is undated, but it must have been composed or last modified no earlier than 1777, since it refers to the result of the work of the Royal Society committee. Cavendish had studied De Luc's work carefully, as documented by Jungnickel and McCormmach 1999, 548, footnote 6.

he thought that "steam must afford a considerably more exact method of adjusting the boiling point than water" (359–360).

De Luc disagreed. Why would the temperature of boiled-off steam be more stable and universal than the temperature of the boiling water itself? In a letter of 19 February 1777 to Cavendish, written in the midst of their collaboration on the Royal Society committee, De Luc commented on Cavendish's "Theory of Boiling" and laid out some of his doubts.[29] The following passage is most significant:

> Setting aside for a moment all theory, it seems that the heat of the vapor of boiling water can be considered only with difficulty as more fixed than that of the water itself; for they are so mixed in the mass before the vapor emerges that they appear to have no alternative but to influence the temperature of each other. So to suppose that the vapor at the moment it emerges has in reality a fixed degree of temperature of its own, it is necessary that it be rigorously true, and demonstrated through some immediate experiments, that the vapor in reality can be *vapor* only at this fixed degree of heat. [But] I do not find that this proceeds from your reasoning.... (De Luc, in Jungnickel and McCormmach 1999, 547 and 550)

As De Luc's doubts underscore, Cavendish's preference for steam rested on an extraordinary belief: that steam emerging from water boiling under a fixed pressure must always have the same temperature, whatever the temperature of the water itself may be. This claim required a defense.

First of all, why would steam (or water vapor)[30] emerging out of water be only at 100°C, if the water itself has a higher temperature? Cavendish's answer was the following:

> These bubbles [of steam] during their ascent through the water can hardly be hotter than the boiling point; for so much of the water which is in contact with them must instantly be turned into steam that by means of the production of cold thereby, the coat of water...in contact with the bubbles, is no hotter than the boiling point; so that the bubbles during their ascent are continually in contact with water heated only to the boiling point. (Cavendish [n.d.] 1921, 359)

The fact that the formation of steam requires a great deal of heat was widely accepted since the work of the celebrated Scottish physician and chemist Joseph Black (1728–1799) on latent heat, and De Luc and Cavendish had each developed

---

[29] De Luc's letter has been published in Jungnickel and McCormmach 1999, 546–551, in the original French and in English translation; I quote from McCormmach's translation except as indicated in square brackets. The essay that De Luc was commenting on must have been an earlier version of Cavendish [n.d.] 1921.

[30] In De Luc's view, as in modern usage, there was no essential difference between steam and water vapor; they were simply different names for the gaseous state of water. Up to the end of the eighteenth century, however, many people saw steam produced by boiling as fundamentally different from vapor produced by evaporation at lower temperatures. A popular idea, which Cavendish favored, was that evaporation was a process in which water became dissolved in air. De Luc was one of those who criticized this idea strongly, for instance by pointing out that evaporation happens quite well into a vacuum where there is no air that can serve as a solvent. For further details, see Dyment 1937, esp. 471–473, and Middleton 1965.

similar ideas, though they recognized Black's priority. However, for a convincing demonstration that the steam in the body of superheated boiling water would always be brought down to the "boiling point," one needed a quantitative estimate of the rate of cooling by evaporation in comparison with the amount of heat continually received by the steam from the surrounding water. Such an estimate was well beyond Cavendish's reach.

De Luc also asked what would prevent the steam from cooling down below the boiling temperature, after emerging from the boiling water. Here Cavendish offered no theoretical account, except a dogmatically stated principle, which he considered empirically vindicated: "[S]team not mixed with air as soon as it is cooled ever so little below the [boiling temperature] is immediately turned back into water" ([n.d.] 1921, 354). De Luc's experience was to the contrary: "[W]hen in a little vessel, the *mouth* and its neck are open at the same time, the vapor, without condensing, becomes perceptibly cooler." (I will return to this point of argument in "A Dusty Epilogue.") He also doubted that steam could realistically be obtained completely free of air, which was necessary in order for Cavendish's principle to apply at all.[31]

Judging from the final report of the Royal Society committee, it is clear that no firm consensus was reached on this matter. While the committee's chief recommendation for obtaining the upper fixed point was to adopt the steam-based procedure advocated by Cavendish, the report also approved two alternative methods, both using water rather than steam and one of them clearly identified as the procedure from De Luc's previous work.[32] De Luc's acquiescence on the chief recommendation was due to the apparent fixedness of the steam temperature, not due to any perceived superiority in Cavendish's theoretical reasoning. In his February 1777 letter to Cavendish mentioned earlier, there were already a couple of indications of De Luc's surprise that the temperature of steam did seem more fixed than he would have expected. De Luc exhorted, agreeably:

> Let us then, Sir, proceed with immediate tests without dwelling on causes; this is our shortest and most certain path; and after having tried everything, we will retain what appears to us the best solution. I hope that we will finally find them, by all the pains that you wish to take. (De Luc, in Jungnickel and McCormmach 1999, 550)

Thus, the use of steam enabled the Royal Society committee to obviate divisive and crippling disputes about theories of boiling. It gave a clear operational procedure that served well enough to define an empirically fixed point, though there was no agreed understanding of why steam coming off boiling water under a given pressure should have a fixed temperature.

A more decisive experimental confirmation of the steadiness of the steam temperature had to wait for sixty-five years until Marcet's work, in which it was shown that the temperature of steam was nearly at the standard boiling point regardless of the temperature of the boiling water from which it emerged. Even when the water temperature was over 105°C, the steam temperature was only a few

---

[31] See De Luc, in Jungnickel and McCormmach 1999, 549–550.
[32] See Cavendish et al. 1777, 832, 850–853, for reference to De Luc's previous practice.

tenths of a degree over 100°C (Marcet 1842, 404–405). That is still not negligible, but it was a reassuring result given that it was obtained with serious superheating in the boiling water. The situation was much ameliorated by the employment of all the various means of converting superheated boiling into common boiling. If these techniques could bring the water temperature down fairly close to 100°C, then the steam temperature would be reliably fixed at 100°C or very near it. Marcet's work closed a chapter in the history of superheating in which it posed a real threat to the fixity of the boiling point, although it did not address the question of whether steam could cool down below the boiling point.

The Understanding of Boiling

The "steam point" proved its robustness. After the challenge of superheating was overcome in the middle of the nineteenth century, the fixity (or rather, the fixability) of the steam point did not come into any serious doubt.[33] The remaining difficulty now was in making sense of the empirically demonstrated fixability of the steam point. This task of understanding was a challenge that interested all but the most positivistic physicists. From my incomplete presentation in the last two sections, the theoretical situation regarding the boiling point would appear to be discordant and chaotic. However, a more careful look reveals a distinct course of advancement on the theoretical side that dovetailed neatly with the practical use of the steam point. Whether this advancement was quite sufficient for an adequate understanding of the phenomenon of boiling and the fixability of the steam point is the question that I will attempt to answer in this section.

Let us start with a very basic question. For anything deserving the name of "boiling" to take place, vapor should form within the body of the liquid water and move out through the liquid. But why should this happen at anything like a fixed temperature? The crucial factor is the relation between the pressure and the temperature of water vapor. Suppose we let a body of water evaporate into an enclosed space as much as possible. In the setup shown in figure 1.4 (left), a small amount of water rests on a column of mercury in a barometer-like inverted glass tube and evaporates into the vacuum above the mercury until it cannot evaporate any more. Then the space is said to be "saturated" with vapor; similarly, if such a maximum evaporation would occur into an enclosed space containing air, the air is said to be saturated. Perhaps more confusingly, it is also said that the vapor itself is saturated, under those circumstances.

It was observed already in the mid-eighteenth century that the density of saturated vapor is such that the pressure exerted by it has a definite value determined by temperature, and temperature only. If one allows more space after saturation is obtained (for instance by lifting the inverted test tube a bit higher), then just enough additional vapor is produced to maintain the same pressure as before; if one reduces the space, enough vapor turns back into water so that the

---

[33]Marcet already stated in 1842 that the steam point was "universally" accepted, and the same assessment can be found at the end of the century, for example in Preston 1904, 105.

 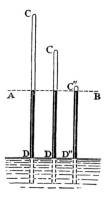

FIGURE 1.4. Experiments illustrating the pressure of saturated vapor (Preston 1904, 395).

vapor pressure again remains the same (see fig. 1.4 [right]). But if the temperature is raised, more vapor per available space is produced, resulting in a higher vapor pressure. It was in fact Lord Charles Cavendish (1704–1783), Henry's father, who first designed the simple mercury-based equipment to show and measure vapor pressures, and the son fully endorsed the father's results and assigned much theoretical significance to them as well.[34] Cavendish's discovery of the exclusive dependence of vapor pressure on temperature was later confirmed by numerous illustrious observers including James Watt (1736–1819), John Dalton (1766–1844), and Victor Regnault (1810–1878). Table 1.2 shows some of the vapor-pressure data obtained by various observers.

As seen in the table, the pressure of saturated vapor ("vapor pressure" from now on, for convenience) is equal to the normal atmospheric pressure when the temperature is 100°C. That observation provided the basic theoretical idea for a causal understanding of boiling: boiling takes place when the water produces vapor with sufficient pressure to overcome the resistance of the external atmosphere.[35] This view gave a natural explanation for the pressure-dependence of the boiling point. It also provided perfect justification for the use of steam temperature to define the boiling point, since the key relation underlying the fixity of that point is the one between the temperature and pressure of saturated steam.

This view, which I will call the *pressure-balance theory of boiling*, was a powerful and attractive theoretical framework. Still, there was a lot of "mopping up" or "anomaly busting" left to do (to borrow liberally from Thomas Kuhn's description of "normal science"). The first great anomaly for the pressure-balance theory of boiling was the fact that the boiling temperature was plainly not fixed even when the external pressure was fixed. The typical and reasonable thing to do was to

---

[34]See Cavendish [n.d.] 1921, 355, and also Jungnickel and McCormmach 1999, 127.
[35]This idea was also harmonious with Antoine Lavoisier's view that a liquid was only prevented from flying off into a gaseous state by the force of the surrounding atmosphere. See Lavoisier [1789] 1965, 7–8.

TABLE 1.2. A comparative table of vapor-pressure measurements for water

Temperature	C. Cavendish (c.1757)[a] Vapor pressure[e]	Dalton (1802)[b] Vapor pressure	Biot (1816)[c] Vapor pressure	Regnault (1847)[d] Vapor pressure
35°F (1.67°C)	0.20 in. Hg	0.221		0.20
40	0.24	0.263		0.25
45	0.28	0.316		0.30
50 (10°C)	0.33	0.375		0.3608
55	0.41	0.443		0.43
60	0.49	0.524		0.52
65	0.58	0.616		0.62
70	0.70	0.721		0.73
75	0.84	0.851		0.87
86 (30°C)		1.21	1.2064	1.2420
104 (40°C)		2.11	2.0865	2.1617
122 (50°C)		3.50	3.4938	3.6213
140 (60°C)		5.74	5.6593	5.8579
158 (70°C)		9.02	9.0185	8.81508
176 (80°C)		13.92	13.861	13.9623
194 (90°C)		20.77	20.680	20.6870
212 (100°C)		30.00	29.921	29.9213
302 (150°C)		114.15		140.993
392 (200°C)				460.1953
446 (230°C)				823.8740

[a]These data are taken from Cavendish [n.d.] 1921, 355, editor's footnote.
[b]Dalton 1802a, 559–563. The last point (for 302°F or 150°C) was obtained by extrapolation.
[c]Biot 1816, 1:531. The French data (Biot's and Regnault's) were in centigrade temperatures and millimeters of mercury. I have converted the pressure data into English inches at the rate of 25.4 mm per inch.
[d]Regnault 1847, 624–626. The entries for Regnault in the 35°–75°F range are approximate conversions (except at 50°F), since his data were taken at each centigrade, not Fahrenheit, degree.
[e]All of the vapor pressure data in this table indicate the height (in English inches) of a column of mercury balanced by the vapor.

postulate, and then try to identify the existence of interfering factors preventing the "normal" operation of the pressure-balance mechanism. An alternate viewpoint was that the matching of the vapor pressure with the external pressure was a necessary, but not sufficient condition for boiling, so other facilitating factors had to be present in order for boiling to occur. The two points of view were in fact quite compatible with each other, and they were used interchangeably sometimes even by a single author: saying that factor x was necessary to enable boiling came to the same thing in practice as saying that the absence of x prevented boiling. There were various competing ideas about the operation of these facilitating or preventative factors. Let us see if any of these auxiliary ideas were truly successful in defending the pressure-balance theory of boiling, thereby providing a theoretical justification for the use of the steam point.

Gay-Lussac (1818, 130) theorized that boiling would be retarded by the adhesion of water to the vessel in which it is heated and also by the cohesion of water within itself. The "adhesion of the fluid to the vessel may be considered as analogous to its viscidity.... The cohesion or viscosity of a fluid must have a considerable effect for its boiling point, for the vapor which is formed in the interior of a fluid has two forces to overcome; the pressure upon its surface, and the cohesion of the particles." Therefore "the interior portions may acquire a greater degree of heat than the real boiling point," and the extra degree of heat acquired will also be greater if the vessel has stronger surface adhesion for water. Gay-Lussac inferred that the reason water boiled "with more difficulty" in a glass vessel than in a metallic one must be because there were stronger adhesive forces between glass and water than between metal and water. Boiling was now seen as a thoroughly sticky phenomenon. The stickiness is easier to visualize if we think of the boiling of a thick sauce and allow that water also has some degree of viscosity within itself and adhesiveness to certain solid surfaces.

Twenty-five years later Marcet (1842, 388–390) tested the adhesion hypothesis more rigorously. First he predicted that throwing in bits of metal into a glass vessel of boiling water would lower the boiling temperature, but not as far down as 100°C, which is where water boils when the vessel is entirely made of metal. This prediction was borne out in his tests, since the lowest boiling temperature he could ever obtain with the insertion of metal pieces was 100.2°C, contrary to Gay-Lussac's earlier claim that it went down to 100°C exactly. More significantly, Marcet predicted that if the inside of the vessel could be coated with a material that has even less adhesion to water than metals do, the boiling temperature would go down below 100°C. Again as predicted, Marcet achieved boiling at 99.85°C in a glass vessel scattered with drops of sulphur. When the bottom and sides of the vessel were covered with a thin layer of *gomme laque*, boiling took place at 99.7°C. Although 0.3° is not a huge amount, Marcet felt that he had detected a definite error in previous thermometry, which had fixed the boiling point at the temperature of water boiling in a metallic vessel:

> It is apparent that previous investigators have been mistaken in assuming that under given atmospheric pressure, water boiling in a metallic vessel had the lowest possible temperatures, because in some cases the temperature could be lowered for a further 0.3 degrees. It is, however, on the basis of that fact, generally assumed to be exactly true, that physicists made a choice of the temperature of water boiling in a metallic vessel as one of the fixed points of the thermometric scale. (Marcet 1842, 391)

Finally, it seemed, theoretical understanding had reached a point where it could lead to a refinement in existing practices, going beyond their retrospective justification.

Marcet's beautiful confirmations seemed to show beyond any reasonable doubt the correctness of the pressure-balance theory modified by the adhesion hypothesis. However, two decades later Dufour (1861, 254–255) voiced strong dissent on the role of adhesion. Since he observed extreme superheating of water drops removed from solid surfaces by suspension in other liquids, he argued that simple adhesion to solid surfaces could not be the main cause of superheating. Instead Dufour stressed the importance of the ill-understood molecular actions at the point of contact between water and other substances:

> For example, if water is completely isolated from solids, it always exceeds 100°C before turning into vapor. It seems to me beyond doubt that heat alone, acting on

water without the joint action of alien molecules, can only produce its change of state well beyond what is considered the temperature of normal ebullition.

Dufour's notion was that the production of vapor would only take place when a sort of equilibrium that maintains the liquid state was broken. Boiling was made possible at the point of pressure balance, but some further factor was required for the breaking of equilibrium, unstable as it may be. Heat alone could serve as the further facilitating factor, but only at a much higher degree than the normal boiling point. Dufour also made the rather subtle point that the vapor pressure itself could not be a cause of vapor production, since the vapor pressure was only a property of "future vapor," which did not yet exist before boiling actually set in. Dufour's critique was cogent, but he did not get very far in advancing an alternative. He was very frank in admitting that there was insufficient understanding of the molecular forces involved.[36] Therefore the principal effect of his work was to demolish the adhesion hypothesis without putting in a firm positive alternative.

There were two main attempts to fill this theoretical vacuum. One was a revival of Cavendish's and De Luc's ideas about the importance of open surfaces in enabling a liquid to boil. According to Cavendish's "first principle of boiling," the conversion of water at the boiling point into steam was only assured if the water was in contact with air or vapor. And De Luc had noted that air bubbles in the interior of water would serve as sites of vapor production. For De Luc this phenomenon was an annoying deviation from true boiling, but it came to be regarded as the definitive state of boiling in the new theoretical framework, which I am about to explain in further detail.

One crucial step in this development was taken by Verdet, whose work was discussed briefly in the last section. Following the basic pressure-balance theory, he defined the "normal" point of boiling as the temperature at which the vapor pressure was equal to the external pressure, agreeing with Dufour that at that temperature boiling was made "possible, but not necessary." Accepting Dufour's view that contact with a solid surface was a key factor promoting ebullition, Verdet also made an interesting attempt to understand the action of solid surfaces along the Cavendish–De Luc line. He theorized, somewhat tentatively, that boiling was not provoked by all solid surfaces, but only by "unwettable" surfaces that also possessed microscopic roughness. On those surfaces, capillary repulsion around the points of irregularity would create small pockets of empty space, which could serve as sites of evaporation. There would be no air or steam in those spaces initially, but it seemed sensible that a vacuum should be able to serve the same role as gaseous spaces in enabling evaporation. If such an explanation were tenable, then not only Dufour's observations but all the observations that seemed to support the adhesion hypothesis could be accounted for.[37]

Verdet's idea was taken up more forcefully by Désiré-Jean-Baptiste Gernez (1834–1910), physical chemist in Paris, who was one of Louis Pasteur's "loyal

---

[36] For this admission, see the last pages of Dufour 1861, esp. 264.
[37] In the exposition of Verdet's view, I follow Gernez 1875, 351–353.

FIGURE 1.5. Gernez's instrument for introducing an open air surface into the interior of water. Courtesy of the British Library.

collaborators" and contributed to various undertakings ranging from crystallography to parasitic etiology.[38] In articles published in 1866 and 1875, Gernez reported that common boiling could always be induced in superheated water by the insertion of a trapped pocket of air into the liquid by means of the apparatus shown in figure 1.5. A tiny amount of air was sufficient for this purpose, since boiling tended to be self-perpetuating once it began. Gernez (1875, 338) thought that at least half a century had been wasted due to the neglect of De Luc's work: "[T]he explanation of the phenomenon of boiling that De Luc proposed was so clear and conformable to reality, that it is astonishing that it was not universally adopted."[39] In Gernez's view a full understanding of boiling could be achieved by a consistent and thorough application of De Luc's idea, a process initiated by Donny, Dufour, and Verdet among others. Donny (1846, 189) had given a new theoretical definition of boiling as *evaporation from interior surfaces*: "[B]oiling is nothing but a kind of extremely rapid evaporation that takes place at interior surfaces of a liquid that surrounds bubbles of a gas."

Gernez (1875, 376) took up Donny's definition, adding two refinements. First, he asserted that such boiling started at a definite temperature, which could be called "the point of normal ebullition." He added that the gaseous surfaces within the liquid could be introduced by hand or produced spontaneously by the disengagement of dissolved gases. (Here one should also allow a possible role of Verdet's

---

[38] The information about Gernez is taken from M. Prévost et al., eds., Dictionnaire de Biographie Française, vol. 15 (1982), and also from comments in G. Geison's entry on Pasteur in the Dictionary of Scientific Biography, 10:360, 373–374.
[39] In Gernez's view, the general rejection of De Luc's ideas had probably been prompted by De Luc's "unfortunate zeal" in opposing Lavoisier's chemistry.

empty spaces created by capillary forces and of internal gases produced by chemical reactions or electrolysis.[40]) Gernez's mopping up bolstered the pressure-balance theory of boiling quite sufficiently; the presence of internal gases was the crucial enabling condition for boiling, and together with the balance of pressure, it constituted a sufficient condition as well. The theoretical foundation of boiling now seemed quite secure.

There were, however, more twists to come in the theoretical debate on boiling. While Verdet and Gernez were busily demonstrating the role of gases, a contrary view was being developed by Charles Tomlinson (1808–1897) in London. Tomlinson believed that the crucial enabling factor in boiling was not gases, but small solid particles. Tomlinson's argument was based on some interesting experiments that he had carried out with superheated liquids. Building on previous observations that inserting solid objects into a superheated liquid could induce boiling, Tomlinson (1868–69, 243) showed that metallic objects lost their vapor-liberating power if they were chemically cleaned to remove all specks of dust. In order to argue conclusively against the role of air, he lowered a small cage made out of fine iron-wire gauze into a superheated liquid and showed that no boiling was induced as long as the metal was clean. The cage was full of air trapped inside, so Tomlinson inferred that there would have been visible production of vapor if air had really been the crucial factor. He declared: "It really does seem to me that too much importance has been attached to the presence of air and gases in water and other liquids as a necessary condition of their boiling" (246). Defying Dufour's warning against theorizing about boiling on the basis of the properties of "future vapor," Tomlinson started his discussion with the following "definition": "A liquid at or near the boiling-point is a supersaturated solution of its own vapour, constituted exactly like soda-water, Seltzer-water, champagne, and solutions of some soluble gases" (242). This conception allowed Tomlinson to make use of insights from his previous studies of supersaturated solutions.

Tomlinson's theory and experiments attracted a good deal of attention, and a controversy ensued. It is not clear to me whether and how this argument was resolved. As late as 1904, the second edition of Thomas Preston's well-informed textbook on heat reported: "The influence of dissolved air in facilitating ebullition is beyond question; but whether the action is directly due to the air itself or to particles of dust suspended in it, or to other impurities, does not seem to have been sufficiently determined" (Preston 1904, 362). Much of the direct empirical evidence cited by both sides was in fact ambiguous: ordinary air typically contained small solid particles; on the other hand, introducing solid particles into the interior of a liquid was likely to bring some air into it as well (as De Luc had noticed when he tried to insert thermometers into his air-free water). Some experiments were less ambiguous, but still not decisive. For example, Gernez acknowledged in his attack on Tomlinson that the latter's experiment with the wire-mesh cage would clearly be negative evidence regarding the role of air; however, he claimed that Tomlinson's result could not be trusted because it had not been replicated by anyone else. Like

---

[40]The latter effect was demonstrated by Dufour 1861, 246–249.

Dufour earlier, Gernez (1875, 354–357, 393) also scored a theoretical point by denigrating as unintelligible Tomlinson's concept of a liquid near boiling as a supersaturated solution of its own vapor, though he was happy to regard a superheated liquid as a supersaturated solution of air.

The Tomlinson-Gernez debate on the theory of boiling is fascinating to follow, but its details were in one clear sense not so important for the question we have been addressing, namely the understanding of the fixity of the steam point. Saturated vapor does obey the pressure–temperature relation, whatever the real cause of its production may be. Likewise, the exact method by which the vapor is produced is irrelevant as well. The pressure–temperature relation is all the same, whether the vapor is produced by steady common boiling, or by bumpy and unstable superheated boiling, or by an explosion, or by evaporation from the external surface alone. After a century of refinement, then, it became clear that *boiling itself was irrelevant to the definition or determination of the "boiling point."*

A Dusty Epilogue

If the determination of the boiling point hinged on the behavior of steam both theoretically and experimentally, we must consider some key points in the physics of steam, before closing this narrative. Most crucial was the definite relationship between the pressure and the temperature of saturated steam. After witnessing the tortuous debates in the apparently simple business of the boiling of water, would it be too rash to bet that there would have been headaches about the pressure–temperature relation of saturated steam, too?

Recall De Luc's worry that saturated steam might cool down below the temperature indicated by the pressure-temperature law without condensing back into water, despite Cavendish's assertion that it could not. In the particular setup adopted by the Royal Society committee for fixing the boiling point, this probably did not happen. However, in more general terms De Luc's worry was vindicated, a whole century later through the late nineteenth-century investigations into the "supersaturation" of steam. This story deserves some brief attention here, not only because supersaturation should have threatened the fixity of the steam point but also because some insights gained in those investigations threw still new light on the understanding of boiling and evaporation.

The most interesting pioneer in the study of supersaturation, for our purposes, was the Scottish meteorologist John Aitken (1839–1919). His work has received some attention from historians of science, especially because it provided such an important stimulus for C. T. R. Wilson's invention of the cloud chamber. Aitken had trained as an engineer, but abandoned the career soon due to ill health and afterwards concentrated on scientific investigations, mostly with various instruments that he constructed himself. According to his biographer Cargill Knott, Aitken had "a mind keenly alive to all problems of a meteorological character," including the origin of dew, glacier motion, temperature measurement, the nature of odorous emanations, and the possible influence of comets on the earth's atmosphere. He was a "quiet, modest investigator" who refused to accept "any theory which seemed to him insufficiently supported by physical reasoning," and studied every problem "in

36 *Inventing Temperature*

his own way and by his own methods." These qualities, as we will see, are amply demonstrated in his work on steam and water.[41]

Aitken was explicit about the practical motivation for his study of supersaturated steam: to understand the various forms of the "cloudy condensations" in the atmosphere, particularly the fogs that were blighting the industrial towns of Victorian Britain (1880–81, 352). His main discovery was that steam could routinely be cooled below the temperature indicated by the standard pressure–temperature relation without condensing into water, if there was not sufficient dust around. Aitken showed this in a very simple experiment (338), in which he introduced invisible steam from a boiler into a large glass receptacle. If the receptacle was filled with "dusty air—that is, ordinary air," a large portion of the steam coming into it condensed into small water droplets due to the considerable cooling it suffered, resulting in a "dense white cloud." But if the receptacle was filled with air that had been passed through a filter of cotton wool, there was "no fogging whatever." He reckoned that the dust particles served as loci of condensation, one dust particle as the nucleus for each fog particle. The idea that dust was necessary for the formation of cloudy condensations obviously had broad implications for meteorology. To begin with:

> If there was no dust in the air there would be no fogs, no clouds, no mists, and probably no rain.... [W]e cannot tell whether the vapour in a perfectly pure atmosphere would ever condense to form rain; but if it did, the rain would fall from a nearly cloudless sky.... When the air got into the condition in which rain falls—that is, burdened with supersaturated vapour—it would convert everything on the surface of the earth into a condenser, on which it would deposit itself. Every blade of grass and every branch of tree would drip with moisture deposited by the passing air; our dresses would become wet and dripping, and umbrellas useless.... (342)

The implication of Aitken's discovery for the fixity of the steam point is clear to me, though it does not seem to have been emphasized at the time. If steam can easily be cooled down below the "steam point" (that is, the temperature at which the vapor pressure of saturated steam equals the external pressure), the steam point is no more fixed than the boiling point of liquid water. Moreover, what allows those points to be reasonably fixed in practice is precisely the same kind of circumstance: the "ordinary" conditions of our materials being full of impurities—whether they be air in water or dust in air. Cavendish was right in arguing that steam would not go supersaturated, but he was right only because he was always dealing with dusty air.[42] Now we can see that it was only some peculiar accidents of human life that gave the steam point its apparent fixity: air on earth is almost always dusty enough, and no one had thought to filter the air in the boiling-point apparatus. (This role of

---

[41] The fullest available account of Aitken's life and work is Knott 1923; all information in this paragraph is taken from that source, and the quoted passages are from xii–xiii. For an instructive discussion of Aitken's work in relation to the development of the cloud chamber, see Galison 1997, 92–97.

[42] What saved Cavendish could actually be the fact that in his setup there was always a water–steam surface present, but that raises another question. If a body of steam is in contact with water at one end, does that prevent supersaturation throughout the body of the steam, however large it is?

serendipity in the production of stable regularities will be discussed further in "The Defense of Fixity" in the analysis part of this chapter.)

Before I close the discussion of Aitken's work, it will be instructive to make a brief examination of his broader ideas. Not only do they give us a better understanding of his work on the supersaturation of steam but also they bear in interesting ways on the debates about boiling that I have discussed in previous sections. His work on the supersaturation of steam came from a general theoretical viewpoint about the "conditions under which water changes from one of its forms to another." There are four such changes of state that take place commonly: melting (solid to liquid), freezing (liquid to solid), evaporation (liquid to gas), and condensation (gas to liquid). Aitken's general viewpoint about changes of state led him to expect that steam must be capable of supersaturation, before he made any observations of the phenomenon.[43] In his own words:

> I knew that water could be cooled below the freezing-point without freezing. I was almost certain ice could be heated above the freezing-point without melting. I had shown that water could be heated above the boiling-point.... Arrived at this point, the presumption was very strong that water vapour could be cooled below the boiling-point... without condensing. It was on looking for some experimental illustration of the cooling of vapour in air below the temperature corresponding to the pressure that I thought that the dust in the air formed 'free surfaces' on which the vapour condensed and prevented it getting supersaturated. (Aitken 1880–81, 341–342)

Changes of state are caused by changes of temperature, but "something more than mere temperature is required to bring about these changes. Before the change can take place, a 'free surface' must be present." Aitken declared:

> When there is no 'free surface' in the water, we have at present no knowledge whatever as to the temperature at which these changes will take place.... Indeed, we are not certain that it is possible for these changes to take place at all, save in the presence of a "free surface." (339)

By a "free surface" he meant, somewhat tautologically, "a surface at which the water is free to change its condition." In an earlier article, Aitken (1878, 252) had argued that a free surface was formed between any liquid and any gas/vapor (or vacuum), which would seem to indicate that he thought the point of contact between any two different states of matter (solid, liquid, or gaseous) constituted a free surface enabling changes between the two states involved.[44] I am not aware whether Aitken ever developed of his concept of "free surface" in a precise way. As it turned out, the

---

[43] Aitken's first observations about the condensation of steam were made in the autumn of 1875. But he had already presented his theoretical paper on the subject, titled "On Boiling, Condensing, Freezing, and Melting," in July 1875 to the Royal Scottish Society of Arts (Aitken 1878).

[44] He also said: "[W]henever a liquid comes in contact with a solid or another liquid, a free surface is never formed." This can only mean that a solid–liquid interface cannot serve as a free surface for the transformation of the liquid into vapor (which would make sense, since the passage occurs in his discussion of boiling in particular). Compare Aitken's notion of free surface with Dufour's idea about the necessity of "alien molecules" for ordinary boiling.

exact mechanism by which dust particles facilitated the condensation of vapor was not a trivial issue, and its elucidation required much theoretical investigation, especially on the effect of surface curvature on vapor pressure.[45] It is unclear whether different kinds of "free surfaces" would have shared anything essential in the way they facilitated changes of state.

When applied to the case of boiling, Aitken's free-surface theory fitted very well with the De Luc–Donny–Dufour line of thought about the role of dissolved air in boiling, which he was quite familiar with. But his ideas did not necessarily go with the pressure-balance theory of boiling, and in fact Aitken actively rejected it: "The pressure itself has nothing to do with whether the water will pass into vapour or not" (1878, 242). Instead, he thought that what mattered for boiling was "the closeness with which the vapour molecules are packed into the space above the water." He redefined the "boiling point" as "the temperature at which evaporation takes place into an atmosphere of its own vapour at the standard atmospheric pressure of 29.905 inches of mercury." This definition is unusual, but may well be quite compatible with Cavendish's operational procedure adopted by the Royal Society committee for fixing the steam point. Aitken recognized that his definition of the boiling point did not require any "boiling" in the sense of vapor rising from within the body of the liquid:

> Where, then, it may be asked, is the difference between boiling and evaporation? None, according to this view. Boiling is evaporation in presence of only its own vapour; and what is usually called evaporation is boiling in presence of a gas. The mechanical bubbling up of the vapour through the liquid is an accident of the boiling.... [W]e may have no free surface in the body of the liquid, and no bubbles rising through it, and yet the liquid may be boiling. (Aitken 1878, 242)

Aitken was clearly working at the frontiers of knowledge. But the fruit of his labors, as far as the theory of boiling was concerned, was only an even more serious disorientation than produced by De Luc's pioneering work on the subject a century earlier.

At the end of his major article on dust and fogs, Aitken expressed his sense that he had only opened this whole subject:

> Much, very much, still remains to be done. Like a traveller who has landed in an unknown country, I am conscious my faltering steps have extended but little beyond the starting point. All around extends the unknown, and the distance is closed in by many an Alpine peak, whose slopes will require more vigorous steps than mine to surmount. It is with reluctance I am compelled for the present to abandon the investigation. (Aitken 1880–81, 368)

Well over a century after Aitken's humble pronouncement, we tend to be completely unaware that boiling, evaporation, and other such mundane phenomena ever constituted "many an Alpine peak" for science. Aitken lamented that he had only been able to take a few faltering steps, but the vast majority of us who have

---

[45]For further details, see Galison 1997, 98–99, and Preston 1904, 406–412.

received today's scientific education are entirely ignorant of even the existence of Aitken's unexplored country.[46]

Already in Aitken's own days, science had gone far enough down the road of specialization that even elementary knowledge became neglected if it did not bear explicitly on the subjects of specialist investigation. In an earlier article Aitken blasted some respectable authors for making inaccurate statements about the boiling and melting temperatures. After citing patently incorrect statements from such canonical texts as James Clerk Maxwell's *Theory of Heat* and John Tyndall's *Heat A Mode of Motion*, Aitken gave a diagnosis that speaks very much to the spirit of my own work:

> Now, I do not wish to place too much stress on statements like these given by such authorities, but would look on them simply as the current coin of scientific literature which have been put in circulation with the stamp of authority, and have been received and reissued by these writers without questioning their value. (Aitken 1878, 252)

Analysis: The Meaning and Achievement of Fixity

> Ring the bells that still can ring.
> Forget your perfect offering.
> There is a crack in everything.
> That's how the light gets in.
>
> Leonard Cohen, "Anthem,"
> 1992

In the preceding narrative, I gave a detailed historical account of the surprising practical and theoretical challenges involved in establishing one particular fixed point for thermometry. But fixing the fixed points was an even less straightforward business than it would have seemed from that narrative. For a fuller understanding of fixed points, some further discussions are necessary. In this part of the chapter, I will start in the first two sections with a philosophical consideration of what it really means for a phenomenon to constitute a fixed point, and how we can judge fixity at all in the absence of a pre-established standard of fixity. Philosophical concerns about standards of fixity did not come into the narrative significantly because the scientists themselves tended not to discuss them explicitly; however, the establishment of fixed points would not have been possible without some implicit standards being employed, which are worth elucidating. Once the standards are clearly identified, we can reconsider the business of assessing whether and to what extent

---

[46]There are, of course, some specialists who know that the boiling and freezing of water are very complicated phenomena and investigate them through sophisticated modern methods. For an accessible introduction to the specialist work, see Ball 1999, part 2.

certain phenomena are actually fixed in temperature. As indicated in the narrative, even the most popular fixed point (the boiling point or steam point) exhibited considerable variations. In "The Defense of Fixity" section, I will discuss the epistemic strategies that can be used in order to defend the fixity of a proposed fixed point, drawing from the salient points emerging in the boiling-point story. Finally, in "The Case of the Freezing Point," I will take a brief look at the freezing point, with the benefit of the broader and deeper perspective given by the foregoing discussions.

The Validation of Standards: Justificatory Descent

Consider, in the abstract, the task of someone who has to come up with a fixed point where none have yet been established. That is not so different from the plight of a being who is hurled into interstellar space and asked to identify what is at rest. Even aside from Einstein saying that there is no such thing as absolute rest, how would our space oddity even begin to make a judgment of what is moving and what is fixed? In order to tell whether something is fixed, one needs something else that is known to be fixed and can serve as a criterion of judgment. But how can one find that first fixed point? We would like to put some nails in the wall to hang things from, but there is actually no wall there yet. We would like to lay the foundations of a building, but there is no firm ground to put it in.

In the narrative, I paid little attention to the question of how it is that the fixity of a proposed fixed point can be assessed because the scientists themselves did not discuss that question extensively. However, a moment's philosophical reflection shows that there must be some independent standard of judgment, if one is going to say whether or not a given phenomenon happens at a fixed temperature. Otherwise all we can have is a chaotic situation in which each proposed fixed point declares itself fixed and all others variable if they do not agree with it. To overcome such chaos, we need a standard that is not directly based on the proposed fixed points themselves. But standards are not God-given. They must be justified and validated, too—but how? Are we stuck with an infinite regress in which one standard is validated by another, that one is validated by yet another, and so forth?

It is helpful to think this issue through by means of a concrete case. Recall Newton's supposed failing in using "blood heat" (human body temperature) as a fixed point of thermometry. It seems that the master instrument-maker Daniel Gabriel Fahrenheit (1686–1736) also used blood heat as one of his three fixed points. Now, we all know that the human body temperature is not constant, even in a given healthy body, which means that using it as a fixed point of thermometry is a mistake.[47] But

---

[47] The modern estimate is that there is variation of around 1°C in the external body temperature of healthy humans, the mean of which is about 37°C (or about 98.5°F); see Lafferty and Rowe 1994, 588. In 1871 the German physician Carl Wunderlich (1815–1877) asserted that in healthy persons the temperature ranged from 98.6 to 99.5°F; see Reiser 1993, 834. But according to Fahrenheit, it was only 96°F. The divergence in early estimates owes probably as much to the differences in the thermometers as to actual variations; that would explain the report in the *Encyclopaedia Britannica*, 3d ed. (1797), 18:500, that the human body temperature ranged from 92 to 99°F. Of course, the temperature also depends on where the thermometer is placed in the body.

how do we "know" that? What most of us in the twenty-first century do is go down to the shop and buy a good thermometer, but that would not have been a possible way of criticizing Newton or Fahrenheit. No such thermometers were available anywhere at that time. Our convenient standard thermometers could not come into being until after scientists settled on good fixed points by ruling out bad ones like blood heat. The issue here is how we exclude blood heat initially as a fixed point, not the vacuous one about how we can obtain measurements showing blood heat not to be fixed, using thermometers based on other fixed points.

The key to resolving this impasse is to recognize that it is sufficient to use a very primitive thermometer to know that blood heat is not fixed. For example, any sealed glass tube filled partway with just about any liquid will do, since most liquids expand with heat. That was the arrangement used by Halley in the experiments reported in his 1693 article discussed in "Blood, Butter, and Deep Cellars" (or one can use an inverted test tube or a flask containing air or some other gas, with the open end plunged into a liquid). To help our observations, some lines can be etched onto the tube, and some arbitrary numbers may be attached to the lines. Many of the early instruments were in fact of this primitive type. These instruments should be carefully distinguished from thermometers as we know them, since they are not graduated by any principles that would give systematic meaning to their readings even when they are ostensibly quantitative. I will follow Middleton in dubbing such qualitative instruments *thermoscopes*, reserving the term *thermometer* for instruments with quantitative scales that are determined on some identifiable method.[48]

In the terminology of standard philosophical theories of measurement, what the thermoscope furnishes is an *ordinal* scale of temperature. An ordinal scale may have numbers attached to them, but those "numbers," or rather numerals, only indicate a definite ordering, and arithmetic operations such as addition do not apply meaningfully to them. In contrast, a proper thermometer is meant to give numbers for which some arithmetical operations yield meaningful results. However, not all arithmetical operations on temperature values yield meaningful results. For instance, a simple sum of the temperatures of two different objects is meaningless; on the other hand, if the temperatures are multiplied by heat capacity, then adding the products gives us total heat content. But this last example also reveals a further subtlety, as that arithmetical operation would be meaningful only for someone who accepts the concept of heat capacity (in the manner of Irvine, to be explained in "Theoretical Temperature before Thermodynamics" in chapter 4). In fact, there are complicated philosophical disputes about just what kind of quantity temperature is, which I will avoid for the time being by vaguely saying that what proper thermometers give us is a *numerical* temperature scale.[49]

---

[48]See Middleton 1966, 4. He thinks that Sanctorius was the first person to attach a meaningful numerical scale to the thermoscope, thereby turning it into a veritable thermometer.

[49]See, for example, Ellis 1968, 58–67, for some further discussion of the classification of scales with some consideration given to the case of temperature.

A thermoscope is exactly what is needed for the establishment of fixed points. It does not measure any numerical quantity, but it will indicate when something is warmer than another thing. If we put a thermoscope in our armpit or some other convenient place at regular intervals, we can observe the indications of the thermoscope fluctuating up and down. If we rule out (without much discussion for the moment) the unlikely possibility that the temperature is actually perfectly still while the thermoscope goes up and down in various ways, we can infer that blood heat is not constant and should not be used as a fixed point. The main epistemic point here is that a thermoscope does not even need to have any fixed points, so that the evaluation of fixed points can be made without a circular reliance on fixed points. Employing the thermoscope as a standard allows an initial evaluation of fixed points.

But we have only pushed the problem of justification one step away, and now we must ask why we can trust the thermoscopes. How do we know that most liquids expand with heat? The thermoscope itself is useless for the proof of that point. We need to bring ourselves back to the original situation in which there were no thermoscopes to rely on. The initial grounding of the thermoscope is in unaided and unquantified human sensation. Although sensation cannot prove the general rule that liquids expand when temperature rises, it does provide a justification of sorts. We get the idea that liquids expand with heat because that is what we observe in the cases that are most obvious to the senses. For example, we put a warm hand on a thermoscope that feels quite cool to the touch and see it gradually rise. We stick the thermoscope into water that scalds our hand and note its rapid rise. We put it into snow and see it plunge. We wet the thermoscope bulb and observe it go down when we blow on it, while remembering that we feel colder if we stand in the wind after getting soaked in the rain. What we see here is that human sensation serves as a prior standard for thermoscopes. The thermoscope's basic agreement with the indications of our senses generates initial confidence in its reliability. If there were clear and persistent disagreements between the indications of a thermoscope and what our own senses tell us, the thermoscope would become subject to doubt. For example, many of us would feel colder when it is 0°C out than when it is 4°C, and question the reliability of the volume changes in water as a standard of temperature when we see it occupy a larger volume at 0°C than at 4°C.

But have we not once again simply pushed the problem around? How do we know that we can trust sensation? From ancient times, philosophers have been well aware that there is no absolute reason for which we should trust our senses. That brings us to the familiar end of foundationalist justification, unsatisfying but inevitable in the context of empirical science. As Ludwig Wittgenstein (1969, 33, §253) puts it: "At the foundation of well-founded belief lies belief that is not founded." The groundlessness cannot be contained: if we follow through the empirical justifications we give for our beliefs, as we have done in the case of thermometric fixed points, we arrive at bodily sensation; if that final basis of justification is seen as untrustworthy, then none of our empirical justifications can be trusted. If we accept that sensations themselves have no firm justification, then we have to reconceptualize the very notion of empirical justification. The traditional notion of justification aspires to the ideal of logical proof, in which the proposition we want to justify is deduced from a set of previously accepted propositions. The trouble arises from the fact that proof only generates a demand for further proof of the previously accepted propositions. That results either in an infinite regress or an unsatisfying stopping point where belief is demanded without proof. But, as the next section will make clearer, justification does not have to consist in proof, and it is not likely that the justification of a standard will consist in proof.

I would like to propose that the justification of a standard is based on a *principle of respect*, which will be shown in action elsewhere in this book, too, and which I will have occasion to define more precisely later. To see how the principle of respect works, consider generally the relationship between human sensation and measuring instruments. Although basic measuring instruments are initially justified through their conformity to sensation, we also allow instruments to augment and even correct sensation. In other words, our use of instruments is made with a respect for sensation as a prior standard, but that does not mean that the verdict of sensation has unconditional authority. There are commonly cited cases to show that sometimes the only reasonable thing to do is to overrule sensations. Put one hand in a bucket of hot water and the other one in cold water; after a while take them out and put them both in a bucket of lukewarm water; one hand feels that the water is cool, and the other one feels it is warm. Our thermoscopes, however, confirm that the temperature of the last bucket of water is quite uniform, not drastically different from one spot to another.[50] Still, these cases do not lead us to reject the evidence of the senses categorically. Why is it that the general authority of sensation is maintained in spite of the acknowledged cases in which it fails?

The reason we accept sensation as a prior standard is precisely because it is *prior* to other standards, not because it has stronger justification than other standards. There is room for the later standard to depart from the prior standard, since the authority of the prior standard is not absolute. But then why should the prior standard be respected at all? In many cases it would be because the prior standard had some recognizable merits shown in its uses, though not foolproof justification. But ultimately it is because we do not have any plausible alternative. As Wittgenstein says, no cognitive activity, not even the act of doubting, can start without first believing something. "Belief" may be an inappropriate term to use here, but it would be safe enough to say that we have to start by accepting and using the most familiar accepted truths. Do I know that the earth is hundreds of years old? Do I know that all human beings have parents? Do I know that I really have two hands? Although those basic propositions lack absolute proof, it is futile to ask for a justification for them. Wittgenstein (1969, 33, §250) notes in his reflections on G. E. Moore's anti-skeptical arguments: "My having two hands is, in normal circumstances, as certain as anything that I could produce in evidence for it." Trusting sensation is the same kind of acceptance.

---

[50]See, for instance, Mach [1900] 1986, §2, for a discussion of this widely cited case.

Exactly what kind of relationship is forged between a prior standard and a later standard, if we follow the principle of respect? It is not a simple logical relation of any kind. The later standard is neither deduced from the prior one, nor derived from it by simple induction or generalization. It is not even a relation of strict consistency, since the later standard can contradict the earlier one. The constraint on the later standard is that it should show sufficient agreement with the prior standard. But what does "sufficient" mean? It would be wrong to try to specify a precise and preset degree of agreement that would count as sufficient. Instead, "sufficient" should be understood as an indication of intent, to respect the prior standard as far as it is plausible to do so. All of this is terribly vague. What the vagueness indicates is that the notion of justification is not rich enough to capture in a satisfactory way what is going on in the process of improving standards. In the following section I will attempt to show that the whole matter can be viewed in a more instructive light if we stop looking for a static logical relation of justification, and instead try to identify a dynamic process of knowledge-building.

The Iterative Improvement of Standards: Constructive Ascent

In the last section I was engaged in a quest of justification, starting with the accepted fixed points and digging down through the layers of grounds on which we accept their fixity. Now I want to explore the relations between the successive temperature standards from the opposite direction as it were, starting with the primitive world of sensation and tracing the gradual building-up of successive standards. This study will provide a preliminary glimpse of the process of multi-stage iteration, through which scientific knowledge can continue to build on itself. If a thermoscope can correct our sensation of hot and cold, then we have a paradoxical situation in which the derivative standard corrects the prior standard in which it is grounded. At first glance this process seems like self-contradiction, but on more careful reflection it will emerge as self-correction, or more broadly, self-improvement.

I have argued that the key to the relation between prior and later standards was the principle of respect. But respect only illuminates one aspect of the relation. If we are seeking to create a new standard, not content to rest with the old one, that is because we want to do something that cannot be achieved by means of the old standard. Respect is only the primary constraint, not the driving force. The positive motivation for change is an *imperative of progress*. Progress can mean any number of things (as I will discuss in more general terms in chapter 5), but when it comes to the improvement of standards there are a few obvious aspects we desire: the consistency of judgments reached by means of the standard under consideration, the precision and confidence with which the judgments can be made, and the scope of the phenomena to which the standard can be applied.

Progress comes to mean a spiral of self-improvement if it is achieved while observing the principle of respect. Investigations based on the prior standard can result in the creation of a new standard that improves upon the prior standard. Self-improvement is possible only because the principle of respect does not demand that the old standard should determine everything. Liberality in respect creates the breathing space for progress. The process of self-improvement arising from the dialectic between respect and progress might be called bootstrapping, but I will not use that term for fear of confusion with other well-known uses of it.[51] Instead I will speak of "iteration," especially with the possibility in mind that the process can continue through many stages.

Iteration is a notion that originates from mathematics, where it is defined as "a problem-solving or computational method in which a succession of approximations, each building on the one preceding, is used to achieve a desired degree of accuracy."[52] Iteration, now a staple technique for computational methods of problem-solving, has long been an inspiration for philosophers.[53] For instance, Charles Sanders Peirce pulled out an iterative algorithm for calculating the cube root of 2 (see fig. 1.6) when he wanted to illustrate his thesis that good reasoning corrects itself. About such processes he observed:

> Certain methods of mathematical computation correct themselves, so that if an error be committed, it is only necessary to keep right on, and it will be corrected in the end.... This calls to mind one of the most wonderful features of reasoning and one of the most important philosophemes [sic] in the doctrine of science, of which, however, you will search in vain for any mention in any book I can think of; namely, that reasoning tends to correct itself, and the more so, the more wisely its plan is laid. *Nay, it not only corrects its conclusions, it even corrects its premises.*... [W]ere every probable inference less certain than its premises, science, which piles inference upon inference, often quite deeply, would soon be in a bad way. Every astronomer, however, is familiar with the fact that the catalogue place of a fundamental star, which is the result of elaborate reasoning, is far more accurate than any of the observations from which it was deduced. (Peirce [1898] 1934, 399–400; emphasis added)

Following the spirit of Peirce's metaphorical leap from mathematical algorithm to reasoning in general, I propose a broadened notion of iteration, which I will call "epistemic iteration" as opposed to mathematical iteration. Epistemic iteration is a process in which successive stages of knowledge, each building on the preceding one, are created in order to enhance the achievement of certain epistemic goals. It differs crucially from mathematical iteration in that the latter is used to approach the correct answer that is known, or at least in principle knowable, by other means. In epistemic iteration that is not so clearly the case.

---

[51]The most prominent example is Glymour 1980, in which bootstrapping indicates a particular mode of theory testing, rather than a more substantial process of knowledge creation.

[52]*Random House Webster's College Dictionary* (New York: Random House, 2000). Similarly, the 2d edition of the *Oxford English Dictionary* gives the following definition: "Math. The repetition of an operation upon its product...esp. the repeated application of a formula devised to provide a closer approximation to the solution of a given equation when an approximate solution is substituted in the formula, so that a series of successively closer approximations may be obtained."

[53]For an introduction to modern computational methods of iteration, see Press et al. 1988, 49–51, 256–258, etc. See Laudan 1973 for further discussion of the history of ideas about self-correction and the place of iteration in that history.

Correct Computation	Sum of Two	Triple	*Erroneous* Computation	Sum of Two	Triple
1				1		
0				0		
1	1	3		1	1	3
4	5	15		4	5	15
15	19	57	*Error!* 16	20	60
58	73	219	61	77	231
223	281	843	235	296	888
858	1081	3243	904	1139	3417
3301	4159	12477	3478	4382	13146
12700			13381		
$1\frac{3301}{12700}$	1.2599213		$1\frac{3478}{13381}$	1.2599208	

Error   +.0000002   Error   −.0000002

FIGURE 1.6. The iterative algorithm for computing the cube root of 2, illustrated by Charles Sanders Peirce ([1898] 1934, 399). Reprinted by permission of The Belknap Press of Harvard University Press, Copyright © 1934, 1962 by the President and Fellows of Harvard College.

Another difference to note is that a given process of mathematical iteration relies on a single algorithm to produce all successive approximations from a given initial conjecture, while such a set mechanism is not always available in a process of epistemic iteration. Rather, epistemic iteration is most likely a process of creative evolution; in each step, the later stage is based on the earlier stage, but cannot be deduced from it in any straightforward sense. Each link is based on the principle of respect and the imperative of progress, and the whole chain exhibits innovative progress within a continuous tradition.

Certain realists would probably insist on having truth as the designated goal of a process like epistemic iteration, but I would prefer to allow a multiplicity of epistemic goals, at least to begin with. There are very few actual cases in which we could be confident that we are approaching "the truth" by epistemic iteration. Other objectives are easier to achieve, and the degree of their achievement is easier to assess. I will discuss that matter in more detail in chapter 5. Meanwhile, for my present purposes it is sufficient to grant that certain values aside from truth can provide the guiding objectives and criteria for iterative progress, whether or not those values contribute ultimately to the achievement of truth.

TABLE 1.3. Stages in the iterative development of thermometric standards

Period and relevant scientists	Standard
From earliest times	Stage 1: Bodily sensation
Early seventeenth century: Galileo, etc.	Stage 2: Thermoscopes using the expansion of fluids
Late seventeenth to mid-eighteenth century: Eschinardi, Renaldini, Celsius, De Luc, etc.	Stage 3a: Numerical thermometers based on freezing and boiling of water as fixed points
Late eighteenth century: Cavendish, The Royal Society committee, etc.	Stage 3b: Numerical thermometers as above, with the boiling point replaced by the steam point

Epistemic iteration will be a recurring theme in later chapters and will be discussed in its full generality in chapter 5. For now, I will only discuss the iteration of standards in early thermometry. Table 1.3 summarizes the stages in the iterative development of temperature standards that we have examined in this chapter.

Stage 1. The first stage in our iterative chain of temperature standards was the bodily sensation of hot and cold. The basic validity of sensation has to be assumed at the outset because we have no other plausible starting place for gaining empirical knowledge. This does not mean that "uneducated" perception is free of theories or assumptions; as numerous developmental and cognitive psychologists since Jean Piaget have stressed, everyday perception is a complex affair that is only learned gradually. Still, it is our starting point in the building of scientific knowledge. After Edmund Husserl (1970, 110–111, §28), we might say that we take the "life world" for granted in the process of constructing the "scientific world."

Stage 2. Building on the commonly observed correlation between sensations of hot and cold and changes in the volume of fluids, the next standard was created: thermoscopes. Thus, thermoscopes were initially grounded in sensations, but they improved the quality of observations by allowing a more assured and more consistent ordering of a larger range of phenomena by temperature. The coherence and usefulness of thermoscope readings constituted an independent source of validation for thermoscopes, in addition to their initial grounding in sensations.

Stage 3a. Once thermoscopes were established, they allowed sensible judgments about which phenomena were sufficiently constant in temperature to serve as fixed points. With fixed points and the division of the interval between them, it became possible to construct a numerical scale of temperature, which then constituted the next standard in the iterative chain. Numerical thermometers, when successfully constructed (see "The Achievement of Observability, by Stages" in chapter 2 for a further discussion of the meaning of "success" here), constituted an improvement upon thermoscopes because they allowed a true quantification of temperature. By means of numerical thermometers, meaningful calculations involving temperature and heat could be made and thermometric observations became possible subjects for mathematical theorizing. Where such theorizing was successful, that constituted another source of validation for the new numerical thermometric standard.[54]

Stage 3b. The boiling point was not as fixed as it had initially appeared (and nor was the freezing point, as I will discuss in "The Case of the Freezing Point"). There were reasonably successful strategies for stabilizing the boiling point, as I will discuss further in "The Defense of Fixity," but one could also try to come up with better fixed points. In fact scientists did find one to replace the boiling point: the steam point, namely the temperature of boiled-off steam, or more precisely, the temperature at which the pressure of saturated steam is equal to the standard atmospheric pressure. The new numerical thermometer using the "steam point" as the upper fixed point constituted an improved temperature standard. Aside from being more fixed than the boiling point, the steam point had the advantage of further theoretical support, both in the temperature-pressure law of saturated steam and in the pressure-balance theory of boiling (see "The Understanding of Boiling" in the narrative). The relation between this new thermometer and the older one employing the boiling point is interesting: although they appeared in succession, they were not successive iterative stages. Rather, they were competing iterative improvements on the thermoscopic standard (stage 2). This can be seen more clearly if we recognize that the steam-point scale (stage 3b) could have been obtained without there having been the boiling-point scale (stage 3a).

The Defense of Fixity: Plausible Denial and Serendipitous Robustness

In the last two sections I have discussed how it was possible at all to assess and establish fixed points. Now I want to address the question of how fixed points are actually created, going back to the case of the boiling point in particular. The popular early judgment reached by means of thermoscopes was that the boiling temperature of water was fixed. That turned out to be quite erroneous, as I have

---

[54] A few qualifications should be made here about the actual historical development of stage 3a, since it was not as neat and tidy as just summarized. My discussion has focused only on numerical thermometers using two fixed points, but there were a number of thermometers using only one fixed point (as explained in note 3) and occasional ones using more than two. The basic philosophical points about fixed points, however, do not depend on the number of fixed points used. Between stage 2 and stage 3a there was a rather long period in which various proposed fixed points were in contention, as already noted. There also seem to have been some attempts to establish fixed points without careful thermoscopic studies; these amounted to trying to jump directly from stage 1 to stage 3 and were generally not successful.

shown in "The Vexatious Variations of the Boiling Point" and "Superheating and the Mirage of True Ebullition." The interesting irony is that the fixity of the boiling point was most clearly denied by the precise numerical thermometers that were constructed on the very assumption of the fixity of the boiling point. Quite inconveniently, the world turned out to be much messier than scientists would have liked, and they had no choice but to settle on the most plausible candidates for fixed points, and then to defend their fixity as much as possible. If defensible fixed points do not occur naturally, they must be manufactured. I do not, of course, mean that we should simply pretend that certain points are fixed when they are not. What we need to do is find, or create, clearly identifiable material circumstances under which the fixity does hold.[55] If that can be done, we can deny the variations plausibly. That is to say, when we say things like "Water always boils at 100°C," that would still be false, but it would be a plausible denial of facts because it would become correct enough once we specify the exceptional circumstances in which it is not true.

There are various epistemic strategies in this defense of fixity. A few commonsensical ones can be gleaned immediately from the boiling-point episode discussed in the narrative.

1. If causes of variation can be eliminated easily, eliminate them. From early on it was known that many solid impurities dissolved in water raised its boiling temperature; the sensible remedy there was to use only distilled water for fixing the boiling point. Another well-known cause of variation in the boiling temperature was the variation in atmospheric pressure; one remedy there was to agree on a standard pressure. There were also a host of causes of variation that were not so clearly understood theoretically. However, at the empirical level what to do to eliminate variation was clear: use metal containers instead of glass; if glass is used, do not clean it too drastically, especially not with sulphuric acid; do not purge the dissolved air out of the water; and so on.
2. If causes of variation cannot be easily eliminated but can be identified and quantified, learn to make corrections. Since it can be tedious to wait for the atmosphere to reach exactly the standard pressure, it was more convenient to create a formula for making corrections for different pressure values. Similarly, the Royal Society committee adopted an empirical formula for correcting the boiling point depending on the depth of water below which the thermometer bulb was plunged. The correction formulas allowed the variations to be handled in a controlled way, though they were not eliminated.
3. Ignore small, inexplicable variations, and hope that they will go away. Perhaps the most significant case in this vein was the variation of temperature depending on the "degree of boiling," which was widely

---

[55]Fixed points can be artificially created in the same way seedless watermelons can be created; these things cannot be made if nature will not allow them, but nonetheless they are our creations.

reported by reputable observers such as Newton, Adams, and De Luc. But somehow this effect was no longer observed from the nineteenth century onward. Henry Carrington Bolton (1900, 60) noticed this curious fact and tried to explain away the earlier observations as follows: "We now know that such fluctuations depend upon the position of the thermometer (which must not be immersed in the liquid), on the pressure of the atmosphere, on the chemical purity of the water, and on the shape of the vessel holding it, so it is not surprising that doubts existed as to the constancy of the phenomenon [boiling temperature]." That explanation is not convincing in my view, as it amounts to sweeping the dirt under the rug of other causes of variation.[56] Other mysterious variations also disappeared in time, including the inexplicable day-to-day variations and the temperature differences depending on the depth of water below the thermometer bulb, both reported by the Royal Society committee. In all those cases the reported variations fell away, for no obvious reason. If the variations do disappear, there is little motivation to worry about them.

In his insightful article on the principles of thermometry, Carl B. Boyer remarked (1942, 180): "Nature has most generously put at the scientist's disposal a great many natural points which serve to mark temperatures with great accuracy." (That is reminiscent of the cryptic maxim from Wittgenstein [1969, 66, §505]: "It is always by favour of Nature that one knows something.") We have now seen that nature has perhaps not been as generous as Boyer imagined, since the fixed points were quite hard won. However, nature had to be generous in some ways if scientists were going to have any success in their attempts to tame the hopeless variations. In "Escape from Superheating" and "A Dusty Epilogue," I have already indicated that it was good fortune, or rather serendipity, that allowed the higher-than-expected degrees of stability that the boiling point and the steam point enjoyed. Serendipity is different from random instances of plain dumb luck because it is defined as "the *faculty* of making fortunate discoveries by accident." The term was coined by Horace Walpole (British writer and son of statesman Robert) from the Persian tale *The Three Princes of Serendip*, in which the heroes possessed this gift.[57] In the following discussion I will use it to indicate a lucky coincidence that results in a tendency to produce desirable results.

The most important serendipitous factor for the fixity of the boiling point is the fact that water on earth normally contains a good deal of dissolved air. For the steam point, it is the fact that air on earth normally contains a good deal of suspended dust. It is interesting to consider how thermometry and the thermal sciences might have developed in an airless and dustless place. To get a sense of the contingency, recall Aitken's speculations about the meteorology of a dustless world,

---

[56]Still, Bolton deserves credit for trying—I know of no other attempts to explain the disappearance of this variation.

[57]*Collins English Dictionary* (London: HarperCollins, 1998).

quoted at more length in "A Dusty Epilogue": "[I]f there was no dust in the air there would be no fogs, no clouds, no mists, and probably no rain...." Possible worlds aside, I will concentrate here on the more sober task of elucidating the implications of the serendipity that we earthbound humans have actually been blessed with.

One crucial point about these serendipitous factors is that their workings are independent of the fortunes of high-level theory. There were a variety of theoretical reasons people gave for using the steam point. Cavendish thought that bubbles of steam rising through water would be cooled down exactly to the "boiling point" in the course of their ascent, due to the loss of heat to the surrounding water. He also postulated that the steam could not cool down below the boiling point once it was released. De Luc did not agree with any of that, but accepted the use of the steam point for phenomenological reasons. Biot (1816, 1:44–45) agreed with the use of the steam point, but for a different reason. He thought that the correct boiling point should be taken from the very top layer of boiling water, which would require the feat of holding a thermometer horizontally right at the top layer of boiling water; thankfully steam could be used instead, because the steam temperature should be equal to the water temperature at the top. Marcet (1842, 392) lamented the wide acceptance of the latter assumption, which was in fact incorrect according to the results of his own experiments. Never mind why—the steam point was almost universally adopted as the better fixed point and aided the establishment of reliable thermometry.

There were also different views about the real mechanism by which the air dissolved in water tended to promote boiling. There was even Tomlinson's view that air was only effective in liberating vapor because of the suspended dust particles in it (in which case dust would be doing a double duty of serendipity in this business). Whatever the theory, it remained a crucial and unchallenged fact that dissolved air did have the desirable effect. Likewise, there were difficult theoretical investigations going on around the turn of the century about the mechanism of vapor formation; regardless of the outcome of such investigations, for thermometry it was sufficient that dust did stabilize the steam point by preventing supersaturation.

Thanks to their independence from high-level theory, the serendipitously robust fixed points survived major changes of theory. The steam point began its established phase in the Royal Society committee's recommendation in 1777, when the caloric theory of heat was just being crafted. The acceptance of the steam point continued through the rise, elaboration, and fall of the caloric theory. It lasted through the phenomenological and then molecular-kinetic phases of thermodynamics. In the late nineteenth century Aitken's work would have created new awareness that it was important not to eliminate all the dust from the air in which the steam was kept, but the exact theoretical basis for that advice did not matter for thermometry. The steam point remained fixed while its theoretical interpretations and justifications changed around. The robustness of the fixed points provided stability to quantitative observations, even as the theoretical changes effected fundamental changes in the very meaning of those observations. The same numbers could remain, whatever they "really meant."

If this kind of robustness is shared in the bases of other basic measurements in the exact sciences, as I suspect it is, there would be significant implications for the persistence and accumulation of knowledge. Herbert Feigl emphasized that the stability of empirical science lies in the remarkable degree of robustness possessed by certain middle-level regularities, a robustness that neither sense-data nor high-level theories can claim: "I think that a relatively stable and approximately accurate basis—in the sense of testing ground—for the theories of the factual sciences is to be located not in individual observations, impressions, sense-data or the like, but rather in the empirical, experimental laws" (Feigl 1974, 8). For example, weight measurements using balances rely on Archimedes's law of the lever, and observations made with basic optical instruments rely on Snell's law of refraction. These laws, at least in the contexts of the measurements they enable, have not failed and have not been questioned for hundreds of years. The established fixed points of thermometry also embody just the sort of robustness that Feigl valued so much, and the insights we have gained in the study of the history of fixed points shed some light on how it is that middle-level regularities can be so robust.

Although it is now widely agreed that observations are indeed affected by the theories we hold, thanks to the well-known persuasive arguments to that effect by Thomas Kuhn, Paul Feyerabend, Mary Hesse, Norwood Russell Hanson, and even Karl Popper, as well as various empirical psychologists, we must take seriously Feigl's point that not all observations are affected in the same way by paradigm shifts or other kinds of major theoretical changes. No matter how drastically high-level theories change, some middle-level regularities may remain relatively unaffected, even when their deep theoretical meanings and interpretations change significantly. Observations underwritten by these robust regularities will also have a fair chance of remaining unchanged across revolutionary divides, and that is what we have seen in the case of the boiling/steam point of water.

The looseness of the link between high-level theories and middle-level regularities receives strong support in the more recent works by Nancy Cartwright (on fundamental vs. phenomenological laws), Peter Galison (on the "intercalation" of theory, experiment, and instrumentation), and Ian Hacking (experimental realities based on low-level causal regularities).[58] Regarding the link between the middle-level regularities and individual sense-observations, James Woodward and James Bogen's work on the distinction between data and phenomena reinforces Feigl's argument; stability is found in phenomena, not in the individual data points out of which we construct the phenomena. In the strategies for the plausible denial of variations in the boiling-point case, we have seen very concrete illustrations of how a middle-level regularity can be shielded from all the fickle variations found in individual observations. This discussion dovetails very nicely with one of Bogen and Woodward's illustrative examples, the melting point of lead, which is a stable phenomenon despite variations in the thermometer readings in individual trials of the experiments for its determination.[59]

---

[58] See Cartwright 1983, Galison 1997, Hacking 1983. I have shown similar looseness in energy measurements in quantum physics; see Chang 1995a.

[59] See Bogen and Woodward 1988; the melting point example is discussed on 308–310. See also Woodward 1989.

The Case of the Freezing Point

Before closing the discussion of fixed points, it will be useful to examine briefly the establishment of the other common fixed point, namely the freezing point of water. (This point was often conceived as the melting point of ice, but measuring or regulating the temperature of the interior of a solid block of ice was not an easy task in practice, so generally the thermometer was inserted into the liquid portion of an ice-water mixture in the process of freezing or melting.) Not only is the freezing-point story interesting in its own right but it also provides a useful comparison and contrast to the boiling-point case and contributes to the testing of the more general epistemological insights discussed in earlier sections. As I will discuss more carefully in "The Abstract and the Concrete" in chapter 5, the general insights were occasioned by the consideration of the boiling-point episode, but they of course do not gain much evidential support from that one case. The general ideas have yet to demonstrate their validity, both by further general considerations and by showing their ability to aid the understanding of other concrete cases. For the latter type of test, it makes sense to start with the freezing point, since there would be little hope for the generalities inspired by the boiling point if they did not even apply fruitfully to the other side of the centigrade scale.

There are some overt parallels between the histories of the boiling point and the freezing point. In both cases, the initial appearance of fixity was controverted by more careful observations, upon which various strategies were applied to defend the desired fixity. In both cases, understanding the effect of dissolved impurities contributed effectively in dispelling the doubts about fixity (this was perhaps an even more important factor for the freezing point than the boiling point). And the fixity of the freezing point was threatened by the phenomenon of supercooling, just as the fixity of the boiling point was threatened by superheating.[60] This phenomenon, in which a liquid at a temperature below its "normal" freezing temperature maintains its liquid form, was discovered in water by the early eighteenth century. Supercooling threatened to make a mockery of the freezing of water as a fixed point, since it seemed that one could only say, "pure water always freezes at 0°C, except when it doesn't."

It is not clear who first noticed the phenomenon of supercooling, but it was most famously reported by Fahrenheit (1724, 23) in one of the articles that he submitted to support his election as a Fellow of the Royal Society of London. De Luc used his airless water (described in the "Superheating and the Mirage of True Ebullition" section) and cooled it down to 14°F (−10°C) without freezing. Supercooling was suspected to happen in mercury in the 1780s, and that gave occasion for important further investigations by Charles Blagden (1748–1820), Cavendish's

---

[60] "Supercooling" is a modern term, the first instance of its use being dated at 1898 in the *Oxford English Dictionary*, 2d ed. In the late nineteenth century it was often referred to as "surfusion" (cf. the French term *surchauffer* for superheating), and in earlier times it was usually described as the "cooling of a liquid below its normal freezing point," without a convenient term to use.

longtime collaborator and secretary of the Royal Society.[61] Research into supercooling continued throughout the nineteenth century. For instance, Dufour (1863) brought small drops of water down to −20°C without freezing, using a very similar technique to the one that had allowed him to superheat water to 178°C as discussed in "Superheating and the Mirage of True Ebullition."

The theoretical understanding of supercooling, up to the end of the nineteenth century, was even less firm than that of superheating, perhaps because even ordinary freezing was so poorly understood. The most basic clue was provided by Black's concept of latent heat. After water reaches its freezing temperature, a great deal more heat has to be taken away from it in order to turn it into ice; likewise, a lot of heat input is required to melt ice that is already at the melting point. According to Black's data, ice at the freezing point contained only as much heat as would liquid water at 140° below freezing on Fahrenheit's scale (if it could be kept liquid while being cooled down to that temperature).[62] In other words, a body of water at 0°C contains a lot more heat than the same amount of ice at 0°C. All of that excess heat has to be taken away if all of the water is to freeze; if just a part of the excess heat is taken away, normally just one part of the water freezes, leaving the rest as liquid at 0°C. But if there is no particular reason for one part of the water to freeze and the rest of it not to freeze, then the water can get stuck in a collective state of indecision (or symmetry, to use a more modern notion). An unstable equilibrium results, in which all of the water remains liquid but at a temperature lower than 0°C, with the heat deficit spread out evenly throughout the liquid. The concept of latent heat thus explained how the state of supercooling could be maintained. However, it did not provide an explanation or prediction as to when supercooling would or would not take place.

How was the fixity of the freezing point defended, despite the acknowledged (and poorly understood) existence of supercooling? Recall, from "Escape from Superheating," one of the factors that prevented superheating from being a drastic threat to the fixity of the boiling point: although water could be heated to quite extreme temperatures without boiling, it came down to much more reasonable temperatures once it started boiling. A similar phenomenon saved the freezing point. From Fahrenheit onward, researchers on supercooling noted a phenomenon called "shooting." On some stimulus, such as shaking, the supercooled water would suddenly freeze, with ice crystals shooting out from a catalytic point. Wonderfully for thermometry, the result of shooting was the production of just the right amount of ice (and released latent heat) to bring up the temperature of the whole to the normal freezing point.[63] Gernez (1876) proposed that shooting from supercooling

---

[61] The supercooling of mercury will be discussed again in "Consolidating the Freezing Point of Mercury" in chapter 3. See Blagden 1788, which also reports on De Luc's work on p. 144.

[62] Measurements by others indicated similar latent heat values, Wilcke giving 130°F and Cavendish 150°F. See Cavendish 1783, 313.

[63] Blagden (1788, 134) noted, however: "If from any circumstances...the shooting of the ice proceeds more slowly, the thermometer will often remain below the freezing point even after there is much ice in the liquor; and does not rise rapidly, or to its due height, till some of the ice is formed close to the bulb."

could actually be used as a more reliable fixed point of temperature than normal freezing.[64]

On the causes and preventatives of shooting there were various opinions, considerable disagreement, and a good deal of theoretical uncertainty. Blagden (1788, 145–146) concluded with the recognition that "the subject still remains involved in great obscurity." However, this was of no consequence for thermometry, since in practice shooting could be induced reliably whenever desired (similarly as superheated "bumping" could be prevented at will). Although the effectiveness of mechanical agitation was seriously debated, from early on all were agreed that dropping a small piece of ice into the supercooled water always worked. This last circumstance fits nicely into a more general theoretical view developed much later, particularly by Aitken, which I discussed in "A Dusty Epilogue." In Aitken's view, fixity of temperature was the characteristic of an equilibrium between two different states of water. Therefore the fixed temperature could only be produced reliably when both liquid and solid water were present together in contact with each other. In addition to supercooling, Aitken reported that ice without a "free surface" could be heated up to 180°C without melting.[65] At a pragmatic level, the importance of equilibrium had been recognized much earlier. De Luc in 1772 argued that the temperature at which ice melted was not the same as the temperature at which water froze and proposed the temperature of "ice that melts, or water in ice" as the correct conception of the freezing/melting point.[66] And Fahrenheit had already defined his second fixed point as the temperature of a water-ice mixture (see Bolton 1900, 70). These formulations specified that there should be both ice and water present in thermal equilibrium, though not on the basis of any general theoretical framework such as Aitken's.

The freezing-point story conforms very well to the general account of standards and their validation and improvement given in "The Validation of Standards" and The Iterative Improvement of Standards." The freezing point was part of the basis on which the dominant form of the numerical thermometer was constructed, respecting and improving upon the thermoscopic standard; most of what I said in that connection about the boiling point applies to the freezing point as well. Likewise, the strategies for the plausible denial of variability discussed in "The Defense of Fixity" also apply quite well to the freezing point. In the freezing-point case, too, causes of variation were eliminated when possible, and small inexplicable variations were ignored. But corrections did not play a big role for the freezing point, since there did not happen to be significant causes of variation that were amenable to corrections.

This episode also reinforces the discussion of serendipity and robustness in "The Defense of Fixity." The robustness of the freezing point was due to the same kind of serendipity that made the boiling/steam point robust. In both cases there was a lack of clear theoretical understanding or at least a lack of theoretical consensus.

---

[64] Gernez actually employed an iterative method in that proposal.
[65] This experiment was attributed to Carnelly; see Aitken 1880–81, 339.
[66] See De Luc 1772, 1:344, §436; and 1:349, §438.

However, pragmatic measures for ensuring fixity were in place long before there was any good theoretical understanding about why they were effective. In fact, they often happened to be in place even before there was a recognition of the problems that they were solving! In the case of the freezing point, it was very easy to disturb the container and break supercooling, if one did not take particular care to maintain tranquility. Black (1775, 127–128) thought that even the imperceptible agitation caused by air molecules spontaneously entering into the water would be sufficient for that effect.[67] Especially if one was taking thermometers in and out of the water to take temperature readings, that would have caused mechanical disturbances and also introduced a convenient solid surface on which ice crystals could start to form. If the experiments were done outdoors on a cold day, which they tended to be in the days before refrigeration, "frozen particles, which in frosty weather are almost always floating about in the air," would have dropped into the water and initiated ice-formation in any supercooled water.[68] But the early experimenters would not have seen any particular reason to cover the vessel containing the water. In short, supercooling is a state of unstable equilibrium, and careless handling of various kinds will tend to break that equilibrium and induce shooting. Therefore, as in the boiling-point story, the theme of the freezing-point story is not the preservation of fixity by random chance, but its protection through a serendipitous meeting of the epistemic need for fixity and the natural human tendency for carelessness.

[67] This was offered as a novel explanation of the puzzling observation that water that had previously been boiled seemed to freeze more easily. Since air would be expelled from the water in boiling, and it would re-enter the water after the boiling ceased, Black reckoned that the molecular agitation involved in that process would be sufficient to prevent supercooling. Therefore the once-boiled water would seem to freeze more readily than ordinary water, which was more liable to supercooling. Blagden (1788, 126–128), however, maintained that boiled water was more susceptible to supercooling, not less; he thought that the purging of air caused by the boiling was responsible for this effect (cf. De Luc's supercooling experiment with airless water, mentioned earlier).

[68] This was a quite popular idea, expressed for example by Blagden (1788, 135), quoted here.

2

Spirit, Air, and Quicksilver

Narrative: The Search for the "Real" Scale of Temperature

> The thermometer, as it is at present construed, cannot be applied to point out the exact proportion of heat.... It is indeed generally thought that equal divisions of its scale represent equal tensions of caloric; but this opinion is not founded on any well decided fact.
>
> Joseph-Louis Gay-Lussac, "Enquiries Concerning the Dilatation of the Gases and Vapors," 1802

In chapter 1, I discussed the struggles involved in the task of establishing the fixed points of thermometry and the factors that enabled the considerable success that was eventually reached. Once the fixed points were reasonably established, numerical thermometers could be created by finding a procedure for assigning numbers to the degrees of heat between the fixed points and beyond them. This may seem like a trivial problem, but in fact it harbored a deep philosophical challenge, which was overcome only after more than a century of debates and experiments.

The Problem of Nomic Measurement

The main subject of this chapter is intimated in a curious passage in *Elementa Chemiae*, the enormously influential textbook of chemistry first published in 1732 by the renowned Dutch physician Herman Boerhaave (1668–1738):

An earlier version of the material in parts of this chapter was published in Chang 2001b. The material in "Comparability and the Ontological Principle of Single Value" and "Minimalism against Duhemian Holism" is partly derived from Chang 2001a.

TABLE 2.1. The discrepancies between thermometers filled with different liquids

Mercury	Alcohol	Water
0 (°C)	0	0
25	22	5
50	44	26
75	70	57
100	100	100

Source: The data are from Lamé 1836, 1:208.

I desired that industrious and incomparable Artist, Daniel Gabriel Fahrenheit, to make me a couple of Thermometers, one with the densest of all Fluids, Mercury, the other with the rarest, Alcohol, which should be so nicely adjusted, that the ascents of the included liquor in the same degree of Heat, should be always exactly equal in both. (Boerhaave [1732] 1735, 87)

The goods were delivered, but Boerhaave found that the two thermometers did not quite agree with each other. Fahrenheit was at a loss for an explanation, since he had graduated the two thermometers in exactly the same way using the same fixed points and the same procedures. In the end he attributed the problem to the fact that he had not made the instruments with the same types of glass. Apparently "the various sorts of Glass made in Bohemia, England, and Holland, were not expanded in the same manner by the same degree of Heat." Boerhaave accepted this explanation and went away feeling quite enlightened.[1]

The same situation was seen in a different light by another celebrated maker of thermometers, the French aristocrat R. A. F. de Réaumur (1683–1757), "the most prestigious member of the *Académie des Sciences* in the first half of the eighteenth century," a polymath known for his works in areas ranging widely from metallurgy to heredity.[2] Roughly at the same time as Boerhaave, Réaumur had noticed that mercury and alcohol thermometers did not read the same throughout their common range (1739, 462). He attributed the discrepancy to the fact that the expansions of those liquids followed different patterns. Réaumur's observation and explanation soon became accepted. It is not a subtle effect, as table 2.1 shows.

There are various attitudes one can take about this problem. A thoroughgoing operationalist, such as Percy Bridgman, would say that each type of instrument defines a separate concept, so there is no reason for us to expect or insist that they should agree.[3] A simple-minded conventionalist would say that we can just choose one instrument and make the others incorrect by definition. As Réaumur put it, one

---

[1] The lesson that Boerhaave drew from this incident was caution against rash assumptions in empirical science: "How infinitely careful therefore ought we to be in our searches after natural knowledge, if we would come at the truth? How frequently shall we fall into mistakes, if we are over hasty in laying down general rules?"

[2] See J. B. Gough's entry on Réaumur in the *Dictionary of Scientific Biography*, 11:334.

[3] See Bridgman 1927, esp. 3–9. Bridgman's views will be discussed in much more detail in "Travel Advisory from Percy Bridgman" and "Beyond Bridgman" in chapter 3.

will calibrate an alcohol thermometer on the standard of a mercury thermometer "when one wishes the alcohol thermometer to speak the language of the mercury thermometer" and vice versa (1739, 462). A more sophisticated conventionalist like Henri Poincaré would say that we ought to choose the temperature standard that makes the laws of thermal phenomena as simple as possible.

Very few scientists making or using thermometers took any of those philosophical positions. Instead, most were realists in the sense that they believed in the existence of an objective property called temperature and persisted in wanting to know how to measure its true values. If various thermometers disagreed in their readings, at most one of them could be right. The question, then, was which one of these thermometers gave the "real temperature" or the "real degree of heat," or most nearly so. This is a more profound and difficult question than it might seem at first glance.

Let us examine the situation more carefully. As discussed in "Blood, Butter, and Deep Cellars" in chapter 1, by the middle of the eighteenth century the accepted method of graduating thermometers was what we now call the "two-point method." For instance, the centigrade scale takes the freezing and boiling points of water as the fixed points. We mark the height of the thermometric fluid at freezing 0°, and the height at boiling 100°; then we divide up the interval equally, so it reads 50° halfway up and so on. The procedure operates on the assumption that the fluid expands uniformly (or, linearly) with temperature, so that equal increments of temperature results in equal increments of volume. To test this assumption, we need to make an experimental plot of volume vs. temperature. But there is a problem here, because we cannot have the temperature readings until we have a reliable thermometer, which is the very thing we are trying to create. If we used the mercury thermometer here, we might trivially get the result that the expansion of mercury is uniform. And if we wanted to use another kind of thermometer for the test, how would we go about establishing the accuracy of that thermometer?

This problem, which I have called the "problem of nomic measurement," is not unique to thermometry.[4] Whenever we have a method of measurement that rests on an empirical law, we have the same kind of problem in testing and justifying that law. To put it more precisely and abstractly:

1. We want to measure quantity X.
2. Quantity X is not directly observable, so we infer it from another quantity Y, which is directly observable. (See "The Validation of Standards" in the analysis part for a full discussion of the exact meaning of "observability.")
3. For this inference we need a law that expresses X as a function of Y, as follows: $X = f(Y)$.
4. The form of this function f cannot be discovered or tested empirically, because that would involve knowing the values of both Y and X, and X is the unknown variable that we are trying to measure.

---

[4]See Chang 1995a, esp. 153–154. The problem first came up in my study of energy measurements in quantum physics.

This circularity is probably the most crippling form of the theory-ladenness of observation. (For further discussions of the problem of nomic measurement, see "Comparability and the Ontological Principle of Single Value" in the analysis part.)

Given this fundamental philosophical puzzle, it should not come as a surprise that there was a complex and protracted fight over the choice of the right thermometric fluid. A bewildering variety of substances had been suggested, according to this or that person's fancy: mercury, ether, alcohol, air, sulphuric acid, linseed oil, water, salt water, olive oil, petroleum, and more. Josiah Wedgwood even used lumps of clay, which actually contracted under high degrees of heat, as in his pottery kilns (see "Adventures of a Scientific Potter" in chapter 3). Just as wonderfully varied as the list of thermometric substances is the list of eminent scientists who concerned themselves seriously with this particular issue: Black, De Luc, Dalton, Laplace, Gay-Lussac, Dulong, Petit, Regnault, and Kelvin, just to mention some of the more familiar names.

Three of the known thermometric fluids became significant contenders for the claim of indicating true temperatures: (1) atmospheric air; (2) mercury, or quicksilver; and (3) ethyl alcohol, most often referred to as "the spirit of wine" or simply "spirit." The rest of the narrative of this chapter charts the history of their contention, ending with the establishment of the air thermometer as the best standard in the 1840s. Throughout the discussion there will be an emphasis on how various scientists who worked in this area attempted to tackle the basic epistemological problem, and these attempts will be analyzed further in broader philosophical and historical contexts in the second part of the chapter.

De Luc and the Method of Mixtures

Thermometry began with no firm principles regarding the choice of thermometric substances.[5] The very early thermoscopes and thermometers of the seventeenth century used air. Those fickle instruments were easily replaced by "liquid-in-glass" thermometers, for which the preferred liquid for some time was spirit. Fahrenheit, working in Amsterdam, was responsible for establishing the use of mercury in the 1710s; small, neat and reliable, his mercury thermometers gained much currency in the rest of Europe partly through the physicians who had received their training in the Netherlands (under Boerhaave, for instance), where they became familiar with Fahrenheit's instruments.[6] Réaumur preferred spirit, and due to his authority spirit thermometers retained quite a bit of popularity in France for some time. Elsewhere mercury came to be preferred by most people including Anders Celsius, who pioneered the centigrade scale.

Initially people tended to assume that whichever thermometric fluids they were using expanded uniformly with increasing temperature. The observations showing

---

[5]For the early history that will not be treated in detail in this chapter, see Bolton 1900, Barnett 1956, and the early chapters of Middleton 1966.

[6]For this account of the dissemination of the mercury thermometer by physicians, see *Encyclopaedia Britannica*, supplement to the 4th, 5th, and 6th editions (1824), 5:331.

the disagreement between different types of thermometers made the need for justification clearer, but it seems that for some time the unreflective habit continued in the form of unsupported assertions that one or another fluid expanded uniformly and others did not. There was even a view that solids expanded more uniformly than liquids and gases, advanced by Thomas Young (1773–1829), the promulgator of the wave theory of light, as late as the beginning of the nineteenth century (Young 1807, 1:647). Jacques Barthélemi Micheli du Crest (1690–1766), Swiss military engineer who spent much of his life in political exile and prison, published an idiosyncratic argument in 1741 to the effect that spirit expanded more regularly than mercury.[7] However, his contemporary George Martine (1702–1741), Scottish physician, stated a contrary opinion: "it would seem, from some experiments, that [spirit] does not condense very regularly" in strong colds; that seemed to go conveniently with Martine's advocacy of the mercury thermometer, which was mostly for practical reasons (Martine [1738] 1772, 26). The German physicist-metaphysician Johann Heinrich Lambert (1728–1777) also claimed that the expansion of spirit was irregular. He believed, as had the seventeenth-century French savant Guillaume Amontons (1663–1738), that air expanded uniformly and liquids did not. Neither Amontons nor Lambert, however, gave adequate arguments in support of that assumption.[8]

Such impulsive and intuitive advocacy, from any side, failed to convince. In the days before the caloric theory there was only one tradition of cogent reasoning and experimentation with a potential to settle the argument. This was the method of mixtures. Mix equal amounts of freezing water (at 0° centigrade, by definition) and boiling water (at 100°, again by definition) in an insulated vessel; if a thermometer inserted in that mixture reads 50°, it indicates the real temperature. Such mixtures could be made in various proportions (1 part boiling water and 9 parts freezing water should give 10° centigrade, and so on), in order to test thermometers for correctness everywhere on the scale between the two fixed points. Given this technique, it was no longer necessary to get into the circular business of judging one type of thermometer against another.

The earliest employment of the method of mixtures intended for the testing of thermometers was probably by Brook Taylor (1685–1731), English mathematician of the Taylor series fame, and secretary of the Royal Society from 1714 to 1718. Taylor (1723) published a brief account reporting that the linseed oil thermometer performed satisfactorily when tested by the method of mixtures. His test was not a very precise one, and he did not even give any numbers in his one-page report. It was also not a test designed to compare the performances of different fluids (which is understandable considering that this was before Boerhaave and Réaumur

---

[7]Du Crest (1741, 9–10) believed that the temperature of the bulk of the Earth (as indicated by the supposedly constant temperature of deep cellars and mines) was fundamentally fixed, and therefore the most extreme temperatures observed on the surface of the earth (in Senegal and Kamchaka, respectively) should be equally far from that median temperature. His spirit thermometers gave readings more in accord with that hypothesis than did his mercury thermometers. See also Middleton 1966, 90–91.

[8]See Lambert 1779, 78, and Amontons 1702. See also Middleton 1966, 108 on Lambert, and 63 on Amontons.

reported the discrepancy between spirit and quicksilver thermometers). A few decades later, in 1760, the method of mixtures was revived by Joseph Black, who carried out similar experiments on the mercury thermometer and obtained a satisfactory verdict regarding its accuracy.[9]

The person who brought the tradition of mixtures to its culmination was Jean-André De Luc (1727–1817), Genevan meteorologist, geologist, and physicist; I have discussed some aspects of his life and work in detail in chapter 1 (see especially "The Vexatious Variations of the Boiling Point" and "Superheating and the Mirage of True Ebullition"). When we left him there, he had just spent four weeks shaking air out of a flask full of water to investigate the boiling behavior of pure water. De Luc had examined almost every conceivable aspect of thermometry in his 1772 treatise, and the choice of fluids was one of his chief concerns. He observed that the choice of thermometric fluids was just that—a matter of choice. However, De Luc insisted that there should be some principle guiding the choice. The "fundamental principle" for him was that the fluid "must measure equal variations of heat by equal variations of its volume" (De Luc 1772, 1:222–223, §§410b–411a). But which fluid, if any, actually satisfied this requirement had not been established.

De Luc's investigations resulted in the conclusion that mercury was the most satisfactory thermometric liquid.[10] What he regarded as the "direct proof" of mercury's superiority and "first reason" for using it in thermometers was the result of the mixing experiments (1:285–314, §422). He attributed the method of mixtures primarily to his mentor and friend George-Louis Le Sage the Younger (1724–1803), a hardly published but highly influential figure in Geneva at this time. Generally speaking, De Luc mixed two samples of water at previously known temperatures and compared the reading given by a thermometer with the calculated temperature. To imagine the simplest case again: equal amounts of water at freezing (0°C) and boiling (100°C) should make a mixture of 50°C; the correct thermometer should read 50°C when inserted into that mixture.[11]

---

[9]See Black 1770, 8–12, and Black 1803, 1:56–59.

[10]As for non-liquid thermometric substances, De Luc dismissed solids relatively quickly and gave many detailed reasons against air, with particular reference to Amontons's air thermometer. See De Luc 1772, 1:275–283, §§420–421.

[11]It should be noted that De Luc did not use water exactly at the boiling and freezing points, so his reasoning was slightly more complicated than Taylor's. This seems to be a fact often ignored by his commentators, friend and foe, but De Luc himself stated well-considered reasons for his practice. Regarding water at the boiling point, he said: "[B]oiling water can be neither measured (in volume) nor weighed [accurately]"; see De Luc 1772, 1:292, §422. One might try to weigh the water before bringing it to boil, but then there would be a significant loss by evaporation once it starts to boil. There was no such problem with the freezing point, but he noted that it was difficult to prepare a large enough volume of liquid water exactly at the freezing point (pp. 298–299). So he was forced to use water that was only nearly boiling and nearly freezing. At first glance it would seem that this saddled De Luc with a vicious circularity, since he first had to use a thermometer to measure the temperatures of his hot and cold waters. He did have a process of correction with which he was satisfied (pp. 299–306), but in any case the basic procedure is not problematic if it is viewed as a test of consistency. *If* the mercury thermometer is correct, and we mix equal amounts of water at temperatures $a°$ and $b°$ *as measured by it*, then the mercury thermometer should give $(a + b)/2°$ for the temperature of the mixture.

TABLE 2.2. Results of De Luc's test of the mercury thermometer by the method of mixtures

	Degree of real heat (calculated)[a]	Reading of the mercury thermometer	Condensation of mercury between last two points
Boiling water	z + 80	80.0	—
	z + 75	74.7	5.3
	z + 70	69.4	5.3
	z + 65	64.2	5.2
	z + 60	59.0	5.2
	z + 55	53.8	5.2
	z + 50	48.7	5.1
	z + 45	43.6	5.1
	z + 40	38.6	5.0
	z + 35	33.6	5.0
	z + 30	28.7	4.9
	z + 25	23.8	4.9
	z + 20	18.9	4.9
	z + 15	14.1	4.8
	z + 10	9.3	4.8
	z + 5	4.6	4.7
Melting ice	z	0.0	4.6

Source: The data are taken from De Luc 1772, 1:301, §422.

[a] All temperatures in this table are on the Réaumur scale. The "z" in the degrees of real heat signifies that the "absolute zero" point of temperature (indicating a complete absence of heat) is not known.

The verdict from De Luc's experiments was unequivocal. The deviation of the mercury thermometer from degrees of real heat was pleasingly small, as shown in table 2.2 (note that De Luc was using the "Réaumur" temperature scale, in which the temperature of boiling water was set at 80° rather than 100°). Even more decisive than this consideration of mercury alone was the comparative view. From De Luc's results juxtaposing the performance of eight different liquids, shown in table 2.3, there was no question that mercury gave the best available approximation to the "real" degrees of heat.

These results were in accord with theoretical considerations as well. De Luc reasoned that the condensation of liquids proceeded uniformly according to temperature until contraction so crowded the molecules that they resisted further condensation.[12] So he inferred that a significant "slowing down" of condensation was a sign that the liquid has entered the crowded phase in which its volume ceases to reflect the true variation in the quantity of heat. Therefore, as temperature goes down "the liquid whose rate of condensation increases in comparison to that of all other liquids is very probably the one in which differences of volume are closest to

---

[12] Here he was certainly not adopting the assumption that particles of matter have inherent mutual attraction only to be counterbalanced by the repulsive action of heat, which would later become a centerpiece of the caloric theory, in the years around 1800.

TABLE 2.3. De Luc's comparison of the readings of various thermometers with the "real" degree of heat

Real degree of heat (calculated)^a	40.0
Mercury thermometer	38.6
Olive oil thermometer	37.8
Camomile oil thermometer	37.2
Thyme oil thermometer	37.0
Saturated salt water thermometer	34.9
Spirit thermometer	33.7
Water thermometer	19.2

Source: Adapted from De Luc 1772, 1:311, §422.

^{a}All temperatures in this table are on the Réaumur scale.

being proportional to differences of heat." On this criterion, too, mercury was shown to be the best choice.[13] De Luc was so confident of his results that he declared that mercury deserved "an exclusive preference" in the construction of the thermometer. Borrowing the words of an acquaintance impressed by his demonstration, he expressed his elation: "[C]ertainly nature gave us this mineral for making thermometers!" (De Luc 1772, 1:330, §426) Unlike his work on the boiling point discussed in chapter 1, De Luc's experiments and arguments in favor of mercury gained wide acceptance, with endorsements from various leading authorities in physics and chemistry throughout Europe. By around 1800 De Luc had created an impressive degree of consensus on this issue, cutting through significant disciplinary, national, and linguistic boundaries.[14]

Caloric Theories against the Method of Mixtures

This consensus on mercury, however, began to crumble just as it was being secured. Trouble developed around De Luc's crucial assumption that the amount of heat needed in heating a given amount of water was simply proportional to the amount of change in its temperature. For instance, in presenting the results listed in table 2.2, De Luc was assuming that it would take the same amount of heat to raise the temperature of a given amount of water by each 5° increment. When applied generally, this amounted to the assumption that the specific heat of water was constant and did not depend on temperature. This was a convenient assumption, and there were no particular reasons for De Luc to doubt it at the time.[15] However,

---

[13]De Luc 1772, 1:284–285, §421. See also his table comparing the "marche" of seven different liquids on p. 271, §418.

[14]For further details of the impressive degree of support De Luc gained, see Chang 2001b, 256–259.

[15]In that sense the status of this assumption was the same as that of the other major assumption in De Luc's experiments (and in all related calorimetric measurements), which was that heat was a conserved quantity.

this assumption was challenged with increasing readiness and confidence as a consequence of the growing sophistication of the caloric theory, which is the main feature in the development of the chemistry and physics of heat in the decades around 1800. For readers unfamiliar with the history of the caloric theory, a few words of background explanation are necessary.[16]

The core of the caloric theory was the postulation of caloric, a material substance that was regarded as the cause of heat or even as heat itself. Most commonly caloric was seen as a subtle fluid (all-penetrating and weightless or nearly so) that was attracted to ordinary matter but self-repulsive (therefore elastic). The self-repulsion of caloric was a crucial quality, since it explained a whole host of effects of heat, ranging from the melting of solids to the increased pressure of gases. There were different versions of the caloric theory developing in competition with each other. I follow Robert Fox (1971) in dividing the caloric theorists (or "calorists") into two broad categories, depending on their views on the meaning of specific and latent heat. I will call these groups "Irvinist" (following Fox) and "chemical."

The Irvinists followed the doctrine of William Irvine (1743–1787), a pupil and collaborator of Black's in Glasgow, who postulated that the amount of caloric contained in a body was the product of its capacity for caloric (heat capacity) and its "absolute temperature" (which would be zero degrees at the point of a total absence of heat). If a body preserved its heat content but its heat capacity was increased for some reason, its temperature would go down; this was explained by an analogy to a bucket that suddenly widens, lowering the level of liquid contained in it. Irvine conceptualized latent heat as the heat required just to keep the temperature at the same level in such a case. (I will discuss Irvinist heat theory further in "Theoretical Temperature before Thermodynamics" in chapter 4.)

In contrast, in the chemical view of caloric, latent heat was seen as a different state of heat, postulated to lack the power of affecting the thermometer. Black had viewed the melting of ice as the combination of ice and caloric to produce liquid water. Though Black himself chose to remain ultimately agnostic about the metaphysical nature of heat,[17] his view on latent heat was taken up and generalized by some chemists into the notion of caloric as a substance that could enter into chemical combinations with ordinary matter. Antoine-Laurent Lavoisier (1743–1794) developed a similar view through the 1770s and went so far as to include caloric (and also light) in the table of chemical elements in his authoritative textbook of the new chemistry, *Elements of Chemistry* (1789).[18] On this chemical view of heat, the latent caloric that entered into combination with particles of matter was the cause of increased fluidity as solids melted into liquids and liquids evaporated into gases; this latent caloric would become sensible again in condensation or congelation.

---

[16]The best source on the history of the caloric theory is still Fox 1971. For a brief yet informative account, see Lilley 1948.

[17]For Black's view on the metaphysical nature of heat, see Black 1803, 1:30–35.

[18]For the development of Lavoisier's view on heat, see Guerlac 1976. His "table of simple substances" can be found in Lavoisier [1789] 1965, 175.

The absorption and emission of heat in ordinary chemical reactions were also explained in the same manner. The notion of the chemical combination of caloric with matter was even incorporated into the terminology of "combined" vs. "free" caloric, which was used alongside the more phenomenological terminology of "latent" and "sensible" caloric (Lavoisier [1789] 1965, 19).

To return to the method of mixtures now: among the Irvinists, the most prominent critic of De Luc was the English Quaker physicist-chemist John Dalton (1766–1844). Dalton's attack was published in his *New System of Chemical Philosophy*, the first part of which (1808) was mostly about heat, although it is now more famous for the statement of his chemical atomic theory. Dalton (1808, 11) confessed that he had been "overawed by the authority of Crawford" initially to trust the mercury thermometer, only to be dissuaded by further considerations. (Here Dalton was referring to the Irish physician Adair Crawford [1748–1795], an Irvinist who had strongly advocated De Luc's method of mixtures in his well-known treatise on animal heat, first published in 1779.) Referring to De Luc's work specifically, Dalton laid the constancy of specific heat open to doubt and declared: "Till this point is settled, it is of little use to mix water of 32° and 212° [in Fahrenheit's scale], with a view to obtain the true mean temperature" (1808, 49–50). According to Dalton, there was an easy argument against the validity of the method of mixtures. The mixing of hot and cold water was observed to result in a slight decrease in overall volume. In Dalton's version of the caloric theory a decrease in volume literally meant less space for caloric to fit in, therefore a decrease in heat capacity. That meant, by basic Irvinist reasoning, that temperature would go up. So Dalton (1808, 3–9) thought that mixtures generally had higher temperatures than those given by De Luc's simple calculations.[19] Although Dalton may not have had any significant following in thermometry, his argument against De Luc would not have been easy to ignore, since it was just the same argument as involved in Dalton's more influential work (1802a) on the explanation of adiabatic heating and cooling by the mechanical compression and decompression of gases.

De Luc's method of mixtures was even more readily questioned by those calorists who inclined toward the chemical view of caloric. Since combined or latent caloric was conceived as the kind of caloric that did not register in thermometers, judging the correctness of thermometers seemed to require knowing the relation between the whole amount of caloric in a body and the amount that was free. That, in turn, required knowing when and how caloric would get bound and unbound to matter, but the exact causes of the transition of caloric between its combined and free states remained under serious dispute. This threw the question of specific heat wide open: specific heat was the amount of total heat input used in raising the temperature of a body by a unit amount, and that would have to include any

---

[19]For a general discussion of Dalton's caloric theory, see Fox 1968. Going beyond mere criticism of De Luc, Dalton (1808, 9ff.) advanced a complex theoretical and experimental argument that the expansion of mercury was quadratic rather than linear with temperature. He even devised a new temperature scale on the basis of this belief, the correctness of which seemed to him confirmed beyond doubt by the way it simplified several empirical laws governing thermal phenomena. See also Cardwell 1971, 124–126.

amount that went into the combined state. Without knowing what the latter amount was, one could hardly say anything theoretically about specific heat.

A most instructive case of the effect of this theoretical worry is the authoritative *Elementary Treatise on Natural Philosophy* (1803) by the renowned mineralogist and divine René-Just Haüy (1743–1794), one of the founders of modern crystallography. This textbook was personally commissioned by Napoleon for use in the newly established *lycées*, and it promptly became a recommended text for the École Polytechnique as well. Thus, it was with Napoleon's authority as well as his own that Haüy had asserted:

> The experiments of De Luc have served...to render evident the advantage possessed by mercury, of being amongst all known liquids,[20] that which approaches the most to the state of undergoing dilatations exactly proportional to the augmentations of heat, at least between zero and the degrees of boiling water. (Haüy [1803] 1807, 1:142)

However, within just three years Haüy withdrew his advocacy of De Luc in the second edition of his textbook, where he gave more highly theoretical treatments. Haüy now emphasized that the expansion of a body and the raising of its temperature were two distinct effects of the caloric that entered the body. He attributed the distinction between these two effects to Laplace (quite significantly, as we shall see in the next section), referring to Lavoisier and Laplace's famous 1783 memoir on heat.

Haüy (1806, 1:86) traced expansion to the part of added caloric that became latent, and the raising of temperature to the part that remained sensible. Then the crucial question in thermometry was the relation between those two amounts: "[I]f the amount of dilatation is to give the measure of the increase in tension,[21] the amount of the caloric that works to dilate the body must be proportional to the amount that elevates the temperature" (1:160). According to Haüy's new way of thinking, De Luc's reasoning was at least slightly negligent. The crucial complication noted by Haüy was that the expansion of water would require more caloric at lower temperatures, since there was stronger intermolecular attraction due to the intermolecular distances being smaller. For that reason, he argued that the real temperature of a mixture would always be lower than the value given by De Luc's simple-minded calculations.[22]

---

[20] As for the air thermometer, Haüy ([1803] 1807, 1:259–260) discussed its disadvantages in a similar vein to De Luc, referring to Amontons's instrument.

[21] Haüy (1806, 1:82) defined temperature as the "tension" of sensible caloric, a notion advanced by Marc-Auguste Pictet in conscious analogy to Volta's concept of electric tension; see Pictet 1791, 9.

[22] Haüy's reasoning is worth following in some detail. Consider the mixing of two equal portions of hot and cold water. In reaching equilibrium, the hot water gives out some heat, which is absorbed by the cold water. One part of the heat given up serves to contract the hot water (call that amount of caloric $C1$), and the rest ($C2$) serves to cool it; likewise, one part of the caloric absorbed by the cold water ($C3$) serves to expand it, and the rest ($C4$) serves to warm it. In order to know the resulting temperature of the mixture, it is necessary to know the quantities $C2$ and $C4$. Since they are not necessarily equal to each other, the temperature of the mixture is not necessarily the arithmetic average of the starting temperatures. Haüy starts with the assumption that the total amount of caloric given out by the hot water should be equal to the amount absorbed by the cold water ($C1 + C2 = C3 + C4$). Then he reasons that $C3$ is

68 *Inventing Temperature*

In summary, it seems that mature theoretical reflections tended to do irrevocable damage to the method of mixtures, by rendering the constancy or variation of the specific heat of water an entirely open question. There is evidence that even De Luc himself recognized this point of uncertainty, actually before Dalton's and Haüy's criticisms were published. Crawford, who was cited by Dalton as the authority who taught him about the method of mixtures, noted in the second edition of his book on animal heat:

> Mr De Luc has, however, himself observed, in a paper, with which he some time ago favoured me on this subject,[23] that we cannot determine with certainty from those experiments, the relation which the expansion of mercury bears to the increments of heat. For when we infer the agreement between the dilatations of mercury and the increments of heat from such experiments, we take it for granted, that the capacity of water for receiving heat, continues permanent at all temperatures between the freezing and boiling points. This, however, should not be admitted without proof. (Crawford 1788, 32–33)

Although Crawford still maintained his belief in the real correctness of mercury thermometers,[24] the statement shows that even the two most important advocates of the method of mixtures came to doubt its theoretical cogency.

---

greater than $C1$, so $C2$ must be greater than $C4$. This is because the thermal expansion of water must require more caloric at lower temperatures, since the molecules of matter would be closer together and therefore offer stronger resistance to the expansive action of caloric; at higher temperatures the intermolecular attraction would be weaker because of the larger distances involved. Therefore the contraction of the hot water would cause less caloric to be given out than the amount taken up by the expansion of the cold water by the same amount ($C1 < C3$), which means that there is more caloric taken away to cool the hot water than that added to heat the cold water ($C2 > C4$). Here Haüy seems to be assuming that the volume of the mixture would be the same as the sum of the initial volumes; this assumption was disputed by Dalton, as we have seen. Apparently also assuming that the specific heat of water is constant if we only consider the part of the caloric that is actually used for raising temperature, Haüy concluded that the temperature of a mixture would always be below the value calculated by De Luc. See Haüy 1806, 1:166–167.

[23] I have not been able to ascertain which paper of De Luc's Crawford is referring to here.

[24] This was on the basis of an additional test of thermometers that Crawford devised. In this experiment he contrived two open metal cylinders containing air at the temperatures of boiling water and melting ice. These two cylinders were put in communication with each other at their open faces, and a mercury thermometer was inserted at that point of contact. Crawford believed that the real temperature of air at that boundary was the arithmetic mean of the two extreme temperatures, and that is what his mercury thermometer indicated. From this result he inferred that the mercury thermometer was indeed correct, and consequently that the method of mixtures must have been quite correct after all. (Incidentally, Crawford believed that the mercury thermometer was almost exactly accurate and disputed De Luc's result that the mercury temperature was appreciably below the real temperature in the middle of the range between the boiling and freezing points of water; cf. the data presented in table 2.2.) See Crawford 1788, 34–54. It is doubtful that anyone would have been persuaded that Crawford's setup was reliably producing the exact mean temperature as intended; I have not come across any discussion of this experiment by anyone else.

The Calorist Mirage of Gaseous Linearity

If the caloric theories rendered the method of mixtures groundless, what alternative did they present in making the choice of thermometric fluids? The answer was not immediately clear. Haüy and Dalton, whom I have discussed as two chief critics of De Luc, only agreed that De Luc was wrong. As for the true temperature of a mixture, Dalton thought it should be higher than De Luc's value, and Haüy thought it should be lower than De Luc's. Was that disagreement ever resolved? I have not seen any evidence of a serious debate on that issue. In fact, Haüy and Dalton were exceptional among calorists in making an attempt at all to theorize quantitatively about the thermal expansion of liquids. The problem was too difficult, and microscopic reasoning rapidly became groundless once it got involved in figuring out the amount of caloric required for effecting the expansion of liquids in opposition to the unspecified forces which material particles exerted on each other.[25] Instead, most caloric theorists were seduced by an apparently easier way out. Caloric theory taught that the action of heat was most purely manifested in gases, rather than liquids or solids. In gases the tiny material particles would be separated too far from each other to exert any nonnegligible forces on each other; therefore, all significant action in gases would be due to the caloric that fills the space between the material particles. Then the theorist could avoid dealing with the uncertainties of the interparticle forces altogether.[26]

Faith in the simplicity of the thermal behavior of gases was strengthened enormously by the observation announced by Joseph-Louis Gay-Lussac (1802) and independently by Dalton (1802a) that all gases expanded by an equal fraction of the initial volume when their temperature was increased by the same amount. This seemed to provide striking confirmation that the thermal behavior of gases had remarkable simplicity and uniformity, and led many calorists to assume that gases expanded uniformly with temperature. A typical instance was the treatment by Louis-Jacques Thenard (1777–1857) in his highly regarded textbook of chemistry, which he dedicated to Gay-Lussac: "[A]ll gases, in contrast [to liquids and solids], expand equally, and their expansion is uniform and equal for each degree— 1/266.67 of their volume at 0°, under atmospheric pressure. The discovery of this law must be attributed to Dalton and Gay-Lussac" (Thenard 1813, 1:37). There was, however, a logical gap in that reasoning, as recognized very clearly by both Dalton and Gay-Lussac themselves despite their works always being cited as evidence for this common opinion. Even if we grant that the thermal expansion of gases is a phenomenon determined exclusively by the effect of temperature, it still does not follow that the volume of a gas should be a linear function of

---

[25] In fact, given the lack of a quantitative estimate, Haüy's reasoning could even have been construed as a further vindication of the mercury thermometer, since in De Luc's mixing experiments mercury did give readings somewhat lower than the calculated real temperatures! In any case Haüy must have thought that the errors in De Luc's calculations were small enough, since he still endorsed De Luc's test when it came to the verdict that mercury was better than alcohol. See Haüy 1806, 1:165.

[26] For more details on this widespread notion, see Fox 1971, ch. 3, "The Special Status of Gases."

temperature. Not all functions of one variable are linear![27] An additional argument was needed for getting linearity, and it was Laplace who took up the challenge most seriously.

The revival of interest in gas thermometers occurred in the context of the ascendancy of what Robert Fox (1974) has famously termed "Laplacian physics," the most dominant trend in French physical science in the years roughly from 1800 to 1815. The mathematician, astronomer, and physicist Pierre-Simon Laplace (1749–1827) worked in close association with the chemist Claude-Louis Berthollet (1748–1822), after each collaborating with Lavoisier. Together they set out a new program for the physical sciences and fostered the next generation of scientists who would carry out the program.[28] Berthollet and Laplace subscribed to a "Newtonian" research program seeking to explain all phenomena by the action of central forces operating between pointlike particles. Renowned for his mathematical refinement of Newtonian celestial mechanics, Laplace aspired to bring its rigor and exactitude to the rest of physics: "[W]e shall be able to raise the physics of terrestrial bodies to the state of perfection to which celestial physics has been brought by the discovery of universal gravitation" (Laplace 1796, 2:198). In the first decade of the nineteenth century Laplace and his followers won wide acclaim by creating new theories of optical refraction, capillary action, and acoustics, based on short-range forces.[29] Heat theory was an obvious next target, since it was already an essential part of Laplace's treatment of the speed of sound, and one of his long-standing interests dating back to his early collaboration with Lavoisier. Besides, with the gradual demise of Irvinism the theoretical lead in heat theory fell to the Lavoisierian chemical tradition, which Laplace transformed in interesting ways as we shall see.

Laplace's early attempt at an argument for the air thermometer, included in the fourth volume of his classic *Treatise of Celestial Mechanics*, was brief and loose (1805, xxii and 270). Laplace said it was "at least very probable" that an air thermometer indicated accurately "the real degrees of heat," but his entire argument consisted in this: "[I]f we imagine the temperature of the air to increase while its volume remains the same, it is very natural to suppose that its elastic force, which is caused by heat, will increase in the same ratio." Then he imagined a relaxation of the external pressure confining the heated gas; if the pressure were brought back to the initial value, the volume of the gas would increase in the same ratio as the pressure had done under constant volume. This last step just follows from assuming

---

[27]Dalton (1808, 9) wrote: "Since the publication of my experiments on the expansion of elastic fluids by heat and those of Gay Lussac, immediately succeeding them... it has been imagined by some that gases expand equally; but this is not corroborated by experience from other sources." Rather, he thought that gases expanded "in geometric progression to equal increments of temperature" (11). See also Gay-Lussac 1802, 208–209, including the passage cited as the epigraph to this chapter. Haüy (1806, 1:263–264), who was clearly aware of this point before he was distracted by Laplacian theorizing, even reported that Gay-Lussac had found the coefficient of thermal expansion of air to vary as a function of temperature.

[28]For a detailed treatment of Laplace and Berthollet's circle, see Crosland 1967.

[29]On the details of these theories, see Gillispie 1997, and also Heilbron 1993, 166–184.

Mariotte's (Boyle's) law.[30] This non-argument, only buttressed by the word "natural," seems to have convinced many people, even the judicious Haüy (1806, 1:167–168). Calorist plausibility combined with Laplacian authority catapulted the air thermometer into the position of the "true thermometer" in the eyes of many active researchers. Thomas Thomson (1773–1852), Regius Professor of Chemistry at the University of Glasgow, granted that "it is at present the opinion of chemists, that...the expansion of all gases is equable," reversing his own earlier view that "none of the gaseous bodies expand equably."[31] It became a general view that the only consolation for the mercury thermometer was that it was practically more convenient to use than the air thermometer, and that its readings agreed closely enough with those of the air thermometer between the freezing and boiling points of water, as shown most clearly by Gay-Lussac (1807).

Meanwhile Laplace himself was not quite satisfied with his 1805 argument for the air thermometer and went on to develop a more detailed and quantitative argument.[32] To make the concept of temperature more precise, he adopted the approach of the Genevan physicist and classicist Pierre Prevost (1751–1839), who had defined temperature through the equilibrium of radiant caloric, conceiving caloric as a "discrete fluid."[33] Extending that kind of view to the molecular level of description, Laplace defined temperature as the density of intermolecular caloric, produced by a continual process of simultaneous emission and absorption between molecules.[34] But why should the caloric contained in molecules be radiated away at all? There would have to be some force that pushes the caloric away from the material core of the molecule that attracts and normally holds it. This force, according to Laplace, was the repulsion exerted by the caloric contained in other molecules nearby. Laplace's model might seem to fit well with the old distinction between free and latent caloric: some of the latent caloric, contained in molecules, would be disengaged by caloric–caloric repulsion and become free caloric. However, that would have conflicted with the Lavoisierian conception that latent caloric was chemically

---

[30] The second step would have been unnecessary for a constant-volume air thermometer, which indicates temperature by pressure, but Laplace was obliged to add it because he was considering a constant-pressure air thermometer.

[31] For the earlier view, see T. Thomson 1802, 1:273; for the later view, T. Thomson 1830, 9–10. In his advocacy of the equable expansion of air, Thomson admitted that "it is scarcely possible to demonstrate the truth of this opinion experimentally, because we have no means of measuring temperature, except by expansion." But he added that "the opinion is founded on very plausible reasons," without actually giving those reasons.

[32] Laplace presented some important derivations on the behavior of gases from general calorist principles to the Paris Academy of Sciences on 10 September 1821. The mathematical article was published promptly (Laplace 1821), and a verbal summary was printed with some delay (Laplace [1821] 1826). Updated versions were included in the fifth volume of *Traité de mécanique céleste* (Laplace [1823] 1825).

[33] This view was initially proposed in Prevost 1791 and elaborated in several subsequent publications.

[34] Here we must note that Laplace's molecules did not touch each other in a gas, unlike Dalton's atoms (each of which consisted of a dense core surrounded by an "atmosphere of caloric") that filled up space even in a gas.

72   Inventing Temperature

bound to matter, incapable of being disengaged from the molecules except through changes of state, chemical reactions, or some unusual physical agitation.

Laplace escaped from this conceptual tangle by taking the extraordinary step of putting free caloric inside molecules, departing considerably from Lavoisier's original picture. The particles of free caloric were bound, but still exerted repulsive forces on each other; this way, free caloric in one molecule was capable of dislodging free caloric from other molecules. On the other hand, latent caloric, also bound in molecules, did not exert repulsive forces and could be ignored in Laplace's force-based derivations.[35] Laplace called the free caloric disengaged from molecules the *free caloric of space*, which was a third state of caloric (very similar to the older notion of radiant caloric), in addition to Lavoisier's latent/combined caloric and free/sensible caloric.[36]

Armed with this refined ontology, Laplace proceeded to argue that there would be a definite correlation between the density of free caloric contained in molecules and the density of free caloric tossing about in intermolecular spaces, because the amount of caloric being removed from a given molecule would clearly be a function of the intensity of the cause of the removal. So the density of free caloric of space could be utilized for the measurement of temperature, even its definition. With this concept of temperature, Laplace's argument that the air thermometer was "the true thermometer of nature" consisted in showing that the volume of air under constant pressure would be proportional to the density of the free caloric of space.[37]

Laplace gave various demonstrations of this proportionality. The most intuitive one can be paraphrased as follows.[38] Laplace took as his basic relations:

$$P = K_1 \rho^2 c^2 \qquad (1)$$

$$T = K_2 \rho c^2 \qquad (2)$$

where P is the pressure, K_1 and K_2 constants, ρ the density of the gas, and c the amount of free caloric contained in each molecule. The first relation follows from regarding the pressure of a gas as resulting from the self-repulsion of caloric contained in it. The repulsive force between any two molecules would be proportional to c^2, and the pressure exerted by a molecular layer of density ρ on a layer of the same density proportional to ρ^2. In favor of the second relation Laplace argued that temperature, the density of free caloric in intermolecular space, would be

---

[35] On this count, curiously, Laplace's mature view was more in agreement with De Luc's than with Haüy's.

[36] See Laplace [1821] 1826, 7, for an explanation of this picture. See also Laplace [1823] 1825, 93, 113, for the emphasis that latent caloric did not enter into his calculations. For the term *la chaleur libre de l'espace*, see Laplace 1821, 335.

[37] Laplace [1821] 1826, 4. The "extreme rarity" of the free caloric of space, due to the high speed at which caloric was transmitted between molecules, guaranteed that its amount would be a negligible fraction of the total amount of free caloric contained in a body. Then the amount of the free caloric of space could serve as a measure of the total free-caloric content, without actually constituting a significant portion of the latter.

[38] This follows the exposition in Laplace [1821] 1826, 3–6, supplemented by insights taken from Laplace [1823] 1825. See Brush 1965, 12–13, for a similar treatment.

proportional to the amount of caloric emitted (and absorbed) by each molecule in a given amount of time. This quantity would be proportional to the intensity of its cause, namely the density of caloric present in its environment, ρc, and also to the amount of free caloric in each molecule available for removal, c. By combining equations (1) and (2) Laplace obtained $P = KT/V$, where V is the volume (inversely proportional to ρ for a given amount of gas), and K is a constant. For fixed P, T is proportional to V; that is, the volume of a gas under constant pressure gives a true measure of temperature.

It now remained to make a truly quantitative derivation and, in the abstract, what a good "Newtonian" had to do was clear: write down the force between two caloric particles as a function of distance, and then perform the appropriate integrations in order to calculate the aggregate effects. Unfortunately, this was a nonstarter. The fact that Laplace (1821, 332–335) did start this derivation and carried it through is only a testimony to his mathematical ingenuity. Laplace had no idea, and nor did anyone else ever, what the intercaloric force function looked like. It was obviously impossible to infer it by making two-particle experiments, and there were few clues even for speculation. In his derivations Laplace simply wrote $f(r)$ for the unknown aspect of that function and kept writing different symbols for its various integrals; the unknown expression in the final formula, a definite integral, was given the symbol K and treated as a constant for a given type of gas, and turned out not to matter for anything important. The real work in the derivation was all done by various other assumptions he introduced along the way.[39] These assumptions make an impressive list. In addition to the basic calorist picture of a gas, Laplace assumed: that the gas would be in thermal equilibrium and uniform in density; that its molecules would be spherical, stationary, and very far away from each other; that each molecule would contain exactly the same amount of caloric; that the force between the caloric particles would be a function of distance and nothing else, and negligible at any sensible distances; that the particles of the free caloric of space moved at a remarkably high speed; and so on.

Since these assumptions were not theoretically defended or empirically testable, it is perhaps not a great surprise that even most French theorists moved away from Laplacian calculations on caloric. Perhaps the sole exception worth noting is Siméon-Denis Poisson (1781–1840), who continued to elaborate the Laplacian caloric theory even after Laplace's death.[40] Not many people bothered to argue against the details of Laplace's caloric theory.[41] Rather, its rejection was made wholesale amid the general decline and rejection of the Laplacian research program.

---

[39] A similar view is given by Heilbron 1993, 178–180, and also Truesdell 1979, 32–33. There was probably a good deal of continuity between this situation and Laplace's earlier, more acclaimed treatments of capillary action and optical refraction, in which he demonstrated that the particular form of the force function was unimportant; see Fox 1974, 101.

[40] See, for instance, Poisson 1835. According to Fox (1974, 127 and 120–121), Poisson "seems to have pursued the [Laplacian] program with even greater zeal than the master himself."

[41] One of those who did was the Scottish mining engineer Henry Meikle, who attacked Laplace's treatment of thermometry directly with a cogent technical argument. See Meikle 1826 and Meikle 1842.

Although generally discredited, the Laplacian treatment remained the only viable theoretical account of the thermal physics of gases until the revival and further development of Sadi Carnot's work in the 1840s and 1850s (see "William Thomson's Move to the Abstract" in chapter 4), and the only viable microphysical account until the maturity of the molecular-kinetic theory in the latter half of the century.

Regnault: Austerity and Comparability

The principles of thermometry thus endured "the rise and fall of Laplacian physics" and returned to almost exactly where they began. The two decades following Laplace's work discussed earlier seem to be mostly characterized by a continuing erosion in the confidence in all theories of heat. The consequence was widespread skepticism and agnosticism about all doctrines going beyond straightforward observations. The loss of confidence also resulted in a loss of theoretical interest and sophistication, with both pedagogic and professional treatments retreating into simpler theoretical conceptions.[42] (I will give a further analysis of post-Laplacian empiricism in "Regnault and Post-Laplacian Empiricism.") An emblematic figure for this period is Gabriel Lamé (1795–1870), renowned mathematician, physicist, and engineer. Lamé was a disciple of Fourier's and also modeled himself after Pierre Dulong and Alexis-Thérèse Petit, who were his predecessors in the chair of physics at the Paris École Polytechnique. He stated his position in no uncertain terms in the preface of his physics textbook for the École:

> Petit and Dulong constantly sought to free teaching from those doubtful and metaphysical theories, those vague and thenceforth sterile hypotheses which used to make up almost the whole of science before the art of experimenting was perfected to the point where it could serve as a reliable guide.... [After their work] it could be imagined that at some time in the future it would be possible to make the teaching of physics consist simply of the exposition of the experiments and observations which lead to the laws governing natural phenomena, without it being necessary to state any hypothesis concerning the first cause of these phenomena that would be premature and often harmful. It is important that science should be brought to this positive and rational state.

For this kind of attitude he won the admiration of Auguste Comte, the originator of "positivism," who had been his classmate at the École Polytechnique.[43]

In his discussion of the choice of thermometric fluids Lamé agreed that gases seemed to reveal, better than other substances, the pure action of heat unadulterated by the effects of intermolecular forces. However, like Dalton and Gay-Lussac (and Haüy before his Laplacian indoctrination), Lamé clearly recognized the limits to the conclusions one could derive from that assumption:[44]

---

[42]See Fox 1971, 261–262, 276–279.
[43]See Fox 1971, 268–270; the quoted passage is from Lamé 1836, 1:ii–iii, in Fox's translation on pp. 269–270.
[44]Lamé 1836, 1:256–258; cf. Haüy [1803] 1807, 1:263–264.

Although the indications of the air thermometer could be regarded as exclusively due to the action of heat, from that it does not necessarily follow that their numerical values measure the energy of that action in an absolute manner. That would be to suppose without demonstrating that the quantity of heat possessed by a gas under a constant pressure increases proportionally to the variation of its volume. If there were an instrument for which such a proportionality actually held, its indications would furnish an absolute measure of temperatures; however, as long as it is not proven that the air thermometer has that property, one must regard its reading as an as yet unknown function of the natural temperature. (Lamé 1836, 1:258)

The blitheness of simply assuming linearity here might have been obvious to Lamé, who is mainly remembered now as the man who introduced the use of curvilinear coordinates in mathematical and physical analysis.

Into this state of resignation entered Henri Victor Regnault (1810–1878), with a solution forged in a most austere version of post-Laplacian empiricism. Regnault's career is worth examining in some detail, since the style of research it shaped is directly relevant to the scientific and philosophical issues at hand. Regnault may be virtually forgotten now, perhaps nearly as much as De Luc, but in his prime he was easily regarded as the most formidable experimental physicist in all of Europe. Regnault's rise was so triumphant that Paul Langevin (1911, 44), though critical of him, drew a parallel with the glory days of Napoleon. Orphaned at the age of 2 and growing up without means, Regnault benefited enormously from the meritocratic educational system that was a legacy of the French Revolution. With ability and determination alone he was able to gain his entry to the École Polytechnique, and by 1840, at the age of 30, succeeded Gay-Lussac as professor of chemistry there. In that same year he was elected to the chemistry section of the Académie des Sciences, and in the following year became professor of experimental physics at the Collège de France. By then he was an obvious choice for a renewed commission by the minister of public works to carry out experimental studies to determine all the data and empirical laws relevant to the study and operation of steam engines.

Thus ensconced in a prestigious institution with ample funds and few other duties, Regnault not only supplied the government with the needed information but also in the course of that work established himself as an undisputed master of precision measurement. Marcelin Berthelot later recalled the strong impression he had received on meeting Regnault in 1849: "It seemed that the very spirit of precision had been incarnated in his person" (Langevin 1911, 44). Young scientists from all over Europe, ranging from William Thomson (later Lord Kelvin) to Dmitri Mendeléeff, visited his fabled laboratory, and many stayed for a while to work and learn as his assistants.[45] Regnault may well have *frightened* the European scientific community into accepting the authority of his results. Matthias Dörries (1998a, 258) notes that it was difficult for other physicists to challenge Regnault's results because they could not afford the apparatus needed to repeat his experiments. The

---

[45] A list of visitors to Regnault's lab is given by Dumas (1885), 178. On Mendeléeff, see Jaffe 1976, 153.

size of his equipment alone might have been enough to overpower potential detractors! Regnault describes in one place a 24-meter tall manometer that he constructed for the measurement of pressure up to 30 atmospheres, later a famous attraction in the old tour of the Collège.[46] The sheer volume and thoroughness of his output would have had the same effect. Regnault's reports relating to the steam engine took up three entire volumes of the *Mémoires* of the Paris Academy, each one numbering 700 to 900 pages, bursting with tables of precise data and interminable descriptions of experimental procedures. In describing the first of these volumes, James David Forbes (1860, 958) spoke of "an amount of minute and assiduous labor almost fearful to contemplate."

But, as I will discuss further in "Minimalism against Duhemian Holism" and "Regnault and Post-Laplacian Empiricism" in the analysis part, it was not mere diligence or affluence that set Regnault apart from the rest. Jean-Baptiste Dumas (1885, 169) asserted that Regnault had introduced a significant new principle to experimental physics, which he regarded as Regnault's service to science that would never be forgotten. To explain this point, Dumas drew a contrast to the methodology exhibited in the classic treatise of physics by Jean-Baptiste Biot. Whereas Biot would employ a simple apparatus to make observations, and then reason clearly through all the necessary corrections, Regnault realized (as Dumas put it): "In the art of experimenting by way of corrections, the only sure procedure is that which does not require any."[47] Dumas summed up Regnault's distinctive style as follows:

> A severe critic, he allows no causes of error to escape him; an ingenious spirit, he discovers the art of avoiding all of them; an upright scholar, he publishes all the elements relevant to the discussion, rather than merely giving mean values of his results. For each question he introduces some characteristic method; he multiplies and varies the tests until no doubts remain about the identity of the results. (Dumas 1885, 174)

Regnault aspired to test all assumptions by measurements. This implied that the measurements would need to be made without relying on any theoretical assumptions: "In establishing the fundamental data of physics one must, as far as possible, only make use of direct methods" (Regnault, quoted in Langevin 1911, 49). Regnault aimed at a puritanical removal of theoretical assumptions in the design of all basic measurement methods. This was, however, easier said than done. It is fine to say that all assumptions should be checked by measurements, but how

---

[46] On that instrument, see Regnault 1847, 349, and Langevin 1911, 53.

[47] For instance, consider the weighing of a given volume of gas, as discussed by Dumas (1885, 174–175). If one puts a sizeable glass balloon containing the gas on one side of the balance and small metal weights on the other, it is necessary to correct the apparent measured weight by estimating exactly the effect of the buoyancy of the surrounding air, for which it is necessary to know the exact pressure and temperature of the air, the exact density and volume of the glass (and its metal frame), etc. Instead of trying to improve that complex and uncertain procedure of correction, Regnault eliminated the need for the correction altogether: he hung an identical glass balloon, only evacuated, on the opposite side of the one containing the gas to be weighed, and thereafter the balance behaved as if it were in a perfect vacuum. In that procedure, the only buoyancy correction to worry about was for the metallic weights balancing the weight of the gas, which would have been quite a negligible effect.

can any measurement instruments be designed, if one can make no assumptions about how the material substances that constitute them would behave? Coming back to thermometry: we have seen that all investigators before Regnault were forced to adopt some contentious assumptions in their attempts to test thermometers. In contrast, Regnault managed to avoid all assumptions regarding the nature of caloric, the constancy or variation of specific heats, or even the conservation of heat.[48] How did he pull off such a feat?

Regnault's secret was the idea of "comparability." If a thermometer is to give us the true temperatures, it must at least always give us the same reading under the same circumstance; similarly, if a type of thermometer is to be an accurate instrument, all thermometers of that type must at least agree with each other in their readings. Regnault (1847, 164) considered this "an essential condition that all measuring apparatuses must satisfy." Comparability was a very minimalist kind of criterion, exactly suited to his mistrustful metrology. All that he assumed was that a real physical quantity should have one unique value in a given situation; an instrument that gave varying values for one situation could not be trusted, since at least some of its indications had to be incorrect. (See "Comparability and the Ontological Principle of Single Value" for further discussion of this "principle of single value.")

The general notion of comparability was not Regnault's invention. It was in fact almost an item of common sense for a long time in thermometry, widely considered a basic requirement for reliability. The term is easier to understand if we go back to its origin, namely when thermometers were so notoriously unstandardized that the readings of different thermometers could not be meaningfully compared with each other. The early difficulty may be illustrated by an exception that proves the rule. In 1714 Fahrenheit astonished Christian Freiherr von Wolff, then professor of mathematics and physics at the University of Halle and later its chancellor, by presenting him with two spirit thermometers that agreed perfectly with each other (a feat he could not manage for Boerhaave when different liquids were involved, as we saw in "The Problem of Nomic Measurement").[49] Comparability was almost a battle cry as the early pioneers of thermometry called for standardization.

Regnault transformed this old notion of comparability into a powerful tool for testing the goodness of each given type of thermometer. The novelty introduced by Regnault was a higher degree of skepticism: in earlier times, once a rigorous method of graduating thermometers was settled on, people tended to assume that all instruments produced by that method would be exactly comparable to each other. For Regnault (1847, 165), this was much too hasty. Methods of standard graduation generally only involved making different thermometers agree with each other at a small number of points on the scale. This gave no guarantee that the thermometers would agree with each other at all other points. The agreement at other points was a hypothesis open to empirical tests, even for thermometers employing the same fluid as long as they differed in any other ways.

---

[48] Regnault did assume heat conservation in some other experiments (especially for calorimetry).
[49] On Fahrenheit's interaction with Wolff, see Bolton 1900, 65–66, and also Van der Star 1983, 5.

78 Inventing Temperature

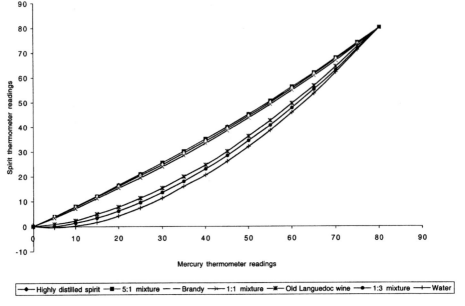

FIGURE 2.1. De Luc's comparison of spirit thermometers. The data and explanations are from De Luc 1772, 1:326, §426.

In the next section we shall see in good detail how fruitful Regnault's use of comparability was. One final note before passing on to that discussion: this innovation in the use of comparability must be credited at least partly to De Luc, although he did not use it nearly as systematically as Regnault did. De Luc was already accustomed to thinking about comparability through his famous work in barometry, and there is also some indication that he regarded comparability as a requirement for measurements in general.[50] In thermometry he used the comparability criterion in order to give an additional argument against the spirit thermometer. De Luc's results, some of which are represented in Figure 2.1, showed that the spirit thermometer was not a comparable instrument, since spirit expanded according to different laws depending on its concentration.[51] But why was it not possible to avoid this difficulty by simply specifying a standard concentration of the

---

[50] See the discussion of barometry in De Luc 1772, vol. 1, part 2, ch. 1, and his attempts to bring comparability into areometry in De Luc 1779, 93ff.

[51] The assumption in De Luc's attack on the spirit thermometer was that mercury thermometers, in contrast, were comparable amongst each other. Since mercury is a homogeneous liquid and does not mix well with anything else, De Luc thought its concentration was not likely to vary; he also believed that any impurities in mercury became readily noticeable by a diminution in its fluidity. For these points, see De Luc 1772, 1:325–326, 330, §426. The comparability of the mercury thermometer seems to have remained the going opinion for decades afterwards; see, for instance, Haüy [1803] 1807, 1:142–143 (and corresponding passages in subsequent editions), and Lamé 1836, 1:219.

spirit to be used in thermometers? That would have created another fundamental difficulty, of having to measure the concentration accurately. This was not easy, as we can see in the extended essay on areometry (the measurement of the specific gravity of liquids), which De Luc (1779) published seven years later.[52]

The Verdict: Air over Mercury

What was the outcome of Regnault's comparability-based tests? Since the spirit thermometer had been discredited beyond rescue (in terms of comparability and in other ways, too), Regnault's main concern was to decide between the air thermometer and the mercury thermometer. This particular issue had also assumed a greater practical urgency in the meantime, thanks to Dulong and Petit's work in regions of higher temperatures that revealed a mercury–air discrepancy reaching up to 10° on the centigrade scale (see "Regnault and Post-Laplacian Empiricism" for further details about this work). Clearly it would not do to use the mercury thermometer and the air thermometer interchangeably for serious scientific work, and a choice had to be made.

With regard to comparability, it was mercury that betrayed clearer signs of trouble. In the course of his work on the comparison between the readings of the air thermometer and the mercury thermometer, first published in 1842, Regnault (1842c, 100–103) confirmed that there was no such thing as "the" mercury thermometer. Mercury thermometers made with different types of glass differed from each other even if they were calibrated to read the same at the fixed points. The divergence was noticeable particularly at temperatures above 100° centigrade. Worse yet, as Regnault (1847, 165) added in his later and more extensive report, samples of the same type of glass that had undergone different thermal treatments did not follow the same law of expansion. Regnault (1847, 205–239) laid to waste the assumed comparability of the mercury thermometer in his painstaking series of experiments on eleven different mercury thermometers made with four different types of glass. As the data in table 2.4 show, there were significant differences, exceeding 5°C in the worst cases. It was as if Fahrenheit's ghost had revisited the scene with a grin: he was correct after all, that the types of glass used made a substantial difference! A further irony is that De Luc's technique originally intended for his advocacy of mercury was now being used to discredit mercury.

It may be objected here that the failure of comparability due to the behavior of glass was merely a practical difficulty, having nothing to do with the thermal expansion of mercury itself. Would it not be easy enough to specify a certain type of glass as the standard glass for making mercury thermometers? But the thermal behavior of glass was complex and not understood very well. Achieving comparability in the standard mercury thermometer would have required the specification

---

[52]See also De Luc 1772, 1:327–328, §426. There is an indication that Fahrenheit, who was an earlier pioneer of areometry, had kept a standard sample of spirit to use as the standard for all his spirit thermometers. It seems that he was aware of the variation in the patterns of expansion, but only as a minor annoyance. See Fahrenheit's letter of 17 April 1729 to Boerhaave, in Van der Star 1983, 161, 163.

TABLE 2.4. Regnault's comparison of mercury thermometers made with different types of glass

Air thermometer	Mercury with "Choisy-le-Roi" crystal	Mercury with ordinary glass (thermometer No. 5)[a]	Mercury with green glass (thermometer No. 10)	Mercury with Swedish glass (thermometer No. 11)
100 (°C)	100.00	100.00	100.00	100.00
150	150.40	149.80	150.30	150.15
200	201.25	199.70	200.80	200.50
250	253.00	250.05	251.85	251.44
300	305.72	301.08	—	—
350	360.50	354.00	—	—

Source: Adapted from Regnault 1847, 239.

[a]Note that the pattern of expansion of ordinary glass happens to match that of mercury quite well up to nearly 300°C, so that the readings of the air thermometer agrees quite well with the readings of the mercury-in-ordinary-glass thermometer.

of the exact chemical composition of the glass, the process of its manufacture (down to the exact manner of blowing the thermometer bulb), and also the conditions of usage. Controlling such specifications to meet the degree of precision wanted by Regnault would have required not only totally impractical procedures but also theoretical and empirical knowledge beyond anyone's grasp at that time. The uncertainties involved would have been enough to defeat the purpose of increased precision. (This is similar to the situation with the failure of comparability in the spirit thermometer due to variations in the concentration of spirit.) In addition, the familiar vicious circularity would also have plagued any attempt to make empirical determinations of the behavior of glass as a function of temperature, since this would have required an already trusted thermometer.

When he announced the mercury thermometer to be lacking in comparability in 1842, Regnault was nearly prepared to endorse the use of the air thermometer as the only comparable type. As the thermal expansion of air was so great (roughly 160 times that of glass), the variations in the expansion of the glass envelope could be made negligible (Regnault 1842c, 103). Still, he was not entirely comfortable. Refusing to grant any special status to gases, Regnault (1847, 167) demanded that the air thermometer, and gas thermometers in general, should be subjected to a rigorous empirical test for comparability like all other thermometers. He had good reason to hesitate. His own work had shown that the average coefficient of expansion was variable according to the density even for a given type of gas. Perhaps the *form* of the expansion law also varied, as in the case of alcohol with different concentrations? The variation in the coefficient was an annoyance, but there was no conceptual problem in graduating each thermometer individually so that it gave 100° at the boiling point of water. On the other hand, variations in the form of the law would have been a more serious matter, resulting in a failure of comparability. Regnault (1847, 172) considered it "absolutely essential" to submit this question to an experimental investigation.

To that end Regnault built constant-volume thermometers filled with gases of various densities, starting with atmospheric air. He rejected air thermometers at

TABLE 2.5. Regnault's comparison of air thermometers filled with different densities of air

Air thermometer A		Air thermometer A'		
Pressure (mmHg)	Temperature reading (°C)	Pressure (mmHg)	Temperature reading (°C)	Temperature difference (A−A')
762.75	0	583.07	0	0
1027.01	95.57	782.21	95.57	0.00
1192.91	155.99	911.78	155.82	+0.17
1346.99	212.25	1030.48	212.27	−0.02
1421.77	239.17	1086.76	239.21	−0.04
1534.17	281.07	1173.28	280.85	+0.22
1696.86	339.68	1296.72	339.39	+0.29

Source: Adapted from Regnault 1847, 181.

constant pressure because they suffered from an inherent lowering of sensitivity at higher temperatures.[53] (The design of these instruments will be discussed further in "The Achievement of Observability, by Stages.") Regnault's typical procedure was to set two such thermometers side by side in a bath of oil, to see how much they differed from each other at each point. Such pairs of temperature readings were taken at various points on the scale ranging from 0° to over 300° centigrade. The results of these tests provided a relief. The data in table 2.5, for instance, give a comparison of the readings of air thermometer A, whose "initial" pressure (that is, pressure at temperature 0°) was 762.75 mm of mercury, with the readings of A', whose initial pressure was 583.07 mm. The divergence between these two thermometers was always less than 0.3° in the range from 0° to 340°, and always below 0.1% of the magnitudes of the measured values. Also attesting to the high comparability of these two thermometers was the fact that the discrepancy between their readings was not systematic, but varied randomly. The results from other similar tests, with initial pressures ranging from 438.13 mm to 1486.58 mm, were similarly encouraging.[54] Regnault (1847, 185) declared: "One can therefore conclude with all certainty from the preceding experiments: the air thermometer is a perfectly comparable instrument even when it is filled with air at different densities."

Regnault also attempted some other experiments to see if the comparability could be extended to the generalized gas thermometer. He found that comparability held well between air and hydrogen, and also between air and carbonic acid gas (carbon dioxide). As with air at different densities, it turned out that these gases had the same form of the law of expansion, though their coefficients of expansion were quite different from each other. However, as shown in table 2.6, there were some serious and systematic discrepancies between air and sulfuric acid gas.[55] So, once again, Regnault exposed a place where the behavior of all gases was not identical

---

[53] See Regnault 1847, 168–171, for an explanation of this effect.
[54] See, for instance, tables in Regnault 1847, 181, 184.
[55] For the other inter-gas comparisons, see the tables on 186–187.

TABLE 2.6. Regnault's comparison of thermometers of air and sulfuric acid gas

Air thermometer A		Sulfuric acid thermometer A'		
Pressure (mmHg)	Temperature reading (°C)	Pressure (mmHg)	Temperature reading (°C)	Temperature difference (A–A')
762.38	0	588.70	0	
1032.07	97.56	804.21	97.56	0.00
1141.54	137.24	890.70	136.78	+0.46
1301.33	195.42	1016.87	194.21	+1.21
1391.07	228.16	1088.08	226.59	+1.57
1394.41	229.38	1089.98	227.65	+1.73
1480.09	260.84	1157.88	258.75	+2.09
1643.85	320.68	1286.93	317.73	+2.95

Source: Adapted from the second series of data given in Regnault 1847, 188.

and showed that the generalized gas thermometer would not be a comparable instrument. Regnault (1847, 259) was happy enough to assert: "[T]he air thermometer is the only measuring instrument that one could employ with confidence for the determination of elevated temperatures; it is the only one which we will employ in the future, when the temperatures exceed 100°."

A couple of questions may be raised regarding this conclusion. First of all, why was air preferred to other kinds of gases? It is not that each of the other gas thermometers had been shown to lack comparability. Having found no explicit discussion of this issue in Regnault's writings, I can only speculate. The practical aspect may have been enough to decide the issue, namely that atmospheric air was the easiest and cheapest gas to acquire, preserve, and control. This may explain why Regnault chose not to produce results on the comparability of other gas thermometers with regard to density, and without such tests he would not have felt comfortable adopting those thermometers for use. It is interesting to note that Regnault was apparently not concerned by the fact that atmospheric air was a mixture of different gases. As long as it satisfied the comparability condition and did not exhibit any overtly strange behavior, he saw no reason to make apologies for air or prefer pure gases to it. This was in line with his antitheoretical bias, which I will discuss further in "Regnault and Post-Laplacian Empiricism."

The second question is whether Regnault thought he had good reason to believe the comparability of air thermometers to be sufficiently proven. Just as his own work had exposed the lack of comparability in the mercury thermometer with the variation in glass-type, was it not possible that there were other parameters of the air thermometer whose variation would destroy its comparability? Again, I can only speculate on what Regnault thought about this. I think that there were no other remaining variations that seemed particularly significant to him, and that he would have made the tests if any had occurred to him. On the other hand, there was no end of parameters to study and subject to doubt, so it is also possible that even Victor Regnault was pragmatically forced to stop somewhere in testing possible variations, against his own principles. In any case, as even Karl Popper would have recommended, the only thing one can do is to adopt and use, for the time being, whatever has not been falsified yet.

Possibly because of these remaining questions, Regnault's final pronouncement in favor of air was muted. His 1842 article on the comparison of mercury and air thermometers had ended on a pessimistic note:

> Such simple laws accepted so far for the expansion of gases had led physicists to regard the air thermometer as a standard thermometer whose indications are really proportional to the increases in the quantities of heat. Since these laws are now recognized as inexact, the air thermometer falls back into the same class as all other thermometers, whose movement is a more or less complicated function of the increases in heat. We can see from this how far we still are from possessing the means of measuring absolute quantities of heat; in our present state of knowledge, there is little hope of finding by experiment simple laws in the phenomena which depend on these quantities. (Regnault 1842c, 103–104)

The later memoir showed no sign that he had become any more optimistic about this issue (Regnault 1847, 165–166). If anything, he expanded further on the scornful remark just quoted about the careless earlier advocates of the gas thermometer. Regnault had no patience with the theoretical arguments trying to show that the thermal expansion of air was uniform, and he was all too aware of the circularity involved in trying to demonstrate such a proposition experimentally. Even when he noted the comparability between the air, hydrogen, and carbonic acid gas thermometers and the deviation of the sulfuric acid gas thermometer from all of them, he was careful not to say that the former were right and the latter was wrong: "Sulfuric acid gas departs notably from the law of expansion which the preceding gases show. Its coefficient of expansion decreases *with temperature as taken by the air thermometer*" (Regnault 1847, 190; emphasis added). He never strayed from the recognition that comparability did not imply truth. In the end, what Regnault managed to secure was only a rather grim judgment that everything else was worse than the air thermometer. Still, that was far from a meaningless achievement. This was the first time ever that anyone had advanced an argument for the choice of the correct thermometric fluid that was based on undisputed principles and unequivocal experimental results.

Regnault's work on thermometry, like most of his experimental work, gained rapid and wide acceptance.[56] His reasoning was impeccable, his technique unmatched, his thoroughness overwhelming. He did not back up his work theoretically, but he succeeded in avoiding theory so skillfully that he left no place open to any significant theoretical criticism. An important phase of the development of thermometry was completed with Regnault's publication on the comparability of gas thermometers in 1847. Ironically, just one year later the basic terms of debate would begin to shift radically and irreversibly, starting with the new theoretical definition of absolute temperature by the same young William Thomson who

---

[56]Statements to that effect are too numerous to cite exhaustively but see, for example, Forbes 1860, 958, and W. Thomson 1880, 40–41. As I will discuss further in chapter 4, Thomson (Lord Kelvin) always relied on Regnault's data in his articles on thermodynamics, with continuing admiration.

had just made his humble pilgrimage to Regnault's laboratory. The conceptual landscape would become unrecognizably altered by the mid-1850s through the promulgation and acceptance of the principle of energy conservation, and subsequently by the powerful revival of the molecular-kinetic theory of heat. What happened to the definition and measurement of temperature through that theoretical upheaval is a story for chapter 4.

Analysis: Measurement and Theory in the Context of Empiricism

> It is not that we propose a theory and Nature may shout NO. Rather, we propose a maze of theories, and Nature may shout INCONSISTENT.
>
> Imre Lakatos, "Criticism and the Methodology of Scientific Research Programmes," 1968–69

In chapter 1 there were no particular heroes in the narrative. In this chapter there is one, and it is Victor Regnault. Hagiography is uninteresting only if it keeps celebrating the tired old saints in the same old way. Regnault's achievement deserves to be highlighted because it has been ignored for no good reason. Today most people who come to learn of Regnault's work tend to find it quite pedestrian, if not outright boring. I hope that the narrative was sufficient to show that Regnault's solution of the thermometric fluid problem was no ordinary success. It is rare to witness such an impeccable and convincing solution to a scientific problem that had plagued the best minds for such a long time. The qualities shown in Regnault's handling of this problem also pervaded his work in general. I will now attempt to elucidate the nature and value of Regnault's achievement further, by analyzing it from various angles: as a step in the extension of observability; as a responsible use of metaphysics; as a solution to the problem of "holism" in theory testing; and as the culmination of post-Laplacian empiricism in French physics.

The Achievement of Observability, by Stages

The improvement of measurement standards is a process contributing to the general expansion and refinement of human knowledge from the narrow and crude world of bodily sensations. The challenge for the empiricist is to make such improvement of knowledge *ultimately on the basis of sense experience*, since empiricism does not recognize any other ultimate authority. In the end I will argue that strict empiricism is not sufficient for the building of scientific knowledge, but it is worthwhile to see just how far it can take us. Regnault is the best guide on that path that we could ever hope for. His rigorous empiricism comes down to an insistence that empirical data will not be acquired by means of measurement procedures that themselves rely on hypotheses that have not been verified by observation.

In order to see whether Regnault was really successful in his aim, and more generally whether and to what extent strict empiricism is viable, we must start with a careful examination of what it means to make observations. This has been a crucial point of debate in philosophical arguments regarding empiricism, especially concerning the viability of scientific realism within an empiricist epistemology. In my view, the key question is just how much we can help ourselves to, in good conscience, in constructing the empirical basis of scientific knowledge. The standard empiricist answer is that we can only use what is observable, but that does not say very much until we specify further what we mean by "observable." Among prominent contemporary commentators, Bas van Fraassen dictates the strictest limitations on what we can count as observable. What van Fraassen (1980, 8–21) means by "observability" is an in-principle perceivability by unaided human senses; this notion forms a cornerstone of his "constructive empiricism," which insists that all we can know about with any certainty are observable phenomena and science should not engage in fruitless attempts to attain truth about unobservable things.

Some realists have attempted to invalidate the observable–unobservable distinction altogether, but I believe that van Fraassen has done enough to show that his concept of observability is coherent and meaningful, despite some acknowledged gray areas. However, I think that his critics are correct when they argue that van Fraassen's notion of observability does not have all that much relevance for scientific practice. This point was perhaps made most effectively by Grover Maxwell, although his arguments were aimed toward an earlier generation of antirealists, namely the logical positivists. Maxwell (1962, 4–6) argued that any line that may exist between the observable and the unobservable was moveable through scientific progress. In order to make this point he gave a fictional example that was essentially not so different from actual history: "In the days before the advent of microscopes, there lived a Pasteur-like scientist whom, following the usual custom, I shall call Jones." In his attempt to understand the workings of contagious diseases, Jones postulated the existence of unobservable "bugs" as the mechanism of transmission and called them "crobes." His theory gained great recognition as it led to some very effective means of disinfection and quarantine, but reasonable doubt remained regarding the real existence of crobes. However, "Jones had the good fortune to live to see the invention of the compound microscope. His crobes were 'observed' in great detail, and it became possible to identify the specific kind of *microbe* (for so they began to be called) which was responsible for each different disease." At that point only the most pigheaded of philosophers refused to believe the real existence of microbes.

Although Maxwell was writing without claiming any deep knowledge of the history of bacteriology or microscopy, his main point stands. For all relevant scientific purposes, in this day and age the bacteria we observe under microscopes are treated as observable entities. That was not the case in the days before microscopes and in the early days of microscopes before they became well-established instruments of visual observation. Ian Hacking cites a most instructive case, in his well-known groundbreaking philosophical study of microscopes:

> We often regard Xavier Bichat as the founder of histology, the study of living tissues. In 1800 he would not allow a microscope in his lab. In the introduction to his *General Anatomy* he wrote that: 'When people observe in conditions of obscurity each sees in his own way and according as he is affected. It is, therefore, observation of the vital properties that must guide us', rather than the blurred imaged provided by the best of microscopes. (Hacking 1983, 193)

But, as Hacking notes, we do not live in Bichat's world any more. Today E. coli bacteria are much more like the Moon or ocean currents than they are like quarks or black holes. Without denying the validity of van Fraassen's concept of observability, I believe we can also profitably adopt a different notion of observability that takes into account historical contingency and scientific progress.

The new concept of observability I propose can be put into a slogan: *observability is an achievement*. The relevant distinction we need to make is not between what is observable and what is not observable to the abstract category of "humans," but between what we can and cannot observe well. Although any basic commitment to empiricism will place human sensation at the core of the notion of observation, it is not difficult to acknowledge that most scientific observations consist in drawing inferences from what we sense (even if we set aside the background assumptions that might influence sensation itself).[57] But we do not count just any inference made from sensations as results of "observation." The inference must be reasonably credible, or, made by a reliable process. (Therefore, this definition of observability is inextricably tied to the notion of reliability. Usually reliability is conceived as aptness to produce correct results, but my notion of observability is compatible with various notions of reliability.) All observation must be based on sensation, but what matters most is what we can infer safely from sensation, not how purely or directly the content of observation derives from the sensation. To summarize, I would define observation as *reliable determination from sensation*. This leaves an arbitrary decision as to just how reliable the inference has to be, but it is not so important to have a definite line. What is more important is a comparative judgment, so that we can recognize an enhancement of observability when it happens.

These considerations of "observation" and "observability" give us a new informative angle on Regnault's achievement. Regnault's contribution to thermometry was to enhance the observability of temperature as a numerical quantity, and to do so without relying on theories. In the philosophical discussions of observability I have just referred to, a very common move is to allow the inferences involved in observation to be validated by scientific theories. For reasons that will be discussed in detail in "Minimalism against Duhemian Holism" and "Regnault and Post-Laplacian Empiricism," Regnault chose a stricter empiricist strategy for the validation of temperature standards. Most of all he wanted to avoid reliance on any quantitative theories of heat, since those theories required verification through readings of

---

[57] See, for instance, Shapere 1982, Kosso 1988, and Kosso 1989 for elaborations on the view that observation consists of a causal chain of interactions conveying information from the observed object to the observer, and a reverse chain of inferences by which the observer traces the flow of information.

an already established numerical thermometer. In establishing the observability of the numerical concept of temperature, he could not use any presumed observations of that very quantity. How, then, did he go about establishing the observability of numerical temperature?

Let us briefly review the overall evolution of temperature standards, as discussed in "The Iterative Improvement of Standards" in chapter 1. If we only have recourse to unaided senses (the "stage 1" standard), temperature is an observable property only in a very crude and limited sense. The invention of the thermoscope (the "stage 2" standard) created a different kind of temperature as an observable property. Numerical thermometers (the "stage 3" standard) established yet another kind of temperature concept that was observable. The single word "temperature" obscures the existence of separate layers of concepts. Now, some people certainly had a theoretical concept of temperature as a numerical quantity before numerical thermometers were actually established. At that point, temperature—stage-3 numerical temperature—existed as an unobservable property. It became observable later. Observability is neither dichotomous nor completely continuous; it progresses, improving continuously in some ways, but also in distinct stages with the successive establishment of distinctly different kinds of standards.

Regnault's air thermometer was the best stage-3 temperature standard ever, to date. In order to establish its reliability (that is to say, to establish the observability of the numerical temperature concept), Regnault used comparability as a nontheoretical criterion of reliability, as I will discuss further in the next two sections. But he also needed other concepts whose observability was already well established, including the ordinal (thermoscope-based) temperature concept. That becomes clearer if we examine more clearly the actual construction and use of Regnault's instrument. Since his air thermometer was the constant-volume type, what it allowed was the determination of temperature from pressure. Such an instrument requires at least a qualitative assurance that the pressure of air varies smoothly with its temperature, which can be verified by means of stage-2 instruments. But how could temperature be determined from pressure in good conscience, when numerical pressure is no more sensory than numerical temperature, and no less theoretical than numerical temperature? That is only thanks to the historical fact that numerical barometers and manometers had already been established to a sufficient degree; therefore, for Regnault pressure was observable as a numerical quantity.[58]

There was another important aid to Regnault's air thermometer, and that was actually the numerical mercury thermometer. This is a difficult matter, which should be considered carefully. Regnault used numerical mercury thermometers in order to measure the temperatures of the air in the tubes connecting the main

---

[58] In a manometer, pressure was determined from the length of the mercury column. The length of the mercury column had been established as an observable quantity before this experiment and prior to the observability of numerical pressure; that is actually not a trivial point, since the precision measurement of length in situations like this was an advanced-stage operation, often done by means of a telescope and a micrometer, in order to prevent disturbances from the observer's own body.

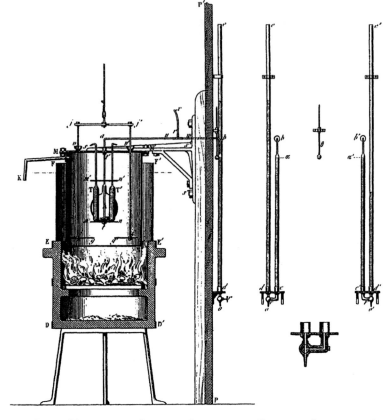

FIGURE 2.2. Regnault's constant-volume air thermometer, illustration from Regnault 1847, 168–171, figs. 13 and 14. Courtesy of the British Library.

section of the air thermometer (the large flask A in Figure 2.2) to the manometer (between point a and the mercury-level α in the right-hand figure).[59] That was necessary because the air in the tubes could not be kept at the same temperature as the air in the large flask in the heat bath, and for practical reasons it was impossible to apply any kind of air thermometer to the tubes. But how was the use of the mercury thermometer allowable, when it had not been validated (worse yet, when Regnault himself had discredited it)? It might seem that even Regnault could not be completely free of unprincipled shortcuts. But the use of the mercury thermometer for this purpose was quite legitimate, for a few different reasons.

---

[59]Regnault's method of obtaining the air-thermometer reading was in fact quite complex. He calculated the temperature in question by equating the expressions for the weight of the air in its "initial" state (at 0°C) and its heated state (at the temperature being measured). For further details, see Chang 2001b, 279–281.

First of all, the amount of air contained in those thin tubes was quite small, so any discrepancy arising from slight errors in the estimate of their temperatures would have been very small; that is a judgment we can reach without consulting any thermometers. Second, although the mercury thermometer was shown to lack comparability generally, the failure of comparability was less severe in lower temperatures (see data in table 2.4); that is helpful since we can expect that the hot air would have cooled down quite a bit in the connecting tubes. Third, earlier comparisons between the mercury thermometer and the air thermometer had shown nearly complete agreement between 0°C and 100°C, and quite good agreement when the temperature did not exceed 100°C too far. Therefore, in the everyday range of temperatures, the reliability of the mercury thermometer stands or falls with the reliability of the air thermometer itself, and the use of the mercury thermometer does not introduce an additional point of uncertainty. Finally, there is a point that ultimately overrides all the previous ones. In the end Regnault did not, and did not need to, justify the details of the design of his thermometers; the test was done by checking comparability in the final temperature readings, not by justifying exactly how those readings were obtained. The use of mercury thermometers in Regnault's air thermometer was only heuristic in the end and did not interfere with establishment of numerical temperature as an observable property by means of the air thermometer.

Comparability and the Ontological Principle of Single Value

Having sharpened our notion of observation and observability, we are now ready to consider again how Regnault solved the greatest problem standing in the way of making numerical temperature observable: the problem of nomic measurement. The solution, as I have already noted, was the criterion of comparability, but now we will be able to reach a deeper understanding of the nature of that solution. Recall the formulation of the problem given in "The Problem of Nomic Measurement." We have a theoretical assumption forming the basis of a measurement technique, in the form of a law that expresses the quantity to be measured, X, as a function of another quantity, Y, which is directly observable: $X = f(Y)$. Then we have a problem of circularity in justifying the form of that function: f cannot be determined without knowing the X-values, but X cannot be determined without knowing f. In the problem of thermometric fluids, the unknown quantity X is temperature, and the directly observable quantity Y is the volume of the thermometric fluid.

When employing a thermoscope (the "stage 2" standard), the only relation known or assumed between X and Y was that they vary in the same direction; in other words, that the function f is monotonic. The challenge of developing a numerical thermometer (the "stage 3" standard) was solved partly by finding suitable fixed points, but fixed points only allowed the stipulation of numerical values of X at isolated points. There still remained the task of finding the form of f, so that X could be deduced from Y in the whole range of values, not just at the fixed points. The usual practice in making numerical thermometers was to make a scale by dividing up the interval between the fixed points equally, which amounted to conjecturing that f would be a linear function. Each thermometric fluid represented the hypothesis that f was a linear function for that substance. (And for other methods of graduating thermometers, there would have been other corresponding hypotheses.) The problem of nomic measurement consists in the difficulty of finding a sufficiently refined standard to be used for testing those hypotheses.

Regnault succeeded first of all because he recognized the starkness of the epistemic situation more clearly than any of his predecessors: the stage-2 standard was not going to settle the choice of stage-3 standards, and no other reliable standards were available. Seeking the aid of theories was also futile: theories verifiable by the stage-2 standard were useless because they provided no quantitative precision; trying to use theories requiring verification by a stage-3 standard created circularities. The conclusion was that each proposed stage-3 standard had to be judged by its own merits. Comparability was the epistemic virtue that Regnault chose as the criterion for that judgment.

But why exactly is comparability a virtue? The requirement of comparability only amounts to a demand for self-consistency. It is not a matter of logical consistency, but what we might call physical consistency. This demand is based on what I have elsewhere called the *principle of single value* (or, single-valuedness): a real physical property can have no more than one definite value in a given situation.[60] As I said in "The Problem of Nomic Measurement," most scientists involved in the debates on thermometric fluids were realists about temperature, in the sense that they believed it to be that sort of a real physical quantity. Therefore they did not object at all to Regnault's application of the principle of single value to temperature.

It is easy enough to see how this worked out in practical terms, but there remains a philosophical question. What kind of criterion is the principle of single value, and what compels our assent to it? It is not reducible to the logical principle of noncontradiction. It would be nonsensical to say that a given body of gas has a uniform temperature of 15°C and 35°C at once, but that nonsense still falls short of the logical contradiction of saying that its temperature is both 15°C and not 15°C. For an object to have two temperature values at once is absurd because of the physical nature of temperature, not because of logic. Contrast the situation with some nonphysical properties, where one object possessing multiple values in a given situation would not be such an absurdity: a person can have two names, and purely mathematical functions can be multiple-valued. We can imagine a fantasy object that can exist in two places at once, but when it comes to actual physical objects, even quantum mechanics only goes so far as saying that a particle can have non-zero probabilities of detection in multiple positions. In mathematical solutions of physical problems we often obtain multiple values (for the simplest example, consider a

---

[60]See Chang 2001a. This principle is reminiscent of one of Brian Ellis's requirements for a scale of measurement. Ellis (1968, 39) criticizes as insufficient S. S. Stevens's definition of measurement: "[M]easurement [is] the assignment of numerals to objects or events according to rule—any rule." Quite sensibly, Ellis (41) stipulates that we have a scale of measurement only if we have "a rule for making numerical assignments" that is "*determinative* in the sense that, provided sufficient care is exercised the same numerals (or range of numerals) would always be assigned to the same things under the same conditions"; to this he attaches one additional condition, which is that the rule should be "non-degenerate" (to exclude non-informative assignments such as "assign the number 2 to everything").

physical quantity whose value is given by the equation $x^2 = 1$), but we select just one of the solutions by considering the particular physical circumstances (if x in the example is, say, the kinetic energy of a classical particle, then it is easy enough to rule out -1 as a possible value). In short, it is not logic but our basic conception of the physical world that generates our commitment to the principle of single value.

On the other hand, it is also clear that the principle of single value is not an empirical hypothesis. If someone would try to support the principle of single value by going around with a measuring instrument and showing that he or she always obtains a single value of a certain quantity at a given time, we would regard it as a waste of time. Worse yet, if someone would try to refute the principle by pointing to alleged observations of multiple-valued quantities (e.g. that the uniform temperature of this cup of water at this instant is 5° and 10°), our reaction would be total incomprehension. We would have to say that these observations are "not even wrong," and we would feel compelled to engage in a metaphysical discourse to persuade this person that he is not making any sense. Any reports of observations that violate the principle of single value will be rejected as unintelligible; more likely, such absurd representations or interpretations of experience would not even occur to us in the first place. Unlike even the most general empirical statements, this principle is utterly untestable by observation.

The principle of single value is a prime example of what I have called *ontological principles*, whose justification is neither by logic nor by experience (Chang 2001a, 11–17). Ontological principles are those assumptions that are commonly regarded as essential features of reality within an epistemic community, which form the basis of intelligibility in any account of reality. The denial of an ontological principle strikes one as more nonsensical than false. But if ontological principles are neither logically provable nor empirically testable, how can we go about establishing their correctness? What would be the grounds of their validity? Ontological principles may be akin to Poincaré's conventions, though I would be hesitant to allow all the things he classified as conventions into the category of ontological principles. Perhaps the closest parallel is the Kantian synthetic a priori; ontological principles are always valid because we are not capable of accepting anything that violates them as an element of reality. However, there is one significant difference between my ontological principles and Kant's synthetic a priori, which is that I do not believe we can claim absolute, universal, and eternal certainty about the correctness of the ontological principles that we hold. It is possible that our ontological principles are false.

This last admission opens up a major challenge: how can we overcome the uncertainties in our ontological principles? Individuals or epistemic communities may be so steeped in some false ontological beliefs that they would be prejudiced against any theories or experimental results that contravened those beliefs. Given that there is notoriously little agreement in ontological debates, is it possible to prevent the use of ontological principles from degenerating into a relativist morass, each individual or epistemic community freely judging proposed systems of knowledge according to their fickle and speculative ontological "principles"? Is it possible to resolve the disagreements at all, in the absence of any obvious criteria of judgment? In short, since we do not have a guarantee of arriving at anything approaching objective certainty in ontology, would we not be better off giving it up altogether?

Perhaps—except that by the same lights we should also have to give up the empiricist enterprise of making observations and testing theories on the basis of observations. As already stressed in "The Validation of Standards" in chapter 1, it has been philosophical common sense for centuries that our senses do not give us certainty about anything other than their own impressions. There is no guarantee that human sense organs have any particular aptitude for registering features of the world as they really are. Even if we give up on attaining objectivity in that robust sense and merely aim at intersubjectivity, there are still serious problems. Observations made by different observers differ, and there are no obvious and fail-safe methods for judging whose observations are right. And the same evidence can be interpreted to bear on theories in different ways. All the same, we do not give up the practice of relying on observations as a major criterion for judging other parts of our systems of knowledge. Instead we do our best to improve our observations. Similarly, I believe that we should do our best to improve our ontological principles, rather than giving up the practice of specifying them and using them in evaluating systems of knowledge. If fallibilist empiricism is allowed to roam free, there is no justice in outlawing ontology because it confesses to be fallible.

Thus, we have arrived at a rather unexpected result. When we consider Regnault's work carefully, what initially seems like the purest possible piece of empiricism turns out to be crucially based on an ontological principle, which can only have a metaphysical justification. What Regnault would have said about that, I am not certain. A difference must be noted, however, between the compulsion to follow an untestable ontological principle and the complacency of relying on testable but untested empirical hypothesis. The former indicates a fundamental limitation of strict empiricism; the latter has no justification, except for practical expediency in certain circumstances. The adherence to an ontological principle satisfies a clear goal, that of intelligibility and understanding. It might be imagined that comparability was wanted strictly for practical reasons. However, I believe that we often want consistency for its own sake, or more precisely, for the sake of intelligibility. It is doubtful that differences of a fraction of a degree or even a few degrees in temperature readings around 300°C would have made any appreciable practical difference in any applications in Regnault's time. For practical purposes the mercury thermometer could have been, and were, used despite its lack of comparability when judged by Regnault's exacting standards. It is not practicality, but metaphysics (or perhaps esthetics) that compelled him to insist that even a slight degree of difference in comparability was a crucial consideration in the choice of thermometric fluids.

Minimalism against Duhemian Holism

Another important way of appreciating Regnault's achievement is to view it as a solution to the problem of "holism" in theory testing, most commonly located in the work of the French physicist-philosopher Pierre Duhem, who summed it up as follows: "An experiment in physics can never condemn an isolated hypothesis but only a whole theoretical group" ([1906] 1962, sec. 2.6.2, 183). That formidable problem is a general one, but here I will only treat it in the particular context of thermometry. Before I can give an analysis of Regnault's work as a solution to this problem of Duhemian holism, some general considerations regarding hypothesis testing are necessary. Take the standard empiricist notion that a hypothesis is tested by comparing its observational consequences with results of actual observations. This is essentially the basic idea of the "hypothetico-deductive" view of theory testing, but I would like to conceptualize it in a slightly different way. What happens in the process just mentioned is the determination of a quantity in two different ways: by deduction from a hypothesis and by observation.

This reconceptualization of the standard notion of theory testing allows us to see it as a type within a broader category, which I will call "attempted overdetermination," or simply "overdetermination": a method of hypothesis testing in which one makes multiple determinations of a certain quantity, on the basis of a certain set of assumptions. If the multiple determinations agree with each other, that tends to argue for the correctness or usefulness of the set of assumptions used. If there is a disagreement, that tends to argue against the set of assumptions. Overdetermination is a test of physical consistency, based on the principle of single value (discussed in the previous section), which maintains that a real physical quantity cannot have more than one value in a given situation. Overdetermination does not have to be a comparison between a theoretical determination and an empirical determination. It can also be a comparison between two (or more) theoretical determinations or two observational ones. All that matters is that some quantity is determined more than once, though for any testing that we want to call "empirical," at least one of the determinations should be based on observation.[61]

Let us now see how this notion of testing by overdetermination applies to the test of thermometers. In the search for the "real" scale of temperature, there was a basic nonobservational hypothesis, of the following form: there is an objectively existing property called temperature, and its values are correctly given by thermometer X (or, thermometer-type X). More particularly, in a typical situation, there was a nonobservational hypothesis that a given thermometric fluid expanded uniformly (or, linearly) with temperature.

De Luc's method of mixtures can be understood as a test by overdetermination, as follows: determine the temperature of the mixture first by calculation, and then by measuring it with the thermometer under consideration. Are the results the same? Clearly not for thermometers of spirit or the other liquids, but much better for mercury. That is to say, the attempted overdetermination clearly failed for the set of hypotheses that included the correctness of the spirit thermometer, and not so severely for the other set that included the correctness of the mercury thermometer instead. That was a nice result, but De Luc's test was seriously weakened by the holism problem because he had to use other nonobservational hypotheses in addition to the main hypothesis he wanted to test. The determination of the final

---

[61] It is in the end pointless to insist on having one theoretical and one observational determination each when we have learned that the line between the theoretical and the observational is hardly tenable. This is, as van Fraassen stressed, different from saying that the observable–unobservable distinction is not cogent. "Theoretical" is not synonymous with "unobservable."

temperature by calculation could not be made without relying on at least two further nonobservational hypotheses: the conservation of heat and the constancy of the specific heat of water. Anyone wishing to defend the spirit thermometer could have "redirected the falsification" at one of those auxiliary assumptions. No one defended the spirit thermometer in that way to my knowledge, but people did point out De Luc's use of auxiliary hypotheses in order to counter his positive argument for mercury, as discussed in "Caloric Theories against the Method of Mixtures." Dalton was one of those who argued that De Luc's successful overdetermination in the mercury case was spurious and accidental: the specific heat of water was not constant; mercury did not expand linearly; according to Dalton, those two errors must have cancelled each other out.

How does Regnault look? The beauty of Regnault's work on thermometry lies in the fact that he managed to arrange overdetermination without recourse to any significant additional hypotheses concerning heat and temperature. Regnault realized that there was already enough in the basic hypothesis itself to support overdetermination. A given temperature could be overdetermined by measuring it with different thermometers of the same type; that overdetermination did not need to involve any uncertain extra assumptions. Regnault's work exemplifies what I will call the strategy of "minimalist overdetermination" (or "minimalism" for short). The heart of minimalism is the removal of all possible extraneous (or auxiliary) nonobservational hypotheses. This is not a positivist aspiration to remove all nonobservational hypotheses in general. Rather, minimalism is a realist strategy that builds or isolates a compact system of nonobservational hypotheses that can be tested clearly. The art in the practice of minimalism lies in the ability to contrive overdetermined situations on the basis of as little as possible; that is what Regnault was so methodically good at.

Minimalism can ameliorate the holism problem, regardless of whether the outcome of the test is positive or negative. Generally the failure of overdetermination becomes a more powerful indictment of the targeted hypothesis when there are fewer other assumptions that could be blamed. If auxiliary hypotheses interfere with the logic of falsification, one solution is to get rid of them altogether, rather than agonizing about which ones should be trusted over which. This Regnault managed beautifully. When there was a failure of overdetermination in Regnault's experiment, the blame could be placed squarely on the thermometer being tested. If that conclusion was to be avoided, there were only two options available: either give up the notion of single-valued temperature altogether or resort to extraordinary skeptical moves such as the questioning of the experimenter's ability to read simple gauges correctly. No one pursued either of these two options, so Regnault's condemnation of the mercury thermometer stood unchallenged.

Successful overdetermination, too, can be more reassuring when there are fewer other hypotheses involved. It is always possible to argue that a given case of successful overdetermination is a result of coincidence, with errors systematically canceling each other out. This is precisely how Dalton criticized De Luc, as I have already mentioned. In contrast, Regnault's experiment was so austere in its logical structure that it left hardly any room for that kind of criticism. Generally speaking, involving a larger number of assumptions would allow more possibilities for explaining away successful overdetermination. Minimalism provides one clear way of fighting this problem.

All of that may sound uncontroversial, but minimalism actually goes against the conventional wisdom, since it recognizes virtue in circularity. What I am calling the conventional wisdom here actually goes back to Duhem. He argued that the physicist had more to worry about than the physiologist regarding the theory-ladenness of observation because laboratory instruments were generally designed on the basis of the principles of physics. Hence, while the physiologist could proceed on the basis of a faith in physics, the physicist was stuck in a vicious circle in which he had to test the hypotheses of physics on the basis of those same hypotheses of physics.[62] There is a widespread impulse to break out of this circle. The minimalist advice, on the contrary, is to tighten the circle.

In the case of negative test outcomes, the lesson from the success of Regnault's minimalism is very clear: we need to lose the unfounded fear that the test of a theory by observations that rely on the same theory will produce a vacuous confirmation of that theory. Whether an apparent confirmation obtained by such circular testing is worthless is an open question. What is certain is that there is no guarantee that observations enabled by a particular theory will always validate that theory. That is a point that was noted at least as early as 1960 by Adolf Grünbaum (1960, 75, 82). Discussing the case of physical geometry in the context of his argument against "Duhem's thesis that the falsifiability of an isolated empirical hypothesis H as an explanans is unavoidably inconclusive," Grünbaum noted: "The initial stipulational affirmation of the Euclidean geometry G_0 in the physical laws P_0 used to compute the corrections [for distortions in measuring rods] in no way assures that the geometry obtained by the corrected rods will be Euclidean." A similar point has also been made more recently by others, including Allan Franklin et al. (1989) and Harold Brown (1993). Therefore we can see that Karl Popper was being somewhat careless when he declared that "it is easy to obtain confirmations, or verifications, for nearly every theory—if we look for confirmations" (1969, 36). In fact it is not always so easy to obtain confirmations. And when a theory is falsified despite being tested by observations made on its basis, it will be very difficult to evade the falsification. The circularity here is a form of the minimalism that renders a negative test result more assuredly damning, as discussed earlier. Therefore there is no clear reason to wish for theory-neutral observations or seek what Peter Kosso (1988, 1989) calls "independence" between the theory of the instrument and the theory to be tested by the observation produced by the instrument.[63]

Even in the case of positive test-outcomes, the comfort provided by independence is illusory. Duhem's physiologist relying on the laws of physics can be comforted only as far as those laws of physics are reliable. Taking observations away

---

[62] See Duhem [1906] (1962), part 2, ch. 6, sec. 1 (pp. 180–183).

[63] Rottschaefer (1976, 499), more than a decade earlier than Kosso's publications, already identified a similar doctrine of "theory-neutrality" as a centerpiece of a "new orthodoxy": "Thus the view that theories are tested by *theory-free* observations is being replaced by the view that theories are tested by *theory-laden* observations, but observations laden with theories neutral to the theory being tested."

from the theory being tested is a good policy only if there are other good theories that can support relevant observations. Minimalism is a plausible strategy in the absence of such alternative theories. As we have seen in the case of De Luc, confirmation is devalued if there is suspicion that it could be a result of unexpected coincidences; minimalism reduces that kind of suspicion, by cutting out as many possible sources of uncertainty in the testing procedure.

Before closing this discussion, I must mention some clear limitations of Regnault's minimalism, as a reminder that I am only admiring it as a creative and effective solution to particular types of problems, not as a panacea. There is no guarantee that a clear winner would emerge through minimalist testing. Fortunately for Regnault, the air thermometer turned out to be the only usable thermometer that survived the test of comparability. But we can easily imagine a situation in which a few different types of thermometers would all pass the comparability test, yet still disagree from each other. It is also imaginable that there might be no thermometers at all that pass the test very well. Minimalism can create a stronger assurance about the verdict of a test when there is a verdict, but it cannot ensure the existence of a clear verdict. Like all strategies, Regnault's strategy worked only because it was applied in appropriate and fortunate circumstances.

Regnault and Post-Laplacian Empiricism

Regnault's empiricism was forged in the context of the empiricist trend dominant in post-Laplacian French science. In order to reach a deeper understanding of Regnault's work, it is important to examine his context further. The direction taken by French physics in the period directly following the end of Laplacian dominance is an important illustration of how science can cope with the failure of ambitious theorizing. The post-Laplacian phase had two major preoccupations: phenomenalistic analysis in theory and precision measurement in experiment. Let us take a closer look at each preoccupation.

The phenomenalist trend, at least in the field of thermal physics, seems to have been a direct reaction against Laplace; more generally, it constituted a loss of nerve in theorizing about unobservable entities. Very symptomatic here was the rise of Jean Baptiste Joseph Fourier (1768–1830). According to Robert Fox (1974, 120, 110), Fourier became "a benign, influential, but rather detached patron of the new generation" of anti-Laplacian rebels including Pierre Dulong (1785–1838), Alexis-Thérèse Petit (1791–1820), François Arago (1786–1853), and Augustin Fresnel (1788–1827). In contrast to the Laplacian dream of the one Newtonian method applied to all of the universe, the power and attraction of Fourier's work lay in a conscious and explicit narrowing of focus. The theory of heat would only deal with what was not reducible to the laws of mechanics: "whatever may be the range of mechanical theories, they do not apply to the effects of heat. These make up a special order of phenomena, which cannot be explained by the principles of motion and equilibrium" (Fourier [1822] 1955, 2; see also p. 23).

Fourier remained noncommittal about the ultimate metaphysical nature of heat, and in theorizing he did not focus on considerations of "deep" causes. The starting point of his analysis was simply that there be some initial distribution of heat, and some specified temperatures on the boundaries of the body being considered; by what mechanisms these initial and boundary conditions would be produced and maintained were not his concerns. Then he produced equations that would predict the observed diffusion of the initial distribution over time, and he hardly made any attempt at a metaphysical justification of his equations. The antimetaphysical bias in Fourier's work had a good deal of affinity to positivist philosophy. As documented by Fox, Fourier attended the lectures of Auguste Comte (1798–1851) on positivism in 1829; Comte for his part admired Fourier's work, so much so that he dedicated his *Course of Positive Philosophy* to Fourier.[64] The affinity of Fourier's work to positivism was also emphasized in Ernst Mach's retrospective appraisal: "Fourier's theory of the conduction of heat may be characterized as an ideal physical theory.... The entire theory of Fourier really consists only in a consistent, quantitatively exact, abstract conception of the facts of conduction of heat—in an easily surveyed and systematically arranged *inventory* of facts" (Mach [1900] 1986, 113).

Fourier represented only one section of heat theory emerging in the twilight of Laplacianism, but the phenomenalistic trend away from microphysics was a broader one, though by no means unanimous. Another important instance of phenomenalism was the work of the engineer and army officer Sadi Carnot (1796–1832), which will be discussed further in "William Thomson's Move to the Abstract" in chapter 4. Carnot's *Reflections on the Motive Power of Fire* (1824) was based on a provisional acceptance of the caloric theory, but it steered away from microphysical reasoning. His analysis of the ideal heat engine only sought to find relations holding between the macroscopic parameters pertaining to a body of gas: temperature, pressure, volume, and the amount of heat contained in the gas; all but the last of these variables were also directly measurable. When the civil engineer Émile Clapeyron (1799–1864) revived Carnot's work in 1834, and even when William Thomson initially took up the Carnot–Clapeyron theory in the late 1840s, it was still in this macroscopic-phenomenalistic vein, though Thomson's later work was by no means all phenomenalistic.

In addition to phenomenalism, experimental precision was the other major preoccupation in the empiricism that increasingly came to dominate nineteenth-century French physics. In itself, the quest for experimental precision was quite compatible with Laplacianism, though it became more prominent as the Laplacian emphasis on microphysical theorizing waned. According to many historians, the drive toward precision measurement was a trend that originated in the "quantifying spirit" originating in the Enlightenment,[65] which continued to develop through and beyond the heyday of Laplace. The highest acclaim for precision in the early nineteenth century went to Dulong and Petit, both identified by Fox as leading rebels against Laplacian physics. The Dulong–Petit collaboration is perhaps best known now for their controversial "law of atomic heat" announced in 1819 (the observation that the product of atomic weight and specific heat is constant for all

---

[64]The dedication was shared with Henri Marie Ducrotay de Blainville, the anatomist and zoologist. See Fox 1971, 265–266, for further discussion of Fourier's relationship with Comte.
[65]See, for example, Frängsmyr et al. 1990 and Wise 1995.

elements, which was taken to imply that all individual atoms had the same heat capacity). However, it was their two earlier joint articles on thermal expansion, the laws of cooling and thermometry (1816 and 1817) that won them undisputed respect at home and abroad, a glimpse of which can be had in Lamé's statement quoted in "Regnault: Austerity and Comparability."[66]

These trends formed the style of science in which Regnault was educated and to which he contributed decisively. Duhem (1899, 392) credited Regnault with effecting "a true revolution" in experimental physics. In Edmond Bouty's estimation (1915, 139): "For at least twenty-five years, the methods and the authority of Regnault dominated all of physics and became imperative in all research and teaching. Scruples for previously unknown degrees of precision became the dominant preoccupation of the young school." But what was so distinctive and powerful about Regnault's work, compared to the works of his important empiricist predecessors such as Fourier, Carnot, Clapeyron, Dulong, Petit, and Lamé?

One point is clear, and banal: Regnault's revolution was a revolution in experimental physics, not in theoretical physics. Regnault made few contributions to theory, and theory could not help the revolution that Regnault was trying to launch. Contrast that to the work of other phenomenalists. Although Fourier and Carnot were empiricists, they did not contribute very much to empirical work; that is not a paradox, only a play on words. Theorizing about observable properties does not necessarily have anything to do with producing actual observations. Fourier's and Carnot's theories did nothing immediately to improve observations. Take, again, the case of thermometry. Because Fourier declined to deal with any mechanical effects of heat (including the thermal expansion of matter), the tradition of heat theory established by him could give no help in elucidating the workings of the thermometer. In fact Fourier displayed a remarkable degree of complacency about thermometry ([1822] 1955, 26–27). Carnot's theory of heat engines only made use of the presumably known relations regarding the thermal expansion of gases and could not make any contributions toward a justification of those relations. Fourier and Carnot were at best consumers of empirical data, and the best that consumers can do is to stimulate production by the demand they create.

Phenomenalists in the tradition of Fourier and Carnot were only antimetaphysical; Regnault was antitheoretical. That is to say, even phenomenological theory was a target of skepticism for Regnault. His experiments subjected wellknown empirical laws to minute scrutiny with corrosive effect. Regnault had initially come to physics from chemistry through the study of specific heats in relation to Dulong and Petit's law.[67] After finding that law to be only approximately true (as many had suspected in any case), he turned to the more trusted regularities

---

[66]The latter paper won a prize competition of the Paris Academy. Fox (1971, 238) lists Comte, Poisson, Lamé, and Whewell as some of the leading authorities who admired this work as a model of experimental method.

[67]See Dumas 1885, 162. The results from this investigation were published in Regnault 1840, as well as two subsequent articles.

regarding the behavior of gases. As Regnault cranked up the precision of his tests, even these laws were shown up as tattered approximations. Already by 1842 Regnault had collected enough data to refute two laws that had been regarded as fundamental truths regarding gases: (1) all types of gases expand to the same extent between the same limits of temperature, which was the conclusion of Gay-Lussac's and Dalton's experiments forty years earlier; (2) a given type of gas expands to the same extent between the same limits of temperature regardless of its initial density, which had been generally believed since Amontons's work in 1702.[68] Regnault's memoir of 1847 repeated the refutation of these laws with further details and also gave results that showed Mariotte's (Boyle's) law to be only approximately and erratically true.[69]

These experiences disillusioned Regnault about the presumed progress that experimental physics had made up to his time. If some of the most trusted empirical regularities were shown to be false, then none of them could be trusted without further assurance. From that point on he eschewed any reliance on presumed laws and set himself the task of establishing true regularities by thorough data-taking through precision measurements. While he was engaged in this enterprise, it is understandable that he did not find himself excited by the new theoretical speculations issuing from the seemingly fickle brains of the likes of Faraday, Ørsted, Joule, and Mayer, whom posterity has praised for their bold and penetrating insights.[70] When De Luc said that "the moral and physical Microscope are equally fit to render men cautious in their theories" (1779, 20), he could not have anticipated the spirit of Regnault's work any better. In Berthelot's estimation, Regnault was "devoted to the search for pure truth, but that search he envisioned as consisting above all in the measurement of numerical constants. He was hostile to all theories, keen to emphasize their weaknesses and contradictions" (Berthelot quoted in Langevin 1911, 44–45). For Regnault, to search for truth meant "to replace the axioms of theoreticians by precise data" (Dumas 1885, 194).

It may seem that the task Regnault set for himself was a laborious yet straightforward one. However, to someone with his intellectual integrity, it was all too obvious that the existing measurement methods relied on theoretical regularities, exactly the kind that he was hoping to test conclusively by measurements. Thus, Regnault came face to face with the fundamental circularity of empiricist theory testing. A complete avoidance of theory would have paralyzed experiment altogether. With the recognition that each experiment had to take some assumptions for granted, Regnault's conscience forced him to engage in further experiments to

---

[68]See Regnault 1842a, 1842b. Part 2 of each memoir deals with the former law, and part 1 of the first memoir deals with the latter.

[69]On the two expansion laws, see Regnault 1847, 91, 119–120. Regarding Mariotte's law, Regnault's results (1847, 148–150, 367–401) showed that it held for carbonic acid at 100°C but not at 0°C, even at low densities; for atmospheric air and nitrogen it was generally not true; the behavior of hydrogen also departed from it, but in the opposite direction from air and nitrogen. Regnault had entertained the belief that the gas laws would be true "at the limit" (that is, when the density of the gas approaches zero), in the conclusion of his earlier investigation (Regnault 1842b). By 1847 he seems to have abandoned even such limited hope.

[70]See Dumas 1885, 191.

test those assumptions. There was no end to this process, and Regnault got caught up in what Matthias Dörries has characterized as a never-ending circle of "experimental virtuosity." It seems that Regnault's original intention was to start with observations cleansed of theory, then to move on to careful theorizing on the basis of indisputable data. However, the task of obtaining indisputable data turned out to have no end, which meant that theoretical activity had to be postponed indefinitely. Regnault himself seems to have felt some frustration at this aspect of his work; in 1862 he referred to a circle that he was not able to get out of: "[T]he more I advanced in my studies, the more the circle kept growing continually..."[71] Such sense of frustration is probably behind much of the lukewarm appraisals that Regnault's work has received from later generations of scientists and historians.[72] For example, Robert Fox is unequivocal in acknowledging Regnault's "monumental achievements," but at the same time he judges that Regnault's "preoccupation with the tedious accumulation of results" was unfortunate especially in view of "the momentous developments in physics taking place outside France during the 1840s" (1971, 295, 299–300).

That seems like a fair assessment in some ways, especially if one focuses on theoretical developments as the most exciting part of scientific progress. However, in some other ways I think Duhem was closer to the mark when he said that Regnault had effected a "true revolution" in physics. To see why, it is important to recognize the ways in which he was more than just an exceptionally careful and skilled laboratory technician. In that respect, it is informative to compare the character of his work with that of his predecessors in the tradition of precision measurement. I will not be able to make that comparison in any comprehensive sense, but as a start it is very instructive to compare Regnault's thermometry with Dulong and Petit's thermometry. Since Dulong and Petit were seen as the unquestioned masters of precision experiments in French physics before Regnault, their work is the best benchmark against which we can assess Regnault's innovations.

In retrospect, Dulong and Petit's decisive contribution to thermometry was to highlight the urgency of the need to make a rational choice of the thermometric fluid. This they achieved in two important ways. First of all they demonstrated that the magnitude of the mercury–air discrepancy was very significant. While confirming Gay-Lussac's earlier result that the mercury thermometer and the air thermometer agreed perfectly well with each other between the freezing and boiling points of water, Dulong and Petit carried out the comparison in high temperatures, where no one before them had been able to make accurate determinations. Their results showed that the discrepancy between mercury and air thermometers increased as temperature went up, reaching about 10 degrees at around 350 degrees centigrade (mercury giving higher numbers).[73] That amount of discrepancy clearly made it impossible to use the two thermometric fluids interchangeably. Second, it was not

---

[71] See Dörries 1998b, esp. 128–131; the quoted passage is from p. 123.
[72] Dörries 1997, 162–164, gives a summary of some of the critical appraisals.
[73] See Dulong and Petit 1817, 117–120, including a summary of results in table 1 on p. 120. See also Dulong and Petit 1816, 250, 252.

plausible to attribute this mercury–air discrepancy to experimental error, given the extraordinary care and virtuosity apparent in Dulong and Petit's experimental procedures. They were justly proud of their achievement in this direction and asserted that they had reached the highest possible precision in this type of experiment.[74] No one credibly challenged their confidence—until Regnault.

Were Dulong and Petit able to reach a definitive verdict on the choice of thermometric fluids, with their superior skills in precision measurement? They certainly set out to do so. Their article began by noting the failure of De Luc and Dalton to provide satisfactory answers to this question. Their chief criticism of Dalton, especially, was that his doctrines were not based on empirical data, and the presumption was that their own precision experiments would provide the needed data.[75] For their own positive contribution, Dulong and Petit (1817, 116) started by stating the requirement of a true thermometer, based on a simple conception of temperature that abandoned all Laplacian sophistication: if additions of equal amounts of heat produce equal increases in the volume of a substance, then that is the perfect thermometric substance. However, they did not consider that condition to be amenable to a direct empirical test, since the quantity of heat was a difficult variable to measure especially at higher temperatures. Instead, their strategy was to start by using the standard mercury thermometer in order to observe the thermal expansion of some candidate substances that were free from obvious factors disturbing the uniformity of expansion, such as gases and metals.

What these observations were meant to allow them to conclude is not absolutely clear, but the thought seems to have been the following (Dulong and Petit 1816, 243). If many candidate substances display the same pattern of thermal expansion, then each of them should be taken to be expanding uniformly. Such agreement is most likely an indication that disturbing factors are not significant, presumably because it would be a very unlikely coincidence if the disturbing factors, which would have to be different for different substances, resulted in exactly the same distortions in the patterns of expansion. Their empirical research revealed that this expectation of uniformity was not fulfilled across different metals, so they concluded that gases were the best thermometric substances.[76] This is an interesting argument but it is ultimately disappointing, first of all because it does not constitute any theoretical advancement over the earlier caloric inference that Gay-Lussac and Dalton had enough perspicuity to distrust. As Dulong and Petit themselves recognized, this was only a plausibility argument—and the plausibility was in fact significantly diminished in the absence of the support by the caloric metaphysics of mutually attracting matter particles held apart by self-repulsive caloric.

[74]Dulong and Petit 1817, 119. A detailed description of their procedures is given in Dulong and Petit 1816, 245–249.
[75]Dulong and Petit 1816, 241–242; Dulong and Petit 1817, 114–115.
[76]Dulong and Petit 1817, 153. For the results on metals, see Dulong and Petit 1816, 263, and Dulong and Petit 1817, 136–150. There is no indication that they carried out any extensive new work showing the uniformity of thermal expansion in different types of gases. Their articles only reported experiments on atmospheric air; aside from that, they merely referred to Gay-Lussac's old results regarding the uniformity across gas types. See Dulong and Petit 1816, 243.

Dulong and Petit failed to solve the problem of thermometric fluids because their work was not framed with philosophical sophistication, not because their experimental technique was insufficient. Only Regnault solved the epistemic problem. Dulong and Petit started on the wrong foot by defining temperature in a nonobservational way; the argument they envisioned for the demonstration of linear expansion was bound to be weak, however good their data might have been. In contrast, as explained in the three previous sections, Regnault devised the strongest possible argumentative strategy and was fortunate enough for that strategy to work. Regnault made the advancement from thermoscopes to numerical thermometers (from stage 2 to stage 3 of temperature standards) about as well as could have been done under the material conditions on earth. In more common scientific terms, he brought practical thermometry to utmost perfection.

Surely, however, there were directions in which Regnault's work left room for further progress. Being completely independent of theory was a clear virtue up to a certain point, but there was a later time when connecting to heat theory was a valuable thing to do. The further development of theoretical thermometry, especially in the hands of William Thomson, is treated in chapter 4. Before we come to that, however, there is another story to tell. Perfect as Regnault's air thermometers were, they were not adapted for extreme temperatures, especially the high end where glass ceased to be robust. In the next chapter we will retrace some key steps in the measurements of very low and very high temperatures, starting again in the middle of the eighteenth century.

3

To Go Beyond

Narrative: Measuring Temperature When
Thermometers Melt and Freeze

> Now, when it is desired to determine the magnitude of some high temperature, the target emissivity is established using a reflected laser beam, the temperature is measured by an infrared-sensing, two-colour pyrometer, information is automatically logged into a computer data bank, and the engineer in charge gives no thought to the possibility that it might not always have been done this way.
>
> J. W. Matousek, "Temperature Measurements in Olden Tymes," 1990

In the last two chapters I examined the establishment of the most basic elements in the measurement of temperature: fixed points and a numerical scale. The narrative in chapter 2 focused on the efforts to establish the numerical scale with increased rigor, continuing into the mid-nineteenth century. But as soon as a reasonable numerical scale was established, a different kind of objective also became apparent: to extend the scale beyond the temperature range in which it was comfortably established. The obvious challenge was that the mercury thermometer physically broke down when mercury froze or boiled. Gaining any definite knowledge of thermal phenomena beyond those breaking-points was very much like mapping a previously uncharted territory. In the narrative part of this chapter I will present a double narrative representing how the challenge was tackled at both ends, focusing on the investigations into the freezing point of mercury, and the master potter Wedgwood's efforts to create a thermometer that could measure the temperature of his kilns. Both are stories of surprising success combined with instructive failure. The analysis will discuss further some crucial issues of justification and meaning raised in these narratives about the extension of empirical knowledge beyond its established domains. There I will use a revitalized version of Percy

Bridgman's operationalist philosophy to shed further light on the process of extending concepts beyond the domains of phenomena in which they were originally formulated.

Can Mercury Be Frozen?

Johann Georg Gmelin (1709–1755), professor of chemistry and natural history at the Imperial Academy in St. Petersburg, had the enormous challenge of leading a team of scholarly observers on a ten-year trek across Siberia starting in 1733.[1] The expedition had been ordered by Empress Anna Ivanovna, who sought to realize a favorite idea of her uncle, Peter the Great, to acquire better knowledge of the vast eastern stretches of the Russian Empire. There was also a plan to make a grand rendezvous at the other end with the sea expedition led by Captains Vitus Bering (1681–1741) and Alexei Chirikov (1703–1748), to explore access to America. Gmelin's party endured a daunting degree of cold. In some places they found that even in the summer the earth had several feet of frozen soil underneath the surface. At Yeniseisk during their second winter, Gmelin recorded (quoted in Blagden 1783, 362–363):

> The air seemed as if it were frozen, with the appearance of a fog, which did not suffer the smoke to ascend as it issued from the chimneys. Birds fell down out of the air as if dead, and froze immediately, unless they were brought into a warm room. Whenever the door was opened, a fog suddenly formed round it. During the day, short as it was, parhelia and haloes round the sun were frequently seen, and in the night mock moons and haloes about the moon.

It was impossible to sense the exact degree of this numbing cold. However, Gmelin noted with satisfaction: "[O]ur thermometer, not subject to the same deception as the senses, left us no doubt of the excessive cold; for the quicksilver in it was reduced to −120° of Fahrenheit's scale [−84.4°C]." This observation astonished the world's scientists, as it was by far the lowest temperature ever recorded anywhere on earth. For example, William Watson (1715–1787), renowned naturalist and "electrician," later to be physician to the London Foundling Hospital, noted that "such an excess of cold can scarcely have been supposed to exist, had not these experiments demonstrated the reality of it." Gmelin's observations were "scarce to be doubted" thanks to the thermometer (Watson 1753, 108–109).

To Charles Blagden (1748–1820) examining this account half a century later, however, Gmelin's mistake was apparent. It did not seem likely at all that even the Siberian winter temperatures would have been so much as almost 100°F lower than the lowest temperatures observed in northern Europe. Blagden (1783, 371) inferred that the mercury in Gmelin's thermometer must have actually frozen and shrunk drastically, indicating a much lower temperature than actual. Gmelin had not

---

[1] The account of Gmelin's Siberian expedition is taken from Blagden 1783, 360–371, and Vladislav Kruta's entry on Gmelin in the *Dictionary of Scientific Biography*, 5:427–429.

considered the possibility of the freezing of mercury, which was commonly considered as essentially fluid at that time.[2] In fact he rejected the idea summarily, when it was brought to his attention forcefully two years later in Yakutsk when one of his colleagues[3] noted that the mercury in his barometer was frozen. Gmelin was shown the solidified mercury, but convinced himself that this was due to the presence of water in the mercury, which had been purified using vinegar and salt. He confirmed this explanation by taking the mercury out of the barometer, drying it well, and seeing that it would not freeze again, "though exposed to a much greater degree of cold, as shown by the thermometer." *By which thermometer?* To Blagden (1783, 364–366), Gmelin's "confirmation" only confirmed that the mercury thermometer could not be trusted. The following winter, Gmelin even observed the evidence of mercury congelation in his own instruments, when he noted that the mercury columns in his thermometer and barometer were broken up by air bubbles. He had some trouble explaining this appearance, but he refused to consider freezing as a possibility (Blagden 1783, 368–369).

The first demonstration that mercury could really be frozen came only twenty-five years later, and that was the work of Joseph Adam Braun (1712?–1768), professor of physics at the St. Petersburg Academy. The winter of 1759–60 in St. Petersburg was very severe, recorded temperatures reaching 212° Delisle in December, which is equivalent to −41.3°C, or −42.4°F. (The Delisle temperature scale, which I will come back to in "Temperature, Heat, and Cold" in chapter 4, was set at 0° at the boiling point of water, and had increasing numbers with increasing cold, with the freezing point of water designated as 150°.) Braun took that opportunity to produce the greatest degree of artificial cold ever observed, using a "freezing mixture" of *aqua fortis* (nitric acid, HNO3) mixed with snow.[4] The mercury thermometer descended beyond 600° Delisle (−300°C), and the mercury in it was quite frozen, which Braun confirmed beyond doubt by breaking the thermometer (not an insignificant material sacrifice at that time) and examining the solid metal (Watson 1761, 158–164). As far as Braun could see, the solidification of mercury was no different from the freezing of any liquid, and hence a mere effect of the "interposition of cold."

---

[2]See Watson 1761, 157, for evidence that this attitude was current well into the middle of the eighteenth century: "[F]or who did not consider quicksilver, as a body, which would preserve its fluidity in every degree of cold?"

[3]This was probably the astronomer Louis de l'Isle de la Croyere—brother of Joseph-Nicolas Delisle, who devised the temperature scale popular in Russia.

[4]"Freezing mixtures" provided the only plausible means of producing extremely cold temperatures at the time. The cooling is produced by the action of a substance (most often an acid) added to snow (or ice), which causes the latter to melt and absorb a great deal of heat (latent heat of fusion) from the surroundings. Fahrenheit is reputed to be the first person to have used freezing mixtures, though it seems quite likely that Cornelius Drebbel had used it much earlier. Before Joseph Black's work on latent heat, the working of freezing mixtures must have seemed quite mysterious. As Watson put it (1761, 169): "That inflammable spirits should produce cold, seems very extraordinary, as rectified spirit seems to be liquid fire itself; and what still appears more paradoxical is, that inflammable spirits poured into water, causes heat; upon snow, cold: and what is water, but melted snow?"

Braun's work was certainly seen as wondrous. William Watson exclaimed in his official report of Braun's work to the Royal Society of London:

> Who, before Mr. Braun's discovery, would have ventured to affirm mercury to be a malleable metal? Who, that so intense a degree of cold could be produced by any means? Who, that the effects of pouring nitrous acid upon snow, should so far exceed those, which result from mixing it with ice...? (Watson 1761, 172)

Afterwards it was confirmed that mercury could be frozen by a natural cold after all, by another German naturalist working in Russia, Peter Simon Pallas (1741–1811). Pallas was invited by Catherine the Great to lead a Siberian expedition, which he did successfully from 1768 to 1774.[5] In December 1772 he observed the freezing of the mercury in his thermometer, and then confirmed the fact more definitely by freezing about a quarter-pound of mercury in a saucer.[6] Those who still found the freezing of mercury difficult to accept tried to explain it away by casting doubts on the purity of Pallas's mercury, but in the end various other experiments convinced the skeptics.

The immediate and easy lesson from the story of freezing mercury is that unexpected things can and do happen when we go beyond the realms of phenomena that are familiar to us. The utilitarian jurist Jeremy Bentham (1748–1832) used this case to illustrate how our willingness to believe is tied to familiarity. When Bentham mentioned Braun's experiment to a "learned doctor" in London, this is the reaction he got: "With an air of authority, that age is not unapt to assume in its intercourse with youth, [he] pronounced the history to be a lie, and such a one as a man ought to take shame to himself for presuming to bring to view in any other character." Bentham compared this with the tale of the Dutch voyagers (reported by John Locke), who were denounced by the king of Siam "with a laugh of scorn" when they told him that in the Netherlands water would become solid in the winter so that people and even wagons could travel on it.[7] Locke's story may be apocryphal, but the philosophical point stands.

Now, this immediate problem of simple prejudice could be solved by sufficient experience, or even be prevented by a healthy does of open-mindedness or theoretical daring. For instance, William Cleghorn of Edinburgh, who developed an early version of the caloric theory, reckoned that "it is not improbable, that by a very great diminution of its heat air itself might become solid," long before any such thing was physically realized.[8] However, Gmelin's predicament also indicated a more profound and difficult question. If we admit, with Blagden, that the mercury thermometer must malfunction as it approaches the freezing point of mercury, what do we propose to use instead for the measurement of such low temperatures? More generally, if we admit that the behavior of matter in new domains may not conform

---

[5]See Urness 1967, 161–168, for a biographical sketch of Pallas.
[6]*Voyages* 1791, 2:237–242.
[7]Bentham 1843, 7:95; I thank Dr. Jonathan Harris for this reference.
[8]This conjecture, he said, was "confirmed by the analogy of other vapours"; see *Encyclopaedia Britannica*, 2d ed., vol. 5 (1780), 3542.

to what we know from more familiar domains, we are also forced to admit that our familiar and trusted observational instruments may cease to function in those new domains. If the observational instruments cannot be trusted, how do we engage in empirical investigations of phenomena in the new domains? We have seen in chapter 2 that the mercury thermometer was the best temperature standard in the eighteenth century. When the best available standard fails, how do we create a new standard, and on what basis do we certify the new standard?

Can Mercury Tell Us Its Own Freezing Point?

The immediate difficulty, after it was admitted that mercury was indeed capable of freezing, was to determine its freezing temperature. Braun admitted that the point seemed to be "at too great a latitude to be exactly determined." His reported that 469° Delisle (−212.7°C) was the warmest at which he observed any congelation of mercury, but the "mean term of the congealation [sic] of mercury" was estimated at 650° Delisle (−333.3°C). These were numbers obtained from the readings of the mercury thermometer. Braun also had thermometers made with "highly rectified spirit of wine" (concentrated ethyl alcohol). He reported that he was unable to freeze the alcohol, and that the alcohol thermometers only indicated 300° Delisle (−100°C) at the degrees of cold that froze mercury. All three alcohol thermometers he employed agreed well with each other and also agreed with mercury thermometers at lesser degrees of cold.[9] Later Pallas reported from Siberia that his frozen mercury was observed to melt at 215° Delisle (−43.3°C).[10] So there was a broad range, spanning nearly 300°C, in which the true freezing point of mercury lay hidden, if one took Braun and Pallas as the best and equally trustworthy authorities on this point. Matthew Guthrie (1732–1807), Scottish physician then working for the army in St. Petersburg, commented in 1785 that nothing at all was certain about the freezing of mercury, except that it was possible. Guthrie (1785, 1–4, 15) concluded from his own experiments, employing an alcohol thermometer, that "the true point of congelation of the purest mercury" was −32° Réaumur (−40°C), and noted the good agreement of that value with Pallas's.

A good sense of the early uncertainties on this matter can be gained from Jean-André De Luc's defense of Braun. In "De Luc and the Method of Mixtures" in chapter 2, we saw that De Luc made a strong argument for mercury as the thermometric fluid of choice, in his 1772 treatise *Inquiries on the Modifications of the Atmosphere*. In the same text he argued that Braun's experiments presented no reason to doubt that the contraction of mercury by cold was quite regular down to its freezing point. Shortly after the announcement of Braun's results, an objection

---

[9]These estimates are cited in Watson 1761, 167–171.
[10]*Voyages* 1791, 2:241. There is a slight puzzle there, since Pallas also seems to have said that this number (215° Delisle) corresponded to 29 degrees below the freezing point (of water) on the Réaumur scale. With the Réaumur scale as commonly understood in the late eighteenth century (De Luc's design), −29°R would have been −36.25°C. But it is not clear how exactly Pallas's "Réaumur" scale was graduated, so it is safer to go with the number cited on the Delisle scale.

had been published in the Paris *Journal des Savants*. At bottom the author, by the name of Anac, was simply incredulous that temperatures like 650° Delisle (−333°C, or −568°F) could have been reached by the quite ordinary cooling methods used by Braun. But De Luc chided Anac for presuming to judge observations about a new domain of phenomena on the basis of what he knew about more familiar cases. Anac also maintained that Braun's temperatures could not be real because they were even lower than absolute zero, which was estimated at 521 and 3/7 degrees Delisle (−247.6°C). Here he was citing the number originating from the work of Guillaume Amontons, obtained by extrapolating the observed temperature–pressure relation of air to the point of vanishing pressure (see "William Thomson's Move to the Abstract" in chapter 4 for more on "Amontons temperature"). De Luc scoffed at this estimate and the very concept of absolute zero, pointing out the shakiness of both the assumption that the observed linear temperature–pressure relation would continue down to zero and the theoretical presumption that temperature was so essentially connected to air pressure (De Luc 1772, 1:256–263, §416).

De Luc also argued that the apparent discrepancy in Braun's own numbers could be reconciled. This argument pointed up a serious question about the apparently sensible practice of using an alcohol thermometer to take temperatures where mercury freezes or presumably begins to behave anomalously near its freezing point. We have seen in "De Luc and the Method of Mixtures" in chapter 2 that De Luc used the method of mixtures to argue that the readings of the standard alcohol thermometer were inaccurate, nearly 8°C below the "real" temperature at the midpoint between the boiling point and the freezing point. De Luc had also demonstrated a serious discrepancy between the readings of the alcohol and the mercury thermometers; the mercury–alcohol discrepancy is real regardless of the cogency of the method of mixtures.

De Luc's data indicated that alcohol had a more "accelerated" expansion than mercury at higher temperatures. If the same pattern continued down to lower temperatures, alcohol would contract considerably less than would be expected from assuming linearity, which means that the linearly graduated alcohol thermometer would show readings that are not as low as the real temperature, or not as low as the readings of a mercury thermometer in any case. De Luc estimated that 300.5° Delisle (−80.75° on De Luc's own scale) on the alcohol thermometer would actually be the same degree of heat as 628° Delisle (−255° De Luc) on the mercury thermometer (see fig. 3.1). That made sense of Braun's puzzling result mentioned earlier, which gave the freezing point of mercury as 650° Delisle on the mercury thermometer and 300° Delisle on the alcohol thermometer. Hence, De Luc argued, there was no mystery there, and no particular reason to distrust the mercury thermometer near its freezing point. A more serious concern was about the accuracy of the alcohol thermometer, which seemed to be at least about 300° Delisle (or 200°C) off the mark near the freezing point of mercury.[11]

---

[11] For this interpretation, see De Luc 1772, 1:255–256, §416. De Luc's extrapolation also gave the result that the condensation of alcohol would stop altogether at 300.5° Delisle (−80.75° De Luc).

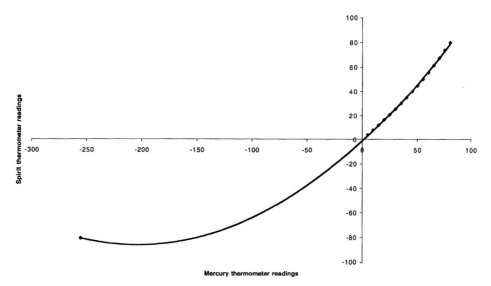

FIGURE 3.1. De Luc's comparison of mercury and alcohol thermometers, and an extrapolation of the results. The same data have been used here as in series 1 in fig. 2.1 in chapter 2. The curve shown here is the best quadratic fit I can make on the basis of De Luc's data and extrapolation.

De Luc's arguments clearly reveal two obstacles in the inquiry concerning the freezing of mercury. First, there were no temperature standards known to be reliable in such extreme degrees of cold. While De Luc shattered the false sense of security that some people might have derived from the alcohol thermometer, he himself could not offer any alternative standard that was more reliable. De Luc only made a wishful and ill-supported conjecture that mercury probably continued its reasonably regular contraction right down to its freezing point. Second, we may note that De Luc's arguments were speculative, even as he chided Anac for his groundless assumptions about the absolute zero. It was difficult for De Luc to avoid speculation, since in the relatively mild climate of Geneva he could not perform his own experiments to clear up the uncertainties that arose in his discussion. At that time, even with the best freezing mixtures, it was impossible to attain low enough temperature unless one started out at very cold natural temperatures found in such places as St. Petersburg.

A development to overcome these obstacles occurred shortly after De Luc's work, with a little help from British imperialism.[12] The Royal Society's desire to arrange new experiments on the freezing of mercury found its fulfillment in the

---

[12] For a concise yet fully informative account of this development, see Jungnickel and McCormmach 1999, 393–400.

person of Thomas Hutchins (?–1790), the governor of Fort Albany, in present-day Ontario.[13] Hutchins had been in North America in the employ of the Hudson's Bay Company since 1766, where he also began to build a reputation as a naturalist in collaboration with Andrew Graham. In 1773, while back in England on leave, he entered into agreement with the Royal Society to make some observations on the dipping needle and the congelation of mercury. Hutchins first made successful experiments in 1775 to freeze mercury (in the relative tranquility of Hudson's Bay, even as Britain started to wage war with the colonial revolutionaries down south). However, like Braun earlier, Hutchins was not able to determine the freezing point with any confidence.

Hearing about Hutchins's experiments, Joseph Black stepped in to pass on some sage advice, via Graham, in a letter of 1779 (reproduced in Hutchins 1783, *305–*306). Black started with a negative assessment: "I have always thought it evident, from Professor Braun's experiments, that this degree of cold [necessary to freeze mercury] cannot be discovered conveniently by congealing the mercury of the thermometer itself." However, he suggested that the mercury thermometer could still be used to determine the freezing point of mercury, as follows. Insert a mercury thermometer into the middle of a wider cylinder filled with mercury, and cool the cylinder gradually from the outside. That way, the mercury outside the thermometer would begin to freeze before the mercury inside the thermometer. As soon as the mercury outside starts to assume the consistency of an amalgam, the reading on the thermometer should be noted. Black predicted confidently: "I have no doubt, that in every experiment, thus made, with the same mercury, the instrument will always point to the same degree."

Directing Hutchins's experiments from London on behalf of the Royal Society was Henry Cavendish. It was a happy enough coincidence that Cavendish, independently of Black, came up with almost precisely the same clever design for Hutchins. Perhaps it was not so much of a coincidence, if we consider that the design hinged crucially on Black's ideas about latent heat, which Cavendish also shared. Since freezing requires a great deal of heat to be taken away from the liquid even after its temperature reaches its freezing point (as explained in "The Case of the Freezing Point" in chapter 1), Black and Cavendish were confident that the larger cylinder full of mercury would take a good amount of time to freeze, and that its temperature would be nearly uniform and constant during that time. Cavendish (1783, 305) had in fact observed a similar pattern of behavior when molten lead and tin cooled and solidified. In short, keeping the thermometer in the middle portion, which would be the last to receive the full effect of the cooling, ensured that the mercury in the thermometer would not itself freeze yet approach the freezing point very closely. Cavendish explained the design of his apparatus as follows:

---

[13]The following information about Hutchins is from Glyndwr Williams's entry on him in the *Dictionary of Canadian Biography*, 4:377–378. I believe he is a different person from a contemporary of the same name, who was geographer to the United States.

If this cylinder is immersed in a freezing mixture till great part of the quicksilver in it is frozen, it is evident, that the degree shewn at that time by the inclosed thermometer is the precise point at which mercury freezes; for as in this case the ball of the thermometer must be surrounded for some time with quicksilver, part of which is actually frozen, it seems impossible, that the thermometer should be sensibly above that point; and while any of the quicksilver in the cylinder remains fluid, it is impossible that it should sink sensibly below it. (Cavendish 1783, 303–304)

Cavendish dispatched to Hudson's Bay the apparatus shown in figure 3.2, with detailed instructions. Hutchins carried out the experiments during the winter of 1781–1782 (just after the surrender of the British Army at Yorktown, Virginia): "The experiments were made in the open air, on the top of the Fort, with only a few deer-skins sewed together, placed to windward for a shelter: there was plenty of snow (eighteen inches deep) upon the works..." (Hutchins 1783, *320). Despite various technical difficulties and interpretive confusions, Hutchins managed to carry out the experiments to Cavendish's satisfaction, and he also performed some experiments of his own design. Hutchins concluded that the freezing point of mercury was −40°F (or −40°C; amusingly, this is exactly the point where the numbers on the Fahrenheit and the centigrade scales coincide). What gave Hutchins and Cavendish confidence about that number was that it was recorded in three different types of circumstances: while the mercury in the cylinder was freezing; while a ball of mercury frozen around the thermometer bulb was melting; and also when the thermometer was inserted into the liquid part of a frozen block of mercury that was melting (Cavendish 1783, 321–322).

The Hutchins-Cavendish result on the freezing point of mercury was accepted very gladly by many scientists. It agreed well with Pallas's earlier estimate (−43°C), and exactly with Guthrie's measurements using alcohol thermometers. Blagden was delighted:[14]

> The late experiments at Hudson's Bay have determined a point, on which philosophers not only were much divided in their opinion, but also entertained, in general, very erroneous sentiments. Though many obvious circumstances rendered it improbable, that the term of mercurial congelation should be 5[00] or 600 degrees below 0 of Fahrenheit's scale...yet scarcely any one ventured to imagine that it was short of 100°. Mr. Hutchins, however, has clearly proved, that even this number is far beyond the truth.... (Blagden 1783, 329)

There was only one slight modification to the result, arising from Cavendish's re-examination of Hutchins's thermometers after they were brought back to London. Hutchins's thermometers were found to be slightly off when compared to the standard thermometers graduated according to the procedures laid out by Cavendish's Royal Society committee in 1777. Accordingly Cavendish (1783, 309 and 321) estimated that the true freezing point of mercury was 38 and 2/3 degrees below

---

[14]According to Jungnickel and McCormmach (1999, 294), Blagden assisted Cavendish in the analysis of Hutchins's data and also in Cavendish's own experiments to freeze mercury in London.

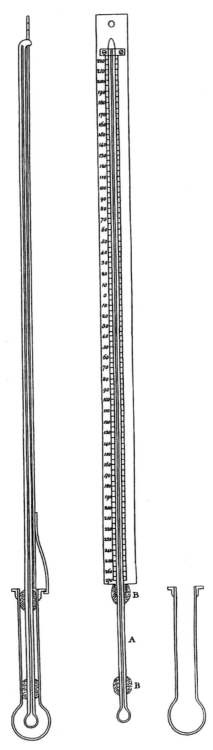

FIGURE 3.2. Cavendish's apparatus for determining the freezing point of mercury, used by Hutchins (1783, tab. 7, facing p. *370). Courtesy of the British Library.

zero, which he rounded up to −39°F. The Hutchins-Cavendish work certainly made a quick impression, as represented in the following small incident. When the Swedish chemist Torbern Bergman published his *Outlines of Mineralogy* in 1782, he had listed the freezing point of mercury as −654°F. William Withering, English physician and member of the Lunar Society of Birmingham, who translated the text into English, felt obliged to correct Bergman's table so that the "melting heat" of mercury read "−39 or −654 degrees Fahrenheit."[15] However, as we will see in the next section, Hutchins's results were not all straightforward.

Consolidating the Freezing Point of Mercury

Cavendish and Blagden spent the winter of 1782–1783 going over Hutchins's results, and the outcome was an article by Cavendish (1783) in which he made a largely successful effort to reach a clear and coherent interpretation of all of Hutchins's observations. From the vantage point of the twenty-first century, it is easy to underestimate the challenges involved in the eighteenth-century experimental work on the freezing of mercury. Most thermometers were unreliable and lacked agreement with each other. Freezing mixtures provided only temporary and variable sources of cold, making it difficult to freeze the mercury either slowly or in large quantities. The degree of cold attained depended crucially on the outside air temperature, over which one had no control. Reading the mercury thermometer near the freezing point of mercury was treacherous, because it was visually difficult to tell whether a thin column of mercury was frozen solid, or still liquid but stationary. Worse yet, frozen mercury would often stick to the inside of the thermometer stem, and then slide off when it got warmer, creating an inexplicable appearance of falling temperature. And the precious thermometers (brought over carefully on long bumpy journeys from St. Petersburg, London, etc. out to the wilderness) routinely broke in the middle of the experiments, because the extreme degree of cold made the glass brittle.

On top of all of those practical problems, there were tricks played by the phenomenon of supercooling. Mercury, like water, was capable of remaining fluid even when it had cooled below its "normal" freezing point. I have quoted Cavendish earlier as saying, "while any of the quicksilver in the cylinder remains fluid, it is impossible that it should sink sensibly below it [the freezing point]." He had to admit the falsity of that statement, in order to explain some "remarkable appearances" in Hutchins's experiments. Cavendish concluded that in some of the experiments the mercury must have been supercooled, remaining liquid below its freezing point. Supercooling created enough confusion when it was detected in water, but one can only imagine the bewilderment caused by the supercooling of mercury, since the "normal freezing point" itself was so disputed in that case. It is also easy to see how supercooling would have wreaked havoc with Cavendish's apparatus. Fortunately, the contemporary understanding of supercooling was just sufficient to

[15]See Jungnickel and McCormmach 1999, 398–399; Bergman 1783, 71, 83.

enable Cavendish to make sense of the most salient anomalies in Hutchins's results. Supercooling in mercury was still a hypothesis, but in fact Cavendish's analysis of Hutchins's results served to confirm the occurrence of supercooling, as well as the value of the freezing point. (This would make a good illustration of what Carl Hempel (1965, 371–374) called a "self-evidencing" explanation.)

In two cases (Hutchins's second and third experiments), the mercury thermometer indicated −43°F while the mercury in the outer cylinder remained fluid; then when the outside mercury started to freeze, the thermometer suddenly went up to −40°F. This was interpreted by Cavendish (1783, 315–316) as the typical "shooting" behavior at the breakdown of supercooling, in which just so much of the supercooled liquid suddenly freezes that the released latent heat brings the whole to the normal freezing temperature (see "The Case of the Freezing Point" in chapter 1 for further information about supercooling and shooting). If so, this observation did provide another piece of evidence that the normal freezing point was indeed −40°F. More challenging was the behavior observed in four other cases (fourth to seventh experiments): each time the thermometer "sank a great deal below the freezing point without ever becoming stationary at −40°" (Cavendish 1783, 317). Not only that but the details recorded by Hutchins showed that there were various other stationary points, some sudden falls, and temperatures down to below −400°F. In Cavendish's view (1783, 318), all of this meant "that the quicksilver in the [outer] cylinder was quickly cooled so much below the freezing point as to make that in the inclosed thermometer freeze, though it did not freeze itself. If so, it accounts for the appearances perfectly well." Guthrie (1785, 5–6), using a similar apparatus by Black's suggestion, reported a similar puzzling observation, in which mercury was still fluid but the thermometer inserted into it indicated −150° Réaumur (−187.5°C, or −305.5°F). He thanked Blagden for providing an explanation of this fact by reference to supercooling (cf. Blagden 1783, 355–359).

There was one remaining reason to be skeptical about the Hutchins-Cavendish value of the freezing point, which was curiously not discussed in any detail in either of their articles, or in Black's communication. The design of the Cavendish-Black apparatus focused on cajoling the mercury in the thermometer just near to the freezing point but not below it, thus avoiding the false readings caused by the significant contraction of mercury that occurs in the process of freezing. However, the design simply assumed that the contraction of mercury would continue regularly until it reached the freezing point. What was the argument for that assumption? Cavendish should have been wary of De Luc's earlier defense of mercurial regularity, since that was geared to defending the now-ludicrous freezing point of −568°F. Guthrie (1785, 5) thought that the assurance was only given by alcohol thermometers, which Hutchins had used alongside the mercury thermometers. I think Guthrie was correct to be suspicious of the mercury thermometer, but what were the grounds of his confidence that alcohol maintained a regular contraction right through the freezing point of mercury? He did not say.

At temperatures near the freezing point of mercury, both the alcohol and the mercury thermometers were on shaky ground. On the one hand, it was a widely accepted view that the thermal expansion of alcohol was generally not linear, as I have discussed in "De Luc and the Method of Mixtures" in chapter 2, and

Cavendish specifically declined to rely on alcohol thermometers.[16] On the other hand, it was difficult not to be wary of the readings of the mercury thermometer near its own freezing point. Even De Luc, the best advocate of the mercury thermometer, had expressed a general view that liquids near their freezing points are not likely to expand linearly (see the end of "De Luc and the Method of Mixtures"). If this was a battle of plausibilities between mercury and alcohol, the result could only be a draw. Hutchins's observations in fact made it clear that there was a discrepancy of about 10°F between the readings of his alcohol and mercury thermometers at the very low temperatures; Cavendish's estimate (1783, 321) was that the freezing point of mercury was between $-28.5°$ and $-30°F$ by the alcohol thermometer.[17]

Cavendish stuck to the mercury thermometer, with characteristic ingenuity. When he commissioned some further experiments in Hudson's Bay, this time carried out by John McNab at Henley House, Cavendish was forced to use alcohol thermometers, since the object was to produce and investigate temperatures clearly lower than the freezing point of mercury by means of more effective freezing mixtures. Still, Cavendish calibrated his alcohol thermometers against the standard of the mercury thermometer. Here is how:

> Mr. McNab in his experiments sometimes used one thermometer and sometimes another; but... I have reduced all the observations to the same standard; namely, in degrees of cold less than that of freezing mercury I have set down that degree which would have been shown by the mercurial thermometer in the same circumstances; but as that could not have been done in greater degrees of cold, as the mercurial thermometer then becomes of no use, I found how much lower the mercurial thermometer stood at its freezing point, than each of the spirit thermometers, and increased the cold shown by the latter by that difference. (Cavendish 1786, 247)

This way the Cavendish-McNab team measured temperatures reaching down to $-78.5°F$ ($-61.4°C$), which was produced by mixing some oil of vitriol (sulphuric acid, H_2SO_4) with snow (Cavendish 1786, 266).

Despite Cavendish's cares and confidence, it seems that temperature measurement at and below the freezing point of mercury remained an area of uncertainty for some time. When Blagden and Richard Kirwan made a suggestion that the freezing point of mercury should be used as the zero point of thermometry, the 3d edition of the *Encyclopaedia Britannica* in 1797 rejected the idea saying that the

---

[16] See Cavendish 1783, 307, where he said: "If the degree of cold at which mercury freezes had been known, a spirit thermometer would have answered better; but that was the point to be determined." I think what Cavendish meant was that knowing that point would have provided a calibration point from which his could extrapolate with some credibility the pattern of alcohol's contraction to lower temperatures, as he did do after the freezing point of mercury was stabilized.

[17] More details can be seen in Hutchins's comparison of the readings of different thermometers (1783, *308ff.); the thermometers are described on p. *307. This, then, throws suspicion on Guthrie's observation (1785, 11, etc.) that mercury was frozen at $-32°$ Réaumur ($-40°F$) by the alcohol thermometer. That would make sense only if his alcohol thermometer was calibrated by comparison with the mercury thermometer, in a procedure explained by Cavendish later.

freezing point of mercury was "not a point well known."[18] In 1808 John Dalton actually constructed a new thermometric scale that took the freezing of mercury as the zero point but, as we saw in "Caloric Theories against the Method of Mixtures" in chapter 2, he disputed the belief that mercury expanded nearly uniformly according to real temperature. Instead, Dalton's temperature scale was based on his own theoretical belief that the expansion of any pure liquid was as the square of the temperature reckoned from the point of greatest density (usually the freezing point, but not for water). Following this law, and fixing the freezing and boiling points of water at 32°F and 212°F, Dalton estimated the freezing point of mercury as −175°F (1808, 8, 13–14). There is no evidence that Dalton's thermometric scale came into any general use, but there were no arguments forthcoming to prove him wrong either. Both the mercury thermometer and the alcohol thermometer remained unjustified, and empirical progress in the realm of frozen mercury was difficult because such low temperatures were only rarely produced and even more rarely maintained steadily, until many decades later.

The most significant advance in the measurement of low temperatures in the first half of the nineteenth century arrived unceremoniously in 1837 and became an unassuming part of general knowledge. The author of this underappreciated work was Claude-Servais-Mathias Pouillet (1790–1868), who was soon to succeed Pierre Dulong in the chair of physics at the Faculty of Sciences in Paris at Dulong's death in 1838. Pouillet held that position until 1852, when he was dismissed for refusing to swear an oath of allegiance to the imperial government of Napoleon III. By the time Pouillet did his low-temperature work he had already taught physics in Paris for two decades (at the École Normale and then at the Faculty of Sciences), and he was also well known as a textbook writer. His early research was in optics under the direction of Biot, and he went on to do important experimental works in electricity and heat. Until Regnault, he was probably the most reliable experimenter on the expansion and compressibility of gases.[19]

The main ingredients of Pouillet's work are quite simple, in retrospect. A. Thilorier had just recently manufactured dry ice (frozen CO_2) and described a paste made by mixing it with sulphuric ether. Thilorier's paste could effect higher degrees of cooling than any freezing mixtures. Pouillet seized this new material as a vehicle to take him further into the low-temperature domain. First of all he wanted to determine the temperature of the cooling paste and concluded that it was −78.8°C.[20] For that purpose he used air thermometers. There is no surprise in Pouillet's use of the air thermometers, since he had worked with them previously and had voiced a clear advocacy of gas thermometers in his 1827 textbook, citing the arguments by

---

[18]See Blagden 1783, 397, and also *Encyclopaedia Britannica*, 3d ed. (1797), 18:496.

[19]See René Taton's entry on Pouillet in the *Dictionary of Scientific Biography*, 11:110–111.

[20]Pouillet 1837, 515, 519. This value is very close to the modern value of the sublimation point of carbon dioxide, which is −78.48°C or −109.26°F.

Dulong and Petit, and Laplace.[21] The more interesting question is why apparently no one had used the air thermometer in the investigation of temperatures near the freezing of mercury, when the advantage would have been obvious that air at these temperatures was not anywhere near liquefying, or changing its state in any other drastic manner. I do not have a convincing answer to that historical question, but the primary reasons may have been practical, particularly that most air thermometers were relatively large in size, and to surround one with a sufficient amount of freezing mercury or other substances to be observed would have required an impractical amount of high-quality cooling material. Pouillet was probably the first person to have sufficient cooling material for the size of the air thermometer available, thanks to Thilorier's paste and his own work on the improvement of air thermometers.

The extension of the air thermometer into the domain of the very cold gave Pouillet important advantages over previous investigators. Pouillet's confidence in the correctness of the air thermometer was probably unfounded (as we have seen in "The Calorist Mirage of Gaseous Linearity" in chapter 2), but at least the air thermometer provided an independent means of testing out the indications of the mercury and alcohol thermometers. Pouillet now set out to confirm the freezing point of mercury. He conceded that it was difficult to use the air thermometer for this experiment. He did not specify the difficulties, but I imagine that the experiment would have required a good deal of mercury for the immersion of the air thermometer, and it would have been quite tricky to cool so much mercury down to such a low temperature slowly enough while keeping the temperature uniform throughout the mass. A direct application of Thilorier's mixture would have produced an overly quick and uneven cooling. Whatever the difficulties were, they must have been severe enough because Pouillet took a laborious detour. This involved the use of a thermocouple, which measures temperature by measuring the electric current induced across the junction of two different metals when heat is applied to it. Pouillet (1837, 516) took a bismuth–copper thermocouple and checked its linearity against the air thermometer in a range of everyday temperatures (17.6°C to 77°C). Then he extended the thermocouple scale by simple extrapolation and was delighted by the "remarkable fact" that it measured the temperature of Thilorier's paste at −78.75°C, within about 0.1° of the air-thermometer readings of that particular point. This close agreement gave Pouillet confidence that the bismuth–copper thermocouple indicated very low temperatures correctly.

The thermocouple was then applied to the freezing mercury and gave the temperature of −40.5°C (or −40.9°F), which was very close to the Cavendish-Hutchins result of −39°F obtained by the mercury thermometer. Now Pouillet constructed alcohol thermometers graduated between the freezing point and the temperature of Thilorier's paste. Note that this scale, assuming the expansion of

---

[21]Pouillet 1827–29, 1:259, 263; see "The Calorist Mirage of Gaseous Linearity" in chapter 2 for the content of the arguments that Pouillet cited.

alcohol to be linear in that sub-zero range, was not the same as the alcohol scale extrapolated linearly from warmer temperatures. Pouillet (1837, 517–519) made six different alcohol thermometers, and noted that they all gave readings within 0.5° of −40.5°C for the freezing point of mercury. He was pleased to note: "These differences are so small as to allow us to conclude that the alcohol thermometer is in perfect accord with the air thermometer from 0°C down to −80°C." What he had established was quite an impressive consistent ring of measurement methods: the expansion of air, the intensity of current in the bismuth–copper thermocouple, and the expansion of alcohol all seemed to be proportional to each other in the range between the freezing point of water and the temperature of Thilorier's paste; moreover, the expansion of mercury also seemed to conform to the same pattern down to the freezing point of mercury itself. Of course, we must admit that Pouillet's work was nowhere near complete, since he had only checked the agreement at one point around the middle of the relevant range, namely the freezing point of mercury, and also did not bring the air thermometer itself to that point. But if we consider how much the expansion of alcohol, air, and mercury differed from each other in higher temperatures,[22] even the degree of consistency shown in Pouillet's results could not have been expected with any complacency. At last, a century after Gmelin's troubles in Siberia, there was reasonable assurance that the freezing point of mercury was somewhere near −40°C.

Adventures of a Scientific Potter

When we shift our attention to scientific work at the other end of the temperature scale, we find a rather different kind of history. The production of temperatures so low as to push established thermometers to their limits was a difficult task in itself. In contrast, extremely high temperatures were well known to humans both naturally and artificially for many centuries; by the eighteenth century they were routinely used for various practical purposes and needed to be regulated. Yet, the measurement of such high temperatures was open to the same kind of epistemic and practical difficulties as those that obstructed the measurement of extremely low temperatures. Therefore *pyrometry*, or the measurement of high temperatures, became an area of research occupying the attention of a wide range of investigators, from the merely "curious" gentlemen to the most hard-nosed industrialists.

In 1797 the much-expanded third edition of the *Encyclopaedia Britannica* was pleased to note the recent progress in thermometry (18:499–500): "We are now therefore enabled to give a scale of heat from the highest degree of heat produced by an air furnace to the greatest degree of cold hitherto known." *Britannica* identified the latest development in pyrometry as the work of "the ingenious Mr Josiah Wedgwood, who is well known for his great improvement in the art of pottery." But

---

[22] See the discussion of De Luc's comparison of mercury and alcohol in "De Luc and the Method of Mixtures" and the discussion of Dulong and Petit's comparison of mercury and air in "Regnault and Post-Laplacian Empiricism" in chapter 2.

the high temperatures reported there are suspect: the melting point of cast iron is put down as 17,977°F, and the greatest heat of Wedgwood's small air-furnace as 21,877°F. Surely these are much too high? Scientists now believe that the surface of the sun has a temperature of about 10,000°F.[23] Can we really believe that Wedgwood's furnace was 10,000 degrees hotter than the sun? A couple of questions arise immediately. What kind of thermometer was Wedgwood using to read temperatures so far beyond where mercury boils off and glass melts away? And how can one evaluate the reliability of a thermometer that claims to work in the range where all other thermometers fail? Wedgwood's thermometer was widely used and his numbers were cited in standard textbooks for some time, but over the course of the first few decades of the nineteenth century they became widely discredited.[24] The rise and fall of the Wedgwood pyrometer is a fascinating story of a pioneering expedition into an entirely unmapped territory in the physics and technology of heat.

By the time Josiah Wedgwood (1730–1795) started in earnest to experiment with pyrometry, he had established himself as the leading manufacturer of porcelains in Britain and indeed one of the most renowned in all of Europe, equaling the long-established masters in Meissen and Sèvres. Wedgwood was proudly called "Potter to Her Majesty" since 1765, a title that he received after fulfilling a high-profile order from Queen Charlotte for a complete tea set (Burton 1976, 49–51). Ever seeking to improve and perfect the many lines of products issuing from his Staffordshire works grandly named "Etruria," Wedgwood was very keen to achieve better knowledge and control of the temperatures in his kilns. In 1768 he wrote exasperatedly to his friend and business partner Thomas Bentley: "Every Vaze [sic] in the last Kiln were [sic] spoil'd! & that only by such a degree of variation in the fire as scarcely affected our creamcolour bisket at all."[25]

For Wedgwood's purposes the standard mercury thermometer was of no help, since mercury was known to boil somewhere between 600°F and 700°F. In his preliminary studies, Wedgwood read an article by George Fordyce (1736–1802), chemist, physician at St. Thomas's Hospital in London, and a recognized authority on medical thermometry.[26] Fordyce asserted that bodies began to be luminous in the dark at 700°F, and Wedgwood noted in his commonplace book: "If mercury boils (or nearly boils) at 600 how did Doctr. Fordyce measure 700 by this instrument." Later he had a chance to buttonhole Fordyce about this, and here is the outcome: "Upon asking the Dr. this question at the Royal societie's rooms in the spring of 1780 he told me he did not mean to be accurate as it was of no great

---

[23] Lafferty and Rowe (1994, 570) give 9980°F or 5530°C.

[24] For a late case of a relatively uncritical citation of Wedgwood's measurements, see Murray 1819, 1:527–529. Pouillet (1827–29, 1:317) used Wedgwood's data, even as he expressed dissatisfaction with the instrument.

[25] Wedgwood to Bentley, 30 August 1768, quoted in Burton 1976, 83; the full text of this letter can be found in Farrer 1903–06, 1:225–226.

[26] The information about Fordyce is taken from the *Dictionary of National Biography*, 19:432–433.

consequence in that instance & he guessed at it as near as he could" (Chaldecott 1979, 74). Air thermometers did not have the same problem, but they generally required other problematic materials, such as glass, and mercury itself (for the measurement of pressure), which would soften, melt, boil, and evaporate.

What was most commonly known as pyrometry at the time was based on the thermal expansion of various metals,[27] but all of those metals melted in the higher degrees of heat found in pottery kilns and, of course, in foundries. Newton ([1701] 1935, 127) had estimated some temperatures higher than the boiling of mercury, but that was done by his unverified "law of cooling," which allowed him to deduce the initial temperatures from the amount of time taken by hot objects to cool down to certain measurable lower temperatures. Newton's law of cooling could not be tested directly in the pyrometric range, since that would have required an independent pyrometer. This was not a problem unique to Newton's method, but one that would come back to haunt every pyrometric scheme. It is very much like the problem of nomic measurement discussed in "The Problem of Nomic Measurement" in chapter 2, but even more difficult because the basis of sensation is almost entirely lacking in the extreme temperatures. (These issues will be considered further in the analysis part of this chapter.)

Initially, in 1780, Wedgwood experimented with certain clay-and-iron-oxide compositions that changed colors according to temperature. However, soon he found a more accurate and extensive measure in a different effect, which is called "burning-shrinkage" in modern terminology:

> In considering this subject attentively, another property of argillaceous [claylike] bodies occurred to me; a property which...may be deemed a distinguishing character of this order of earths: I mean, the *diminution of their bulk by fire*....I have found, that this diminution begins to take place in a low red-heat; and that it proceeds regularly, as the heat increases, till the clay becomes vitrified [takes a glassy form]. (Wedgwood 1782, 308–309)

These pyrometric pieces were extremely robust, and their shrinkage behavior answered the purpose even better than Wedgwood had initially hoped.[28] Their final sizes were apparently only a function of the highest temperatures they had endured, depending neither on the length of time for which they were exposed to the heat,[29] nor on the lower temperatures experienced before or after the peak; "in three minutes or less, they are perfectly penetrated by the heat which acts upon them, so

---

[27]See, for example, Mortimer [1735] 1746–47, Fitzgerald 1760, and De Luc 1779.

[28]Years later Guyton de Morveau, whose work I will be examining in detail later, expressed similar surprise at how well the clay pieces worked. Guyton (1798, 500) tried two pyrometer pieces in the same intense heat for half an hour, and was impressed to see that their readings differed very little (160° and 163.5° on Wedgwood's scale): "I confess I did not expect to be so completely successful in this verification." See the end of "Ganging Up on Wedgwood," for Guyton's further defense of the comparability of the Wedgwood pyrometer.

[29]Chaldecott (1979, 79) notes that we now know this claim to be false, but that the proof of its falsity was not forthcoming until 1903.

as to receive the full contraction which that degree of heat is capable of producing." So the pieces could be left in the hot places for any reasonable amount of time, and taken out, cooled, and measured at leisure afterwards (Wedgwood 1782, 316–317). This was an entirely novel method of pyrometry, and at that time the only one usable at extremely high temperatures. It was an achievement that propelled Wedgwood, already the master "artist," also into the top ranks of the (natural) "philosophers." Not only was Wedgwood's first article on the contraction pyrometer published in the *Philosophical Transactions* of the Royal Society in 1782, communicated by no less than Joseph Banks, the president, but it was also well enough received to facilitate Wedgwood's prompt election as a Fellow of the Royal Society.

In that article, Wedgwood made no secret of his initial motivations:

> In a long course of experiments, for the improvement of the manufacture I am engaged in, some of my greatest difficulties and perplexities have arisen from not being able to ascertain the heat to which the experiment-pieces had been exposed. A red, bright red, and white heat, are indeterminate expressions...of too great latitude....

In the absence of a thermometer, he had relied on very particular benchmarks:

> Thus the kiln in which our glazed ware is fired furnishes three measures, the bottom being of one heat, the middle of a greater, and the top still greater: the kiln in which the biscuit ware is fired furnishes three or four others, of higher degrees of heat; and by these I have marked my registered experiments. (Wedgwood 1782, 306–307)

But these measures were inadequate, and clearly unusable to anyone not baking clay in Etruria. In contrast, using the contraction of standard clay pieces had sufficient promise of quantifiability and wider applicability.

Wedgwood (1782, 309–314) gave detailed instructions on the preparation of the clay, to be formed in small rectangular shapes (0.6 inch in breadth, 0.4 inch deep, and 1 inch long), and the brass gauge for measuring the sizes of the shrunken pieces. He attached a numerical scale assigning 1 degree of heat to contraction by 1/600 of the width of a clay piece. He acknowledged that "the divisions of this scale, like those of the common thermometers, are unavoidably arbitrary." However, Wedgwood was confident that the procedures specified by him would ensure comparability (as defined in "Regnault: Austerity and Comparability" in chapter 2):

> By this simple method we may be assured, that thermometers on this principle, though made by different persons, and in different countries, will all be equally affected by equal degrees of heat, and all speak the same language: the utility of this last circumstance is now too well known to need being insisted on. (314–315)

With this instrument, Wedgwood succeeded in attaching numbers to a vast range of high temperatures where no one had confidently gone before (318–319). The pyrometer quickly taught him many interesting things because it enabled clear comparisons of various degrees of heat. Wedgwood found that brass melted at 21 degrees on his scale (which I will write as 21°W), while the workmen in brass

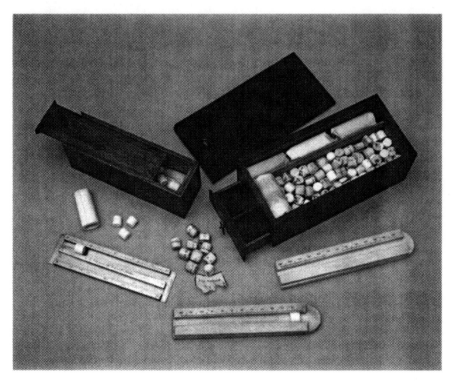

FIGURE 3.3. The pyrometer that Wedgwood presented to King George III (inventory no. 1927-1872). Science Museum/Science & Society Picture Library.

foundries were in the habit of carrying their fires to 140°W and above; clearly this was a waste of fuel, for which he saw no purpose.[30] He also found that an entire range from 27°W upwards had all been lumped together in the designation "white heat," the vagueness of which was obvious considering that the highest temperature he could produce was 160°W. Regarding his own business, Wedgwood (1784, 366) confessed: "Nor had I any idea, before the discovery of this thermometer, of the extreme difficulty, not to say impracticability, of obtaining, in common fires, or in common furnaces, an uniform heat through the extent even of a few inches."

Wedgwood's pyrometer was an instant success, popular among technologists and scholars alike. For example, the Scottish chemist John Murray wrote in his textbook of chemistry (1819, 1:226): "The pyrometer which has come into most

---

[30]Wedgwood's enlightenment on this point may have been a shallow one, however. As noted by Daniell (1830, 281): "When metals are melted for the purposes of the arts, they of course require to be heated very far beyond their fusing points, that they may flow into the minutest fissures of the moulds in which they are cast, notwithstanding the cooling influences to which they are suddenly exposed. In some of the finer castings of brass, the perfection of the work depends upon the intensity to which the metal is heated, which in some cases is urged even beyond the melting point of iron."

general use is that invented by Mr Wedgwood." Although Guyton de Morveau (1811b, 89) was critical of Wedgwood's instrument as we shall see later, he was clear about its proven utility and further potential: "I stressed the benefits that one could draw from this instrument, as already testified by the routine use of it by most of the physicists and chemists who make experiments at high temperatures." Lavoisier's collaborator Armand Séguin wrote to Wedgwood about the "greatest use" and "indispensability" of the Wedgwood pyrometer for their experiments on heat. Numerous other positive appraisals can be found very easily. There is an impressive list of chemists and physicists who are on record as having received pyrometer sets from Wedgwood, including Black, Hope, Priestley, Wollaston, Bergman, Crell, Guyton de Morveau, Lavoisier, Pictet, and Rumford. Wedgwood was proud enough to present a pyrometer to George III, which is the very instrument now preserved in the Science Museum in London (figure 3.3).[31]

It Is Temperature, but Not As We Know It?

Toward the end of his article Wedgwood indicated the next step clearly (1782, 318): "It now only remains, that the language of this new thermometer be understood, and that it may be known what the heats meant by its degrees really are." This desire was certainly echoed by others who admired his invention. William Playfair[32] wrote Wedgwood on 12 September 1782:

> I have never conversed with anybody on the subject who did not admire your thermometer...but I have joined with severall [sic] in wishing that the scale of your thermometer were compared with that of Fahrenheit... [so] that without learning a new signification [or] affixing a new idea to the term degree of heat we might avail ourselves of your useful invention. (Playfair, quoted in McKendrick 1973, 308–309)

Wedgwood did his best to meet this demand, apparently helped by a useful suggestion from Playfair himself. Wedgwood opened his next communication to the Royal Society as follows:

> This thermometer...has now been found, from extensive experience, both in my manufactories and experimental enquiries, to answer the expectations I had conceived of it as a measure of all degrees of common fire above ignition: but at present it stands in a detached state, not connected with any other, as it does not begin to take place till the heat is too great to be measured or supported by mercurial ones. (Wedgwood 1784, 358)

To connect up his pyrometric scale and the mercury-based Fahrenheit scale, Wedgwood looked for a mediating temperature scale that would overlap with both. For that purpose he followed the familiar pyrometric tradition of utilizing the expansion of metals, and he chose silver. Wedgwood's patching-up strategy is

---

[31] For Séguin's letter, the list of those who received pyrometer sets from Wedgwood, and other records of the esteem that Wedgwood's pyrometer enjoyed among his contemporaries, see Chaldecott 1979, 82–83, and McKendrick 1973, 308–309.

[32] This was probably William Playfair (1759–1823), Scottish publicist who later participated in the French Revolution, and brother of John (1748–1819), the geologist and mathematician.

124 *Inventing Temperature*

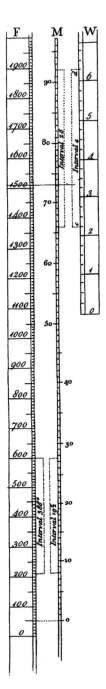

FIGURE 3.4. Wedgwood's patching of three temperature scales; the upper portion of the Fahrenheit measure is not known to begin with. Source: Wedgwood 1784, figs. 1, 2, and 3, from plate 14. Courtesy of the Royal Society.

represented graphically in figure 3.4, which he originally presented to the Royal Society "in the form of a very long roll."

The basis of this Wedgwood-Fahrenheit comparison was quite straightforward (Wedgwood 1784, 364–369). The expansion of the silver piece was measured in a gauge very much like the one he used to measure the clay pyrometer pieces. The low end of the silver scale overlapped considerably with the mercury scale. He set the zero of the silver scale at 50°F and found that the expansion of silver from that point to the heat of boiling water (212°F) amounted to 8 (arbitrary) units on the gauge; so an interval of 8° on the silver scale corresponded to an interval of 162° on the mercury-Fahrenheit scale. Assuming linearity, Wedgwood concluded that each silver degree "contained" 20.25 mercury-Fahrenheit degrees, simply by dividing 162 by 8. Putting this result together with a similar one obtained between 50°F and the boiling point of mercury, Wedgwood arrived at an approximate conversion ratio of 20 mercury-Fahrenheit degrees for each silver degree. Similar calculations at the high end of the silver scale gave the conversion ratio between the Wedgwood scale and the silver scale as 6.5 silver degrees for each Wedgwood degree.[33] Putting the two conversion factors together, Wedgwood found that each Wedgwood degree was worth $20 \times 6.5 = 130$ Fahrenheit degrees. From that conversion ratio and the information that the zero of the silver scale was set at 50°F, he also located the start of his scale (red heat) at 1077.5°F.[34] Now Wedgwood had achieved his stated aim of having "the whole range of the degrees of heat brought into one uniform series, expressed in one language, and comparable in every part" (1784, 358). He presented a double series of temperatures in Fahrenheit and Wedgwood degrees, some of which are shown here in table 3.1.

Wedgwood's measurements were the first concrete temperature values with any sort of credibility in the range above the upper limit of the mercury thermometer. However, much as his work was admired, it also drew increasingly sharp criticism. First of all, there were enormous difficulties in reproducing his clay pyrometer pieces exactly. His initial desire was that others would be able to make their own Wedgwood thermometers following his published instructions, but that did not turn out to be the case.[35] As Wedgwood himself acknowledged (1786, 390–401), the properties of the clay pieces depended on intricate details of the process by which they were made, and producing uniformly behaving pieces turned out to be very difficult. The making of the standard clay pieces required "those peculiar niceties and precautions in the manual operations, which theory will not suggest, and which practice in the working of clay can alone teach" (quoted in Chaldecott 1979, 82). There was also the need to use exactly the same kind of clay as used for the original pieces. Initially Wedgwood was sanguine about this problem (1782, 309–311), thinking

---

[33]In a furnace, when the Wedgwood pyrometer indicated 2.25°, the silver thermometer gave 66°; in another instance, 6.25° Wedgwood corresponded to 92° silver. Hence an interval of 4° on Wedgwood's thermometer was equivalent to an interval of 26° on the silver thermometer, which gives 6.5 silver degrees for 1 Wedgwood degree.

[34]Wedgwood made the calculation as follows: $2.25°W = 66°$ silver $= (50°F + 66 \times 20°F) = 1{,}370°F$; so $0°W = (1{,}370°F - 2.25 \times 130°F) = 1077.5°F$.

[35]This was noted particularly in France; see Chaldecott 1975, 11–12.

TABLE 3.1. Some high temperatures measured by Wedgwood, with conversion to Fahrenheit degrees

Phenomenon	°Wedgwood	°Fahrenheit
Greatest heat of Wedgwood's air-furnace	160	21877
Cast iron melts	130	17977
Welding heat of iron, greatest	95	13427
Welding heat of iron, least	90	12777
Fine gold melts	32	5237
Fine silver melts	28	4717
Swedish copper melts	27	4587
Brass melts	21	3807
Red-heat fully visible in daylight	0	1077
Red-heat fully visible in the dark	−1	947
Mercury boils	−3.673	600

Source: Wedgwood 1784, 370.

that there was sufficient uniformity in the clays found in various places at equal depths. If problems arose, he thought that all the pyrometer pieces ever needed could be made with clay from one particular bed in his own possession in Cornwall: "[T]he author offers to this illustrious Society [Royal Society of London], and will think himself honoured by their acceptance of, a sufficient space in a bed of this clay to supply the world with thermometer-pieces for numerous ages." But to his great chagrin, Wedgwood (1786, 398–400) later found out that different samples of clay even from that same area of Cornwall differed from each other in their thermometric behavior in uncontrollable ways. In short, even Wedgwood himself had trouble reproducing the "standard clay" pieces that he had initially used. The shrinkage behavior became more controllable when he employed a slightly artificial preparation, a mixture of alum and natural clay, but in the end Wedgwood had to resort to the use of several fixed points to ensure sufficient comparability (0°W at red heat as before, 27°W at the melting point of silver, 90°W at the welding heat of iron, and 160°W at the greatest possible heat of a good air-furnace).[36]

The other major difficulty, which Wedgwood never addressed in print, concerned the connection with ordinary scales of temperature. Many of Wedgwood's critics did not believe that he had worked out the Wedgwood-Fahrenheit conversion correctly, in three different ways. These points will be discussed in more detail in the next section:

1. His estimate of the temperature of red heat (the beginning and zero point of his scale) was too high.
2. His estimate of the number of Fahrenheit degrees corresponding to one degree of his scale was also too high.
3. There was no positive reason to believe that the contraction of clay was linear with temperature.

---

[36] For a description of the alum mixture, see Wedgwood 1786, 401–403; for the fixed points, p. 404.

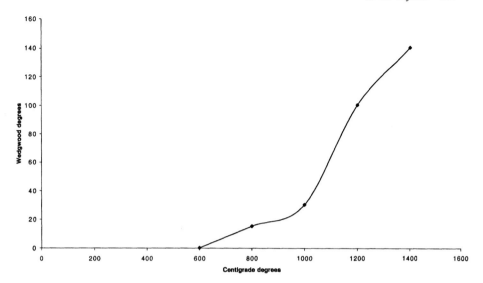

FIGURE 3.5. Late nineteenth-century comparison of Wedgwood and centigrade degrees. The data represented in the figure are as reported in Le Chatelier and Boudouard 1901, 164.

These points are supported by later appraisals, summarized by the physical chemist Henri Louis Le Chatelier (1850–1936) in the late nineteenth century, shown graphically in figure 3.5.

If all the points of criticism mentioned earlier are indeed correct, the most charitable light in which we can view Wedgwood pyrometry is that it gave some rough indication of high temperatures, but without conceptual or quantitative accuracy. Some later commentators have used this apparent failure of pyrometry as evidence that Wedgwood was scientifically unsophisticated. But the first set of problems, namely those concerning the lack of uniformity in the behavior of clay, cannot be held against Wedgwood, since he recognized them clearly and devised very reasonable methods for dealing with them. The second set of problems is a different matter. When Neil McKendrick (1973, 280, 310) disparages Wedgwood's pyrometry for "its obvious scientific shortcomings" and "its lack of scientific sophistication and lack of command of theory," the chief fault that he finds is Wedgwood's "failure to calibrate its scale with that of Fahrenheit." McKendrick surely could not be missing the fact that Wedgwood did calibrate his thermometric scale with the Fahrenheit scale, so what he means must be that Wedgwood did it incorrectly.

But how can we be so sure that Wedgwood was wrong? And, more pertinently, how can we be sure at all that any of the proposed alternatives to Wedgwood pyrometry were any better? It is quite true that Wedgwood had not produced any direct empirical justification for his assumption that the expansion and contraction of his clay and silver pieces depended only on temperature and linearly on temperature. But these assumptions could not be tested without an independent method of measuring the temperatures involved, and there were none available to

Wedgwood. He was striking out into virgin territory, and no previous authority existed to confirm, or contradict, his reports. On what grounds did his opponents declare his numbers false? That is the great epistemic puzzle about the downfall of the Wedgwood pyrometer. We need to examine with some care the process by which physicists, chemists, and ceramic technologists came to rule against Wedgwood, and the character of the alternative standards with which they replaced Wedgwood's pyrometer. (The doubts were raised strongly only after Wedgwood's death in 1795, so we can only speculate on how the master artist himself would have responded.)

Ganging Up on Wedgwood

In discussing the work of Wedgwood's critics, I will depart slightly from the chronological order and organize the material in terms of the alternative pyrometric methods they proposed and developed. I will discuss the alternatives one by one, and then assess their collective effect. To anticipate the conclusion: I will show that each of the temperature standards favored by Wedgwood's critics was as poorly established as Wedgwood's own. Their main strength was in their agreement with each other. What exactly such a convergence of standards was capable of underwriting will be discussed fully in the analysis part.

The Expansion of Platinum

This alternative to Wedgwood pyrometry was conceptually conservative but materially innovative. It hinged on a new material, platinum. Although platinum was known to Europeans since the mid-eighteenth century, it was only at the beginning of the nineteenth century that William Hyde Wollaston (1766–1828), English physician and master of "small-scale chemistry," managed to come up with a secret process to render it malleable so that it could be molded into useful shapes and drawn into fine wires.[37] As platinum was found to withstand higher degrees of heat than any previously known metals, it was naturally tempting to enlist it in the service of pyrometry. The simplest scheme was to use the thermal expansion of platinum, in the familiar manner in which pre-Wedgwood pyrometry had used the expansion of various metals.

The clear pioneer in platinum pyrometry was Louis-Bernard Guyton de Morveau (1737–1816), who had probably one of the most interesting scientific and political careers through the French Revolution and Empire. A prominent lawyer in Dijon whose reforming zeal had drawn Voltaire's praise, Guyton became gradually swept up in revolutions both in chemistry and politics. Having collaborated with Lavoisier on the new chemical nomenclature, Guyton threw himself into the political revolution that would later claim the life of his esteemed colleague. He moved to Paris

---

[37]This process gave Wollaston a comfortable income for the rest of his life. Much further useful information can be found in D. C. Goodman's entry on Wollaston in the *Dictionary of Scientific Biography*, 14:486–494.

as a member of the National Assembly and then the Convention, dropped the aristocratic-sounding "de Morveau" from his name, voted for the execution of the king, and served as the first president of the Committee of Public Safety until he was removed as a moderate with the coming of the Reign of Terror. He did return to the committee briefly after the fall of Robespierre, but soon retired from politics and concentrated on his role as a senior statesman of science. Guyton was one of the first members of the new French Institute at its founding, and the president of its First Class (mathematical and physical sciences) in 1807. At the École Polytechnique he was a professor for nearly twenty years and director twice.[38]

The chemist formerly known as De Morveau started working with Wedgwood pyrometers in his research on the properties of carbon published in 1798 and 1799, which reported that charcoal was an effective insulator of extreme heat and that diamond burned at 2765°F according to Wedgwood's pyrometer and conversion table. He announced at that time that he was engaged in research toward improving the pyrometer.[39] In 1803 he presented his platinum pyrometer to the French Institute and promised to present further research on its relation to the Wedgwood pyrometer. Guyton's comparison of the platinum pyrometer with the Wedgwood pyrometer led him to propose a serious recalibration of the Wedgwood scale against the Fahrenheit scale, bringing 0°W down to 517°F (from Wedgwood's 1077°F), and estimating each Wedgwood degree as 62.5°F (not 130°F as Wedgwood had thought).[40] The overall effect was to bring Wedgwood's temperature estimates down considerably; for instance, the melting point of cast iron was adjusted from 17,327°F down to 8696°F (see the first two columns of data in table 3.2). Guyton does not specify explicitly how exactly his recalibration was done, but judging from the data it seems as though he fixed the clay scale to agree with the platinum scale at the melting points of gold and silver. Guyton was well aware that there was no guarantee that the expansion of platinum at high temperatures was linear with temperature, and his reasons for trusting the platinum pyrometer over the Wedgwood pyrometer rested on the agreement of the former with a few other types of pyrometers, as we will see shortly.

No significant progress on platinum pyrometry seems to have been made after Guyton's work until John Frederic Daniell (1790–1845) made an independent invention of the platinum pyrometer in 1821, nearly two decades after Guyton. At that point Daniell was best known as a meteorologist, although later he would achieve more lasting fame in electrochemistry largely with the help of the "constant battery" that he invented himself. He spent the last fifteen years of his life as the first professor of chemistry at the newly established King's College in London, widely respected for his "lofty moral and religious character" as well as his successes

---

[38]For these and many further details, see W. A. Smeaton's entry on Guyton de Morveau in the *Dictionary of Scientific Biography*, 5:600–604.
[39]See also the note Guyton attached (pp. 171–172) in communicating Scherer 1799 to the *Annales de chimie*.
[40]See Guyton 1811b, 90–91, and also table 3.

TABLE 3.2. A comparative table of data produced by various pyrometric methods, published up to the first half of the nineteenth century

	Clay °W[a]	Conversion into °F	Mercury	Metal	Ice	Water	Air	Cooling	Current values
Melting point of tin			481 (N) 415 (B)[b]	441 (Da) 442 (G)			383 (C/D)[G] 410 (Pa) 455 (Pc)		449 [L]
Melting point of bismuth			537 (N) 494 (B)[b]	462 (Da) 476 (G)			662 (C/D)[G] 493 (Pa) 518 (Pc)		521 [L]
Melting point of lead			631 (N) 595 (B)[b]	609 (Da) 612 (G)			617 (C/D)[G] 500 (Pa) 630 (Pc)		621 [L]
Melting point of zinc	3 (G)	705 (G)		699 (B)[b] 680 (G) 648 (Da) 773 (Di)			932 (C/D)[G] 680 (Pa) 793 (Pc)		787 [L]
Red heat visible in the dark				947 (W) 977 (Dr)				743 (N)	
Melting point of antimony	7 (G)	955 (G)		809 (B)[b] 810 (G)			847 (C/D)[G] 810 (Pa)	942 (N)	1167 [L]
Red heat visible in daylight	0, by definition	1077 (W)		1050 (B)[b] 517 (G) 980 (Da)		1272 (C/D)[G]	977 (Pb) 1200 (Pr)	1036 (N)	
Melting poin of brass	21 (W) 21 (G)	3807 (W) 1836 (G)		1869 (Da)					1706–1913 [R]

Melting point of silver	28 (W) 22 (G)	4717 (W) 1893 (G)		1000 (B)^b 1893 (G) 2233 (Da) 1873 (Db) 1682 (Di)	1000 (Pa) 1832 (Pb) 1830 (Pr)	1763 [C] 1761 [R] 1763 [L]
Melting point of copper	27 (W) 27 (G) 27 (Pa)	4587 (W) 2205 (G)	2295 (C/D)[G]	1450 (B)^b 2313 (G) 2548 (Da) 1996 (Db)		1984 [C] 1981 [R] 1984 [L]
Melting point of gold	32 (W) 32 (G) 32 (Pa)	5237 (W) 2518 (G)		1301(B)^b 2518 (G) 2590 (Da) 2016 (Db) 1815 (Di)	2192 (Pb) 2282 (Pc)	1948 [C] 1945 [R] 1948 [L]
Welding heat of iron, least	90 (W) 95 (G)	12777 (W) 6504 (G)				1922 [R]
Welding heat of iron, greatest	95 (W) 100 (G)	13427 (W) 6821 (G)				2192 [R]
Red hot iron	88 (C/D)[G]	12485 (C/D)[G]	2732 (C/D)[G]			
White hot iron	100 (C/D)[G]	14055 (C/D)[G]	3283 (C/D)[G]			
Melting point of cast iron	130 (W) 130 (G)	17977 (W) 8696 (G)	3164 (C/D)[G]^c	1601 (B)^b 3479 (Da) 2786 (Db)	1922–2192 (Pb)	2100–2190 [R]
melting point of soft iron	174 (C/D)[G] 175 (G)^d	23665 (C/D)[G] 11455 (G)	3988 (C/D)[G]		3902 (C/D)[G] 2700–2900 (Pb)	

(Continued)

TABLE 3.2. (Continued)

	Clay °W[a]	Conversion into °F	Mercury	Metal	Ice	Water	Air	Cooling	Current values
Melting point of steel	160 [R] 154 [R]								
Greatest heat, air furnace	160 (W) 170 (G)	21877 (W)					~2370—2550 (Pb)		
Melting point of platinum	unknown (G)			over 3280 (Db)					3215 [L]

Source code:
B: Bergman 1783, 71, 94
C: Chaldecott 1979, 84
C/D: Clément and Desormes
Da: Daniell 1821, 317–318, by platinum
Db: Daniell 1830, 279, by platinum, corrected for non-linearity
Di: Daniell 1830, 279, by iron
Dr: Draper 1847, 346
G: Guyton 1811b, 90, table 3; 117, table 5; and 120, table 7
L: Lide and Kehiaian 1994, 26–31
N: Newton [1701] 1935, 125–126; data converted assuming the highest boiling heat of water at 212° F.
Pa: Pouillet 1827–29, 1:317
Pb: Pouillet 1836, 789
Pc: Pouillet 1856, 1:265
Pr: Prinsep 1828, 94
R: Rostoker and Rostoker 1989, 170
W: Wedgwood 1784, 370

[a] All degrees are on the Fahrenheit scale, except in the first column of data, which gives Wedgwood degrees. The last column gives the currently accepted values, for comparison. The code in parentheses indicates the authorities cited, and the code in square brackets indicates my sources of information, if they are not the original sources.
[b] Bergman's 1783 text does not generally indicate how his melting points were determined. But on p. 94 he notes that the number for the melting point of iron was based on Mortimer's work on the expansion of metals, so I have put all of his values above the boiling point of mercury into the "metal" column. I assume that he used the mercury thermometer for lower temperatures, but that is only a conjecture.
[c] This point is described as *fonte de fer prête à couler* by Guyton.
[d] This is described as *fusion de fer doux, sans cément*.

in research and teaching.[41] Apparently unaware of Guyton's work, Daniell (1821, 309–310) asserted that in pyrometry "but one attempt has ever been made, with any degree of success," which was Wedgwood's. He lamented the fact that Wedgwood's measurements were still the only ones available, although the Wedgwood pyrometer had "long fallen into disuse" for good reasons (he cited the difficulty of making clay pieces of uniform composition, and the observation that the amount of their contraction depended on exposure time as well as temperature).

Daniell (1821, 313–314) graduated his instrument on two fixed points, on the model of the standard thermometers. He put the zero of his pyrometer where the mercury thermometer gave 56°F, and then set 85° of his scale at the boiling point of mercury, which he took as 656°F. Simple comparison gave about 7°F per each Daniell degree. That was an estimate made assuming the linearity of the thermal expansion of both mercury and platinum, but Daniell did test that assumption to some extent by comparing the two thermometers at various other points. The result was reasonably reassuring, as the differences were well within about 3 Daniell degrees through the range up to the boiling point of mercury. However, on the most crucial epistemic point regarding the validity of extending that observed trend beyond the boiling point of mercury, he made no progress beyond Guyton. The following non-argument was all that Daniell (1821, 319) provided for trusting the expansion of platinum to remain linear up to its melting point: "[T]he equal expansion of platinum, with equal increments of heat, is one of the best established facts of natural philosophy, while the equal contraction of clay, is an assumption which has been disputed, if not disproved." After taking our lessons from Regnault in chapter 2, we may be pardoned if we cannot help pointing out that the "equal expansion of platinum" was an "established fact" only in the temperature range below the boiling point of mercury, and even then only if one assumes that mercury itself expands linearly with temperature.

By 1830 Daniell had discovered Guyton's work, and he had some interesting comments to make. All in all, Daniell's work did constitute several advances on Guyton's. Practically, he devised a way of monitoring the expansion of the platinum bar more reliably.[42] In terms of principles, Daniell chastised Guyton for going along with Wedgwood's assumption of linearity in the contraction of clay (1830, 260): "Guyton, however, although he abundantly proves the incorrectness of Mr. Wedgwood's estimate of the higher degrees of temperature, is very far indeed from establishing the point at which he so earnestly laboured, namely, the regularity of the contraction of the clay pieces." Daniell made much more detailed comparisons between temperature readings produced by the various methods in question (1830, 260–262). His general conclusion regarding Guyton's correction of the Wedgwood scale was that he had not shrunk it sufficiently, while his lowering of the zero point (red heat visible in daylight) went too far. In Daniell's view, the best correction of the Wedgwood scale was obtained by actually raising Wedgwood's

---

[41] See the *Dictionary of National Biography*, 14 (1888), 33.
[42] See Daniell 1821, 310–311, for his basic design, and Daniell 1830, 259, for a critique of Guyton's design.

estimate of the zero point a little bit, and shrinking each Wedgwood degree drastically down to about 20°F. However, he thought that no simple rescaling of the Wedgwood scale would bring one to true temperatures, since the contraction of clay was not regular.

Not regular when judged by the platinum pyrometer, that is. Daniell (1830, 284–285) admitted that the expansion of platinum was also unlikely to be linear and took Dulong and Petit to have shown that "the dilatability of solids, referred to an airthermometer, increases with the heat." Extrapolating Dulong and Petit's results, Daniell arrived at some corrections to his pyrometer readings obtained with the assumption of linearity. (Daniell's corrected results are labeled *Db* in table 3.2, which summarizes various pyrometric measurements.) These corrections were nontrivial. Dulong and Petit had indicated that a thermometer of iron graduated between 0°C and 100°C assuming linearity would read 372.6°C when the air thermometer gave 300°C; the deviation from linearity was not so dire for platinum, but even a platinum thermometer would give 311.6°C when the air thermometer gave 300°C (1817, 141; 1816, 263). Daniell deserves credit for applying this knowledge of nonlinearity to correct metallic thermometers, but there were two problems with his procedure. First of all, Dulong and Petit's observations went up to only 300°C (572°F), probably since that was the technical limit of their air thermometer. So, even in making corrections to his linear extrapolation, Daniell had to extrapolate an empirical law far beyond the domain in which it was established by observation. Daniell carried that extrapolation to about 1600°C, covering several times as much as Dulong and Petit's entire range. Second, Daniell's correction of the platinum pyrometer made sense only if there was assurance that the air thermometer was a correct instrument in the first place; I will return to that issue later.

Ice Calorimetry

Given the futility of relying on unverifiable expansion laws for various substances at the pyrometric range, it seems a sensible move to bring the measurements to the easily observable domains. The chief methods for doing so were calorimetric: the initial temperature of a hot object can be deduced from the amount of ice it can melt, or the amount of temperature rise it can produce in a body of cold water. The latent and specific heats of water being so great, a reasonable amount of ice or water sufficed to cool small objects from very high temperatures down to sensible temperatures. Different calorimetric techniques rested on different assumptions, but all methods were founded on the assumption of conservation of heat: the amount of heat lost by the hot body is equal to the amount of heat gained by the colder body, if they reach equilibrium in thermal isolation from the external environment. In addition, inferring the unknown initial temperature of a body also requires further assumptions about the specific heat of that body, a more problematic matter on which I will comment further shortly.

Calorimetry by means of melting ice was a well-publicized technique ever since Lavoisier and Laplace's use of it, described in their 1783 memoir on heat. Wedgwood (1784, 371–372) read a summary of the Lavoisier-Laplace article after his initial publication on the pyrometer, with excitement: "The application of this important discovery, as an intermediate standard measure between Fahrenheit's thermometer and mine, could not escape me." However, Wedgwood (1784, 372–384) was greatly disappointed in his attempt to use the ice calorimeter to test the soundness of his pyrometer. Ice calorimetry rested on the assumption that all the melted water would drip down from the ice so that it could be collected and weighed up accurately. Wedgwood noted that even a solid block of ice imbibed and retained a considerable amount of melted water, and the problem was much worse for pounded ice, used by Lavoisier and Laplace. Wedgwood's confidence in the ice calorimeter was further shaken by his observation that even while the melting of ice was proceeding as designed, there was also fresh ice forming in the instrument, which led him to conclude that the melting point of ice and the freezing point of water, or water vapor at any rate, were probably not the same.

The French savants, however, were not going to give up the invention of their national heroes so easily. Claude-Louis Berthollet, Laplace's close associate and the dean of French chemistry since Lavoisier's demise at the guillotine in 1794, defended the ice calorimeter against Wedgwood's doubts in his 1803 textbook of chemistry. The problem of ice retaining the melted water could be avoided by using ice that had already imbibed as much water as it could. As Lodwig and Smeaton (1974, 5) point out, if Wedgwood had read Lavoisier and Laplace's memoir in full he would have realized that they had considered this factor but thought that their crushed ice was already saturated with water to begin with.[43] As for the refreezing of ice that takes place simultaneously with melting, it did not in itself interfere with the functioning of the instrument. Citing Berthollet, Guyton (1811b, 102–103) argued that the calorimeter was "the instrument best suited for verifying or correcting Wedgwood's pyrometric observations."

Guyton never had the opportunity to carry out this test of the Wedgwood pyrometer by the ice calorimeter, not having met with the right weather conditions (low and steady temperatures). But he cited some relevant results that had been obtained by two able experimenters, Nicolas Clément (1778/9–1841) and his father-in-law Charles-Bernard Desormes (1777–1862), both industrial chemists; Desormes was for a time an assistant in Guyton's lab at the École Polytechnique.[44] Their data showed that temperatures measured by ice calorimetry were generally much lower than those obtained with the Wedgwood pyrometer (not only by Wedgwood but also by Clément and Desormes themselves). All of their results cited by Guyton are included in table 3.2. As one can see there, Clément and Desormes measured four very high temperatures with an ice calorimeter, and the same temperatures were also measured by a Wedgwood pyrometer. If one adopted Wedgwood's own conversion of Wedgwood degrees into Fahrenheit degrees (the second column in

---

[43]Lodwig and Smeaton also note that Wedgwood's criticism was quite influential at least in England and discuss various other criticisms leveled against the Laplace–Lavoisier calorimeter.

[44]See Guyton 1811b, 104–105, and the data in table 7. As far as I can ascertain, Clément and Desormes's pyrometric work was not published anywhere else; the reference that Guyton gives in his article seems misplaced. The biographical information is cited from Jacques Payen's entry on Clément in the *Dictionary of Scientific Biography*, 3:315–317.

the table), the resulting temperatures were about 10,000 to 20,000 higher than the numbers given by ice calorimetry. Even using Guyton's revised conversion gave numbers that were thousands of degrees higher than those obtained by ice calorimetry.

Clément and Desormes's ice-calorimetry results were scant and lacked independent confirmation. But the real problem lay in the theoretical principles, as mentioned briefly earlier. We have already seen the same kind of problem in De Luc's method of mixtures for testing thermometers (see "Caloric Theories against the Method of Mixtures" in chapter 2), which is a calorimetric technique (though it is more akin to water calorimetry, to be discussed shortly). Ice calorimetry measures the amount of heat lost by the hot object in coming down to the temperature of melting ice, not the temperature initially possessed by the hot object. Ice calorimetry at that time relied on the assumption that the specific heat of the hot object was constant throughout the temperature range. That assumption was clearly open to doubt, but it was difficult to improve on it because of the circularity that should be familiar by now to the readers of this book: the only direct solution was to make accurate measurements of specific heats as a function of temperature, and that in turn required an accurate method of temperature measurement, which is precisely what was missing in the pyrometric range. In a work that I will be discussing later, Dulong and Petit (1816, 241–242) regarded the "extreme difficulty" of determining the specific heat of bodies with precision, especially at high temperatures, as one of the greatest obstacles to the solution of the thermometry problem.

Water Calorimetry

How about the other major calorimetric method, using the temperature changes in water? Water calorimetry had fewer problems than ice calorimetry in practice, but it was open to the same problem of principle, namely not knowing the specific heat of the object that is being cooled down. In fact the theoretical problem was worse in this case, since there was also a worry about whether the specific heat of water itself varied with temperature. The history of water calorimetry was long, but it seems that Clément and Desormes were the first people to employ the method in the pyrometric range. As shown in table 3.2, they only obtained three data points by this method. The melting point of copper by this method was in good agreement with Guyton's result by platinum pyrometry and also in rough agreement with the two later results from Daniell by the same method. The number for the melting point of soft iron was drastically lower than the value obtained by the Wedgwood method, and in quite good agreement with the result by ice calorimetry. The estimate of "red heat" was nearly 200 degrees higher than Wedgwood's, and over 700 degrees higher than Guyton's, and not much of anything could be concluded from that disagreement.

Time of Cooling

Another method of avoiding the taking of data in the pyrometric range was to estimate the temperature of a hot object from the amount of time taken for it to cool down to a well-determined lower temperature. As I mentioned earlier, this method had been used by Newton for temperatures above the melting point of lead. Dulong and Petit revived the method and employed it with more care and precision than Newton, but the fundamental problem remained: verifying the law of cooling would have required an independent measure of temperature. For Dulong and Petit, that independent measure of temperature was the air thermometer, which made sense for them, since they regarded the air thermometer as the true standard of temperature, as discussed in "Regnault and Post-Laplacian Empiricism" in chapter 2.

Air Pyrometry

The thermometer in general theoretical favor by the beginning of the nineteenth century was the air thermometer, so it might have made sense to compare the readings of the various pyrometers with the readings of the air thermometer as far as possible. However, as discussed at length in chapter 2, no conclusive argument for the superiority of the air thermometer to the mercury thermometer was available until Regnault's work in the 1840s. And Regnault never claimed that his work showed that air expanded linearly with temperature. Moreover, Regnault's painstaking work establishing the comparability of air thermometers was only carried out in relatively low temperatures, up to about 340°C (644°F). In short, there was no definite assurance that comparison with the air thermometer was an absolutely reliable test for the accuracy of pyrometers. Even so, the air thermometer certainly provided one of the most plausible methods of measuring high temperatures.

In practical terms, if one was going to rely on the expansion of anything in the pyrometric range, air was an obvious candidate as there were no conceivable worries at that time about any changes of state (though later the dissociation of air molecules at very high temperatures would become an issue). But that was illusory comfort: an air thermometer was good only as long as the container for the air remained robust. Besides, air thermometers were usually large and very unwieldy, especially for high temperatures. Clément and Desormes took the air-in-glass thermometer to its material limits, using it to measure the melting point of zinc, which they reported as 932°F (Guyton 1811b, table 7). Dulong and Petit's work with the air thermometer was more detailed and precise, but did nothing to extend its range; in fact, in order to ensure higher precision they restricted the range in which they experimented, going nowhere beyond 300°C (572°F). Even Regnault only managed to use air thermometers credibly up to temperatures around 400°C (around 750°F).

The obvious solution was to extend the range of the air thermometer by making the reservoir with materials that were more robust in high temperatures.[45] The first

---

[45] As an alternative (or additional) solution, Pouillet had the idea that the range of the air thermometer might be extended in the high-temperature domain by the employment of the constant-pressure method, which had the advantage of putting less strain on the reservoirs. Regnault agreed with this idea (1847, 260–261, 263), but he criticized Pouillet's particular setup for having decreasing sensitivity as temperature increased (170), and also expressed worries about the uncertainty arising from the lack of knowledge in the law of expansion of the reservoir material (264–267).

credible step in that direction was taken by the Anglo-Indian antiquarian James Prinsep (1799–1840), who was also the assay master at the Calcutta mint.[46] He began with a bold condemnation (1828, 79–81): "If all the experiments had been recorded, which at different times must undoubtedly have been made on the subject of Pyrometry... the catalogue would consist principally of abortive attempts, if not of decided failures." Dulong and Petit's work was valuable, but only for relatively low temperatures. The Wedgwood pyrometer was the only instrument applicable in the higher heats produced by furnaces, but "a slight practical acquaintance with metals and crucibles" was sufficient to teach one that Wedgwood's results were not reliable. Prinsep thought that Daniell's more recent work was much more promising, but still he saw some problems in the design of Daniell's instrument, which he thought were manifested in the lack of "a desirable accordance in the result of different trials."[47]

Prinsep first tried to construct an air thermometer with a cast-iron reservoir. After experiencing various technical difficulties, Prinsep finally opted for a much more expensive solution (1828, 87–89): "a retort or bulb of pure gold, weighing about 6,500 grains troy [about 420 g], containing nearly ten cubic inches of air." This gold-based instrument was robust, but he recognized two problems of principle. First, the thermal expansion of gold was not well known, so it was difficult to correct for the errors arising from the expansion of the vessel. Second, he was not so convinced about the correctness of "the absolute law of gaseous expansion" in the pyrometric range, either. There were also practical difficulties, most of which were common to all air thermometers. Nonetheless, Prinsep (1828, 95) carried out some elaborate measurements and concluded that his results were unequivocal on certain important points, particularly on the melting point of silver: "[T]hese experiments... are sufficiently trustworthy to warrant a reduction in the tabular melting point of pure silver of at least 400 degrees [Fahrenheit] below the determination of Mr. Daniell, while they indisputably prove the superiority of that gentleman's thermometric table as contrasted with that of Mr. Wedgwood."

When Prinsep discarded iron and went for gold, he was not only compromising on economy but the range as well, as iron could withstand a much higher degree of heat than gold. Given what people knew about metals at that time, there was only one hope: platinum. But, as noted earlier, the handling of platinum was still a very difficult art in the early nineteenth century. In fact J. G. Schmidt (1805) in Moldavia had already proposed making a pyrometer with air enclosed in a platinum container, but there is no indication that he ever executed this idea; Guyton (1811b, 103–104) could not imagine making such a contraption without soldering platinum plates,

---

[46]See *Encyclopaedia Britannica*, 11th ed., for brief biographical information about Prinsep.

[47]There is some irony in Prinsep's attack on Wedgwood. The prime example he gave of the unreliability of the Wedgwood pyrometer was the overly high melting point of silver, particularly the fact it was placed above the melting point of copper. As Prinsep noted, Wedgwood had put forward these erroneous silver and copper melting points "on the authority of Mr. Alchorne," who had performed the experiments for Wedgwood (see also Wedgwood 1782, 319). What kind of authority was Alchorne? He was the assay master at the Tower of London, a man with a bit more than "a slight practical acquaintance with metals and crucibles"!

which would make the instrument only as robust as the soldering material. Similarly, according to Prinsep (1828, 81), Andrew Ure had recommended "an air thermometer made of platina," and even got such an instrument made for sale, but no reports of any experiments done with them could be found.

By 1836, however, Pouillet managed to construct an air pyrometer with the reservoir made out of a single piece of platinum. I have already discussed Pouillet's work at the low temperature end in "Consolidating the Freezing Point of Mercury," and in fact one of the air thermometers he used for that work had originally been constructed for pyrometric purposes. The platinum-based air thermometer was capable of recording temperatures well over 1000°C (about 1830°F). As shown in table 3.2, the melting points of metals that Pouillet (1836, 789) obtained by this means were mostly quite consistent with values obtained by Daniell with his platinum pyrometer. Having the air thermometer readings available to such high temperatures also aided the development of calorimetry because it enabled specific heat measurements at high temperatures. Pouillet (1836, 785–786) reported that the specific heat of platinum increased steadily, going from 0.0335 around 100°C (212°F) to 0.0398 around 1600°C (2912°F). This knowledge allowed him to estimate the melting point of iron by water calorimetry, by putting a piece of platinum in the same heat that melted iron, and then performing calorimetry on the platinum piece. The resulting value was 1500–1600°C (roughly 2700–2900°F) for the melting point of iron.

After that whirlwind tour of early pyrometry, we can now come back to the question that we set out to answer: on what grounds did people decide that Wedgwood's temperature values were incorrect? For quite a while after the Wedgwood pyrometer was generally rejected, none of the available alternative pyrometric methods were clearly superior to Wedgwood's, either in principle or in practice. Le Chatelier's harsh retrospective judgment on nineteenth-century pyrometry is not entirely an exaggeration:

> Since Wedgwood, many have undertaken the measurement of high temperatures, but with varying success. Too indifferent to practical requirements, they have above all regarded the problem as a pretext for learned dissertations. The novelty and the originality of methods attracted them more than the precision of the results or the facility of the measurements. Also, up to the past few years, the confusion has been on the increase. The temperature of a steel kiln varied according to the different observers from 1,500° to 2,000°; that of the sun from 1,500° to 1,000,000°. First of all, let us point out the chief difficulty of the problem. Temperature is not a measurable quantity in the strict sense of the term.... It is evident that the number of thermometric scales may be indefinitely great; too often experimenters have considered it a matter of pride for each to have his own. (Le Chatelier and Boudouard 1901, 2–3)

An inspection of the data collected in table 3.2 shows that by the middle of the nineteenth century the comparability of each of these methods had not been established: either the data were too scant for comparisons to take place or the measurements of the same phenomena obtained by the same method differed considerably from each other. In fact, in terms of comparability, it could easily be argued that the Wedgwood clay pyrometer was superior to alternative methods because the results obtained by this method by Wedgwood, Guyton, Clément and Desormes, and Pouillet were in close agreement with each other for most phenomena (see the first column of data in table 3.2). Guyton (1811a, 83–84) very ably defended the Wedgwood pyrometer and its comparability. Fourmi had just published an argument that the contraction of the Wedgwood pyrometric pieces was a function of the exposure time, as well as the temperature to which it was exposed. Guyton took Fourmi's own data and argued quite convincingly that they actually showed a remarkable degree of comparability between different trials with very different amounts of exposure time.[48]

The non-Wedgwood methods did not agree all that well between themselves, either. However, as we can see at a glance in table 3.2, it was still very clear that the numbers produced by them tended to agree much more with each other than with Wedgwood's. As Guyton put the matter already in 1811:

> I believe that we can conclude that the values assigned by Wedgwood to the degrees of his pyrometric scale ought to be reduced considerably, and that all the known means of measuring heat contribute equally toward the establishment of that result, from the zero of the thermometer to the temperature of incandescent iron. (Guyton 1811b, 112)

Beyond the melting point of gold, the clay pyrometer readings, even as recalibrated by Guyton, were distinctly far away from the range where the numbers produced by other methods tended to cluster.

That is where matters stood for quite some time. The transition into the kinds of pyrometry that would be recognizable at all to modern physicists and engineers did not occur until the last decades of the nineteenth century.[49] The most important basis of modern pyrometry is a quantitative knowledge of the radiation of heat and light from hot bodies, and of the variation of the electrical properties of matter with temperature. Such knowledge could not be gained without basing itself on previous knowledge gained by the types of pyrometry discussed in this chapter. Modern pyrometry was the next stage in the saga of the extension of the temperature

---

[48]Fourmi exposed various Wedgwood clay pieces to very high degrees of heat, around or beyond the melting point of cast iron, for repeated periods of 30 to 40 hours each. For instance, one piece shrank to the size corresponding to 146°W, after one period of exposure to a heat estimated by Fourmi at 145°W; two more exposures each at 145°W brought the piece only down to the size of 148°W; another exposure estimated at 150–151°W brought it to 151°W. Another piece (no. 20), on the other hand, contracted to 151°W after just one exposure at 150–151°W. Later commentators, however, have sided with Fourmi's verdict. Daniell (1821, 310) voiced the same opinion and that is also in line with the modern view, as indicated by Chaldecott (1975, 5).

[49]Matousek (1990, 112–114) notes that electrical-resistance pyrometry was only proposed in 1871, by William Siemens; radiation pyrometry started with the Stefan-Boltzmann law, discovered around 1880; optical pyrometry was pioneered by Le Chatelier in 1892; thermoelectric pyrometry did not become reliable until the 1880, although its basic idea can be traced back to Seebeck's work in the 1820s. (We have seen that Pouillet began to gain confidence in the thermoelectric method in the low-temperature range in the 1830s; Melloni used it in the same period with great effect in his study of radiant heat, but not as a pyrometer.)

scale, and I expect that its development involved the same kind of epistemic challenges as we are examining here.

The Wedgwood pyrometer was discredited long before the establishment of the methods we now trust. All in all, it seems that the Wedgwood pyrometer met its demise through a gradual convergence of a host of other methods all lined up against it. But does such epistemic ganging up prove anything? The Wedgwood pyrometer continued to be used with practical benefit well into the nineteenth century. Even if we disregard Le Chatelier's retrospective hyperbole that the Wedgwood pyrometer was "the only guide in researches at high temperatures" for "nearly a century," we cannot dismiss the estimate in E. Péclet's 1843 textbook on the practical applications of heat that the instrument of "Vedgwood" was the most generally employed pyrometer, even sixty years after its invention.[50] There are various other reports showing the uses of the Wedgwood pyrometer later in the century, too.[51] What exactly was gained by declaring it to be incorrect, on the basis of a convergence of various other methods that were each insecure in themselves? These questions will be addressed more systematically in the analysis part.

Analysis: The Extension of Concepts beyond Their Birth Domains

> [Physics] has come to see that thinking is merely a form of human activity...with no assurance whatever that an intellectual process has validity outside the range in which its validity has already been checked by experience.
>
> P. W. Bridgman, "The Struggle for Intellectual Integrity," 1955

To make and describe scientific observations and measurements, we must make use of certain concepts and material instruments. These concepts and instruments embody certain regularities. In the first two chapters we have seen how the great difficulties involved in establishing such regularities can be overcome, at least to some extent. However, there are new challenges in extending these regularities to new domains of phenomena, so that the concepts can function usefully and meaningfully there. The double narrative given in the first part of this chapter gives

---

[50] See Le Chatelier and Boudouard 1901, 1; Péclet 1843, 1:4.
[51] According to Rostoker and Rostoker 1989, the *Ordnance Manual for the Use of Officers in the United States* (1841) gave the melting point of steel as 160°W; H. C. Osborn, also in America, used the Wedgwood pyrometer to help the manufacture of "blister" steel, as he reported in 1869. In France, Alphonse Salvétat (1857, 2:260), the chief of chemistry at the porcelain works at Sèvres, criticized the Wedgwood thermometer but still reported that the Wedgwood degrees for the melting points of gold, silver, and cast iron were sufficiently exact.

ample illustrations of those challenges. Now I will give a more thorough and general analysis of this problem of the extension of concepts, in their measurement and in their meaning. I will start in "Travel Advisory from Percy Bridgman" with a more careful characterization of the challenge of extension, with the help of Percy Bridgman's ideas on operational analysis. In "Beyond Bridgman" I will argue that Bridgman's ideas need to be modified in order to avoid the reduction of meaning to measurement, which makes it impossible to question the validity of proposed conceptual extensions. After these preliminary steps, "Strategies for Metrological Extension" and "Mutual Grounding as a Growth Strategy" will present "mutual grounding" as a strategy of extension that can help knowledge grow in the absence of previously established standards.

Travel Advisory from Percy Bridgman

The scientists discussed in the narrative part of this chapter were explorers into unknown territories—some of them literally, and all of them metaphorically. There were clear dangers and mirages awaiting them in the new lands. Their journeys would have been much easier with a travel advisory from a knowledgeable authority, but there were no such authorities then. However, there is no reason why we should not retrace, analyze, and reconsider their steps, thinking about how they could have avoided certain pitfalls, where else they might have gone, or how they might have reached the same destinations by more advisable routes. There will be fresh understanding and new discoveries reached by such considerations.

In our own journey we should seek the help of any guides available, and I cannot imagine a better one than Percy Williams Bridgman (1882–1961) (fig. 3.6), the reluctant creator of "operationalism,"[52] whose pioneering work in the physics of high pressures was rewarded with a Nobel Prize in 1946. His chief scientific contribution was made possible by technical prowess: in his Harvard laboratory Bridgman created pressures that were nearly 100 times higher than anyone else had reached before him and investigated the novel behavior of various materials under such high pressures. But as Gerald Holton has noted, Bridgman was placed in a predicament by his own achievements: at such extreme pressures, all previously known pressure gauges broke down; how was he even to know what level of pressures he had in fact reached?[53] That was just the same sort of pitfall as Braun fell into by his success with freezing mixtures, which made the mercury thermometer

---

[52]Bridgman denied that he ever intended to create a rigid and systematic philosophy. In a conference session devoted to the discussion of his ideas in 1953, he complained: "As I listened to the papers I felt that I have only a historical connection with this thing called 'operationalism.' In short, I feel that I have created a Frankenstein, which has certainly got away from me. I abhor the word *operationalism* or *operationism*, which seems to imply a dogma, or at least a thesis of some kind. The thing I have envisaged is too simple to be dignified by so pretentious a name; rather, it is an attitude or point of view generated by continued practice of operational analysis" (Bridgman in Frank 1954, 74–75).

[53]See the entry on Bridgman in *The Dictionary of Scientific Biography*, 2:457–461, by Edwin C. Kemble, Francis Birch, and Gerald Holton. For a lengthier treatment of Bridgman's life and work viewed within broader contexts, see Walter 1990.

FIGURE 3.6. Percy Williams Bridgman. Courtesy of the Harvard University Archives.

inoperable. The situation was even starker for Bridgman because he kept breaking his own pressure records, creating the need to establish pressure measures fit for a succession of higher and higher pressures. Therefore it is no surprise that Bridgman thought seriously about the groundlessness of concepts where no methods were available for their measurement.

With the additional stimulus of pondering about the lessons from Albert Einstein's definition of simultaneity in his special theory of relativity, Bridgman formulated the philosophical technique of operational analysis, which always sought to ground meaning in measurement. The operational point of view was first set out systematically in his 1927 *Logic of Modern Physics* and became very influential among practicing physicists and various thinkers inspired by the tradition of American pragmatism or the new philosophy of logical positivism. In the rest of this chapter we shall see that Bridgman had much careful warning to offer us about extending concepts beyond the domains in which they were born. Bridgman's concern about the definition and meaning of scientific concepts was also forged in the general climate of shock suffered by the early twentieth-century physicists from a barrage of phenomena and conceptions that were entirely alien to everyday expectations, culminating with quantum mechanics and its "Copenhagen" interpretation. In a popular article, Bridgman (1929, 444) wrote: "[I]f we sufficiently extend our range we shall find that nature is intrinsically and in its elements neither understandable nor subject to law."

But as we have seen in the narrative, the challenges of the unknown were amply present even in much more prosaic circumstances. Bridgman was aware of the ubiquity of the problem and chose to open his discussion of operational analysis with the example of the most mundane of all scientific concepts: length. He was both fascinated and concerned by the fact that "essential physical limitations" forced us to use different operations in measuring the same concept in different realms of phenomena. Length is measured with a ruler only when we are dealing with dimensions that are comparable to our human bodies, and when the objects of measurement are moving slowly relative to the measurer. Astronomical lengths or distances are measured in terms of the amount of time that light takes to travel, and that is also the procedure taken up in Einstein's theorizing in special relativity; "the space of astronomy is not a physical space of meter sticks, but is a space of light waves" (Bridgman 1927, 67).

For even larger distances we use the unit of "light-year," but we cannot actually use the operation of sending off a light beam to a distant speck of light in the sky and waiting for years on end until hopefully a reflected signal comes back to us (or our descendants). Much more complex reasoning and operations are required for measuring any distances beyond the solar system:

> Thus at greater and greater distances not only does experimental accuracy become less, but the very nature of the operations by which length is to be determined becomes indefinite.... To say that a certain star is 10^5 light years distant is actually and conceptually an entire different *kind* of thing from saying that a certain goal post is 100 meters distant. (17–18; emphasis original)

Thus operational analysis reveals that the length is not one homogeneous concept that applies in the whole range in which we use it:

> In *principle* the operations by which length is measured should be *uniquely* specified. If we have more than one set of operations, we have more than one concept, and strictly there should be a separate name to correspond to each different set of operations. (10; emphases original)

In practice, however, scientists do not recognize multiple concepts of length, and Bridgman was willing to concede that it is allowable to use the same name to represent a series of concepts, if the different measurement operations give mutually consistent numerical results in the areas of overlap:

> If we deal with phenomena outside the domain in which we originally defined our concepts, we may find physical hindrances to performing the operations of the original definition, so that the original operations have to be replaced by others. These new operations are, of course, to be so chosen that they give, within experimental error, the same numerical results in the domain in which the two sets of operations may be both applied. (23)

Such numerical convergence between the results of two different operations was regarded by Bridgman (16) as merely "the practical justification for retaining the same name" for what the two operations measure.

Even in such situations, we have to be wary of the danger of slipping into conceptual confusion through the use of the same word to refer to many operations. If our thoughts are not tempered by the operationalist conscience always referring us back to concrete measurement operations, we may get into the sloppy habit of using one word for all sorts of different situations without checking for the required convergence in the overlapping domains. Bridgman warned (1959, 75): "[O]ur verbal machinery has no built-in cutoff." In a similar way, we could be misled by the representation of a concept as a number, into thinking that there is naturally an infinitely extendable scale for that concept, the way the real-number line continues on to infinity in both directions. Similarly it would be easy to think that physical quantities must meaningfully exist down to infinite precision, just because the numerical scale we have pinned on them is infinitely divisible. Bridgman reminded us:

> Mathematics does not recognize that as the physical range increases, the fundamental concepts become hazy, and eventually cease entirely to have physical meaning, and therefore must be replaced by other concepts which are operationally quite different. For instance, the equations of motion make no distinction between the motion of a star into our galaxy from external space, and the motion of an electron about the nucleus, although physically the meaning in terms of operations of the quantities in the equations is entirely different in the two cases. The structure of our mathematics is such that we are almost forced, whether we want to or not, to talk about the inside of an electron, although physically we cannot assign any meaning to such statements. (Bridgman 1927, 63)

Bridgman thus emphasizes that our concepts do not automatically extend beyond the domain in which they were originally defined. He warns that concepts in far-out domains can easily become meaningless for lack of applicable measurement operations. The case of length in the very small scale makes that danger clear. Beyond the resolution of the eye, the ruler has to be given up in favor of various micrometers and microscopes. When we get to the realm of atoms and elementary particles, it is not clear what operations could be used to measure length, and not even clear what "length" means any more. Bridgman asked:

146 Inventing Temperature

What is the possible meaning of the statement that the diameter of an electron is 10^{-13}cm? Again the only answer is found by examining the operations by which the number 10^{-13} was obtained. This number came by solving certain equations derived from the field equations of electrodynamics, into which certain numerical data obtained by experiment had been substituted. The concept of length has therefore now been so modified as to include that theory of electricity embodied in the field equations, and, most important, assumes the correctness of extending these equations from the dimensions in which they may be verified experimentally into a region in which their correctness is one of the most important and problematical of present-day questions in physics. To find whether the field equations are correct on a small scale, we must verify the relations demanded by the equations between the electric and magnetic forces and the space coördinates, to determine which involves measurement of lengths. But if these space coördinates cannot be given an independent meaning apart from the equations, not only is the attempted verification of the equations impossible, but the question itself is meaningless. If we stick to the concept of length by itself, we are landed in a vicious circle. As a matter of fact, the concept of length disappears as an independent thing, and fuses in a complicated way with other concepts, all of which are themselves altered thereby, with the result that the total number of concepts used in describing nature at this level is reduced in number.[54]

Such a reduction in the number of concepts is almost bound to result in a corresponding reduction in the number of relations that can be tested empirically. A good scientist would fight against such impoverishment of empirical content in new domains.

Before closing this section, I would like to articulate more clearly a new interpretation of Bridgman's operationalism, which will also be helpful in framing further analysis of the problem of measuring extreme temperatures. Operationalism, as I think Bridgman conceived it, is a philosophy of extension. To the casual reader, much of Bridgman's writing will seem like a series of radical complaints about the meaninglessness of the concepts we use and statements we make routinely without much thinking. But we need to recognize that Bridgman was not interested in skepticism about established discourses fully backed up by well-defined operations. He started getting worried only when a concept was being extended to new situations where the familiar operations defining the concept ceased to be applicable.[55] His arguments had the form of iconoclasm only because he was exceptionally

---

[54]Bridgman 1927, 21–22. Similarly he asked (1927, 78): "What is the meaning, for example, in saying that an electron when colliding with a certain atom is brought to rest in 10^{-18} seconds? . . . [S]hort intervals of time acquire meaning only in connection with the equations of electrodynamics, whose validity is doubtful and which can be tested only in terms of the space and time coordinates which enter them. Here is the same vicious circle that we found before. Once again we find that concepts fuse together on the limit of the experimentally attainable."

[55]This was not only a matter of scale, but all circumstances that specify the relevant measurement operations. For example, if we want to know the length of a moving object, such as a street car, how shall it be measured? An obvious solution would be to board the car with a meter stick and measure the length of the car from the inside just as we measure the length of any stationary everyday-size object. "But here

good at recognizing where a concept had in fact been extended to new domains, especially when the extension was made unthinkingly and most people were not even aware that it had been made. He felt that all physicists, including himself, had been guilty of such unthinking extension of concepts, especially on the theoretical side of physics. Einstein, through his special theory of relativity, taught everyone what dangerous traps we can fall into if we step into new domains with old concepts in an unreflective way. At the heart of the special theory of relativity was Einstein's recognition that judging the simultaneity of two events separated in space required a different operation from that required for judging the simultaneity of two events happening at the same place. Fixing the latter operation was not sufficient to determine the former operation, so a further convention was necessary. But anyone thinking operationally should have recognized from the start that the meaning of "distant simultaneity" was not fixed unless an operation for judging it was specified.[56]

In Bridgman's view, Einstein's revolution should never have been necessary, if classical physicists had paid operational attention to what they were doing. He thought that any future toppling of unsound structures would become unnecessary if the operational way of thinking could spread and quietly prevent such unsound structures in the first place. Operational awareness was required if physics was not to be caught off guard again as it was in 1905: "We must remain aware of these joints in our conceptual structure if we hope to render unnecessary the services of the unborn Einsteins" (Bridgman 1927, 24). As with Descartes, skepticism for Bridgman was not an end in itself, but a means for achieving a more positive end. Bridgman was interested in advancing science, not in carping against it. Operational analysis is an excellent diagnostic tool for revealing where our knowledge is weak, in order to guide our efforts in strengthening it. The Bridgmanian ideal is always to back up concepts with operational definitions, that is, to ensure that every concept is independently measurable in every circumstance under which it is used. The operationalist dictum could be phrased as follows: increase the empirical content of theories by the use of operationally well-defined concepts. In the operationalist ethic, extension is a duty of the scientist but unthinking extension is the worst possible sin.

---

there may be new questions of detail. How shall we jump on to the car with our stick in hand? Shall we run and jump on from behind, or shall we let it pick us up in front? Or perhaps does now the material of which the stick is composed make a difference, although previously it did not? All these questions must be answered by experiment" (Bridgman 1927, 11).

[56]See Bridgman 1927, 10–16. This lesson from Einstein was so dear to Bridgman that he did not shrink from attacking Einstein himself publicly when he seemed to betray his own principle in the general theory of relativity: "[H]e has carried into general relativity theory precisely that uncritical, pre-Einsteinian point of view which he has so convincingly shown us, in his special theory, conceals the possibility of disaster" (Bridgman 1955, 337). The article in which this argument occurs was initially published in the collection entitled *Albert-Einstein: Philosopher-Scientist*, edited by Paul A. Schilpp, in the Library of Living Philosophers series. Einstein replied briefly with bemused incomprehension, much the same way in which he responded to Heisenberg's claim that he was following Einstein in treating only observable quantities in his matrix mechanics (see Heisenberg 1971, 62–69).

148 Inventing Temperature

Beyond Bridgman: Meaning, Definition, and Validity

There is one stumbling block to clear away if we are to construe Bridgman's operationalism as a coherent philosophy of conceptual extension. That obstacle is an overly restrictive notion of meaning, which comes down to the reduction of meaning to measurement, which I will refer to as *Bridgman's reductive doctrine of meaning*. It is a common opinion that operationalism failed as a general theory of meaning, as did its European cousin, namely the verification theory of meaning often attributed to the logical positivists.[57] I do not believe that Bridgman was trying to create a general philosophical theory of meaning, but he did make remarks that certainly revealed an impulse to do so. The following two statements are quite significant, and representative of many other remarks made by Bridgman:

> In general, we mean by any concept nothing more than a set of operations; the concept is synonymous with the corresponding set of operations. (Bridgman 1927, 5)
>
> If a specific question has meaning, it must be possible to find operations by which an answer may be given to it. (28)

The reductive doctrine of meaning indicated by these remarks is not only untenable in itself but unhelpful for understanding the extension of concepts.

Bridgman reminded us so forcefully that measurement operations did not have unlimited domains of application and that our conceptual structures consequently had "joints" at which the same words might continue to be used but the actual operations for measuring them must change. But there can be no "joints" if there is no continuous tissue around them at all. Less metaphorically: if we reduce meaning entirely to measurement operations, there are no possible grounds for assuming or demanding any continuity of meaning where there is clear discontinuity in measurement operations. Bridgman recognized that problem, but his solution was weak. He only proposed that there should be a continuity of numerical results in the overlapping range of two different measurement operations intended for the same concept. Such numerical convergence is perhaps a necessary condition if the concept is to have continuity at all, but it is not a positive indication of continuity, as Bridgman recognized clearly. If we are to talk about a genuine extension of the concept in question, it must be meaningful to say whether what we have is an entirely accidental convergence of the measured values of two unrelated quantities, or a convergence of values of a unified concept measured by two different methods. In sum: a successful extension of a concept requires some continuity of meaning, but reducing meaning entirely to measurement operations makes such continuity impossible, given that measurement operations have limited domains of application.

Moreover, if we accept Bridgman's reductive doctrine of meaning, it becomes unclear why we should seek extensions of concepts at all. That point can be illustrated very well through the case of Wedgwood pyrometry. It would seem that Wedgwood had initially done exactly what would be dictated by operationalist

---

[57]For this and other various important points of criticism directed against operationalism, see Frank 1954.

conscience. He created a standard of temperature measurement that applied successfully to pyrometric phenomena. As the new instrument did not operate at all in the range of any trustworthy previous thermometers, he attached a fresh numerical scale to his own thermometer. Why was that not the honest thing to do, and quite sufficient, too? Why did everyone, including Wedgwood himself, feel compelled to interpret the Wedgwood clay scale in terms of the mercury scale? Why was a continuous extension desired so strongly, when a disjointed set of operations seemed to serve all necessary practical purposes? It is difficult to find adequate answers to these questions, if we adhere to Bridgman's reductive doctrine of meaning.

The key to understanding the urge for conceptual extension, in the Wedgwood case, lies in seeing that there was a real and widespread sense that a property existed in the pyrometric range that was continuous in its meaning with temperature in the everyday range. Where did that feeling come from? If we look closely at the situation in pyrometry, numerous connections that are subtle and often unspoken do emerge between pyrometric temperature and everyday temperature. In the first place, we can bring objects to pyrometric domains by prolonged heating—that is to say, by the continuation of ordinary processes that cause the rise of temperature within the everyday domain. Likewise, the same causes of cooling that operate in the everyday domain, if applied for longer durations or with greater intensity, bring objects from pyrometric temperatures down to everyday temperatures; that is precisely what happens in calorimetric pyrometry (or when we simply leave very hot things out in cold air for a while). These are actually concrete physical operations that provide a continuity of meaning, even operational meaning, between the two domains that are not connected by a common measurement standard.

The connections listed earlier rest on very basic qualitative causal assumptions about temperature: fire raises the temperature of any ordinary objects on which it acts directly; if two objects at different temperatures are put in contact with each other, their temperatures tend to approach each other. There are semi-quantitative links as well. It is taken for granted that the consumption of more fuel should result in the generation of more heat, and that is based on a primitive notion of energy conservation. It is assumed that the amount of heat communicated to an object is positively correlated with the amount of change in its temperature (barring changes of state and interfering influences), and that assumption is based on the rough but robust understanding of temperature as the "degree of heat." So, for example, when a crucible is put on a steady fire, one assumes that the temperature of the contents in the crucible continues to rise, up to a certain maximum. That is exactly the kind of reasoning that Daniell used effectively to criticize some of Wedgwood's results:

> Now, any body almost knows, how very soon silver melts after it has attained a bright red heat, and every practical chemist has observed it to his cost, when working with silver crucibles. Neither the consumption of fuel, nor the increase of the air-draught, necessary to produce this effect, can warrant us in supposing that the fusing point of silver is 4 1/2 times higher than a red heat, fully visible in daylight. Neither on the same grounds, is it possible to admit that a full red-heat being 1077°[F], and the welding heat of iron 12,777°, that the fusing point of cast iron can be more than 5000° higher. The welding of iron must surely be considered as incipient fusion. (Daniell 1821, 319)

Similar types of rough assumptions were also used in the study of lower temperatures, as can easily be gleaned from the narrative in "Can Mercury Tell Us Its Own Freezing Point?" and "Consolidating the Freezing Point of Mercury."

These cases illustrate that concepts can and do get extended to fresh new domains in which experiences are scant and observations imprecise, even if no definite measurement operations have been worked out. I will use the phrase *semantic extension* to indicate any situation in which a concept takes on any sort of meaning in a new domain. We start with a concept with a secure net of uses giving it stable meaning in a restricted domain of circumstances. The extension of such a concept consists in giving it a secure net of uses credibly linked to the earlier net, in an adjacent domain. Semantic extension can happen in various ways: operationally, metaphysically, theoretically, or most likely in some combination of all those ways in any given case. One point we must note clearly, which Bridgman did not tend to emphasize in his philosophical discourses, is that not all concrete physical operations are measurement operations (we may know how to make iron melt without thereby obtaining any precise idea of the temperature at which that happens). Therefore even operational meaning in its broader sense is not exhausted by operations that are designed to yield quantitative measurement results.[58] What I would call *metrological extension*, in which the measurement method for a concept is extended into a new domain, is only one particular type of *operational extension*, which in itself is only one aspect of *semantic extension*. What I want to argue, with the help of these notions, is that the justification of a metrological extension arises as a meaningful question only if some other aspects of semantic extension (operational or not) are already present in the new domain in question.

Now, before I launch into any further discussion of semantic extension, I must give some indication of the conception of meaning that I am operating on, although I am no keener to advance a general theory of meaning than Bridgman was. One lesson we can take from Bridgman's troubles is that meaning is unruly and promiscuous. The kind of absolute control on the meaning of scientific concepts that Bridgman wished for is not possible. The most control that can be achieved is by the scientific community agreeing on an explicit definition and agreeing to respect it. But even firm definitions only regulate meaning; they do not exhaust it. The entire world can agree to define length by the standard meter in Paris (or by the wavelength of a certain atomic radiation), and that still does not come close to exhausting all that we mean by length. The best common philosophical theory of meaning for framing my discussion of conceptual extension is the notion of "meaning as use," which is often traced back to the later phase of Ludwig

---

[58]Bridgman himself recognized this point, at least later in his career. In the preface to *Reflections of a Physicist*, we read (1955, vii): "This new attitude I characterized as 'operational'. The essence of the attitude is that the meanings of one's terms are to be found by an analysis of the operations which one performs in applying the term in concrete situations or in verifying the truth of statements or in finding the answers to questions." The last phrase is in fact much too broad, embodying the same kind of ambiguity as in Bridgman's notion of "paper and pencil operations," which threatened to take all the bite out of the operational attitude.

Wittgenstein's work.[59] If we take that view of meaning, it is easy to recognize the narrowness of Bridgman's initial ideas. Since measurement is only one specific context in which a concept is used, the method of measurement is only one particular aspect of the concept's meaning. That is why Bridgman's reductive doctrine of meaning is inadequate.

In fact, there are some indications that even Bridgman himself did not consistently subscribe to the reductive doctrine of meaning. Very near the beginning of *The Logic of Modern Physics* (1927, 5) as he was trying to motivate the discussion about the importance of measurement operations, Bridgman asserted: "We evidently know what we mean by length if we can tell what the length of any and every object is, and for the physicist nothing more is required." It would have been better if Bridgman had stuck to this weaker version of his ideas about meaning, in which possessing a measurement operation is a sufficient condition for meaningfulness, but not a necessary condition. Even more significant is Bridgman's little-known discussion of "mental constructs" in science (53–60), particularly those created in order "to enable us to deal with physical situations which we cannot directly experience through our senses, but with which we have contact indirectly and by inference." Not all constructs are the same:

> The essential point is that our constructs fall into two classes: those to which no physical operations correspond other than those which enter the definition of the construct, and those which admit of other operations, or which could be defined in several alternative ways in terms of physically distinct operations. This difference in the character of constructs may be expected to correspond to essential physical differences, and these physical differences are much too likely to be overlooked in the thinking of physicists.

They were very easily overlooked in the thinking of philosophers who debated his ideas, too. What Bridgman says here is entirely contrary to the common image of his doctrines. When it came to constructs, "of which physics is full," Bridgman not only admitted that one concept could correspond to many different operations but even suggested that such multiplicity of operational meaning was "what we mean by the reality of things not given directly by experience." In an illustration of these ideas, Bridgman argued that the concept of stress within a solid body had physical reality, but the concept of electric field did not, since the latter only ever manifested itself through force and electric charge, by which it was defined. To put it in my terms, Bridgman was saying that a mental construct could be assigned physical reality only if its operational meaning was broader than its definition.

That last thought gives us a useful key to understanding how metrological validity can be judged: validity is worth debating only if the meanings involved are not exhausted by definitions. If we accept the most extreme kind of operationalism, there is no point in asking whether a measurement method is valid; if the measurement method defines the concept and there is nothing more to the meaning of the concept, the measurement method is automatically valid, as a matter of

---

[59]See, for instance, Hallett 1967.

convention or even tautology. In contrast, validity becomes an interesting question if the concept possesses a broader meaning than the specification of the method of its measurement. Then the measurement method can be said to be valid if it coheres with the other aspects of the concept's meaning. Let us see, in the next section, how this general view on validity can be applied to the more specific matter of metrological extension.

Strategies for Metrological Extension

I am now ready to consider what the validity of a metrological extension consists in and to apply that consideration to the case of extending temperature measurements to extreme domains. We start with a concept that has a well-established method of measurement in a certain domain of phenomena. A metrological extension is made when we make the concept measurable in a new domain. By definition, a metrological extension requires a new standard of measurement, and one that is connected to the old standard in some coherent way. In order for the extension to be valid, there are two different conditions to be satisfied:

> *Conformity.* If the concept possesses any pre-existing meaning in the new domain, the new standard should conform to that meaning.
>
> *Overlap.* If the original standard and the new standard have an overlapping domain of application, they should yield measurement results that are consistent with each other. (This is only a version of the comparability requirement specified in chapter 2, as the two standards are meant to measure the same quantity.)

As we have seen in the last section, the second condition is stated plainly by Bridgman, and the first is suggested in his discussion of constructs.[60]

With that framework for considering the validity of metrological extension, I would now like to return to the concrete problem of extending temperature measurement to the realms of the very cold and the very hot. In the rest of this section I will attempt to discern various strategies that were used in making the extension in either direction, each of which was useful under the right circumstances. As we have seen in chapter 2, by the latter part of the eighteenth century (when the main events in the narratives of the present chapter began), the widespread agreement was that the mercury-in-glass thermometer was the best standard of temperature measurement. From around 1800 allegiances distinctly started to switch to the air thermometer. Therefore the extensions of temperature measurement that we have been considering were made from either the mercury or the air thermometer as the original standard.

---

[60]The target of Bridgman's critique of constructs was what we might call *theoretical* constructs, generally defined by mathematical relations from other concepts that have direct operational meaning. But concepts like pyrometric temperature are also constructs. The only direct experience we can have of objects possessing such temperatures would be to be burned by them, and nothing in our experience corresponds to various magnitudes of such temperatures.

Disconnected Extension

It is a somewhat surprising consequence of the viewpoint I have taken, that the original Wedgwood pyrometric scale, without the conversion to the Fahrenheit scale, constituted a valid extension of the mercury standard. What Wedgwood did was to create an entirely new measurement standard that was not connected directly to the original standard. As there was no direct area of overlap between the mercury thermometer and the Wedgwood pyrometer, the overlap condition was irrelevant. The conformity condition was met in quite a satisfactory way. The connections between Wedgwood temperatures and various aspects of ceramic chemistry and physics were amply testified by Wedgwood's increased success in the art of pottery achieved with the help of the pyrometer, and the stated satisfaction of numerous others who put the Wedgwood pyrometer to use. It is true, as noted earlier, that Daniell made a compelling critique of Wedgwood's melting point of silver on the basis of pre-existing meanings, but that only amounted to a correction of an isolated data point, rather than the discrediting of the Wedgwood pyrometric standard on the whole.

The Wedgwood Patch

The only obvious shortcoming of the original Wedgwood extension was that it left a considerable stretch of the scale without a measurement standard, as the starting point of the pyrometric scale was already quite a bit higher than the endpoint of the mercury scale. Although that gap does not make the Wedgwood pyrometer invalid in itself, it is easy enough to understand the desire for a continuous extended temperature scale, particularly for anyone doing practical work in the range between the boiling point of mercury and red heat. As we have seen in "It Is Temperature, but Not As We Know It," Wedgwood's solution to this problem was to connect up the new standard and the original standard by means of a third standard bridging the two. The intermediate silver scale connected with the Wedgwood scale at the high end, and the mercury scale at the low end. In principle, this strategy had the potential to satisfy both the conformity and the overlap conditions.

Wedgwood's implementation of the patching strategy, however, left much to be desired. He did not check whether the pattern of expansion of silver was the same as the pattern of expansion of mercury in the range up to mercury's boiling point, or whether the expansion of silver at higher temperatures followed a congruent pattern with the contraction of his clay pieces. Instead, Wedgwood simply picked two points and calculated the silver–clay conversion factor assuming linearity. In the case of the silver–mercury comparison he did make two different determinations of the conversion factor and saw that they agreed well with each other, but that was still not nearly enough. Therefore Wedgwood's patched-up scale was only as good as a bridge made of three twisted planks held together with a few nails here and there. This bridge was not good enough to pass the Bridgman test (that is, to satisfy the overlap condition). However, Wedgwood's failure should not be taken as a repudiation of the general strategy of patching.

Whole-Range Standards

Instead of patching up disconnected standards, one could seek a single standard to cover the entire range being considered. This was a popular strategy in the history that we have examined, but its success depended entirely on whether a suitable physical material could be found. For extension into the pyrometric range, measurement methods based on ice calorimetry, water calorimetry, cooling rates, and metallic expansion were all candidates for providing a whole-range standard. For extension into the very cold domain, the alcohol thermometer was a clear candidate but there was a problem with satisfying the overlap condition: it was well known that the alcohol thermometer disagreed significantly with the mercury thermometer and the air thermometer in the everyday temperature range, which is their area of overlap. In the end the best solution was found in Pouillet's air-in-platinum thermometer, which in fact covered the entire range from the lowest known temperatures at the time up to near the melting point of platinum (see the last parts of "Consolidating the Freezing Point of Mercury" and "Ganging Up on Wedgwood").[61]

But there is ultimate futility in this strategy. No matter how broadly a standard is applicable, it will have its limits. Even platinum melts eventually; air liquefies at the cold end and dissociates at the hot end. There are also more mundane limits, for example, in how hot an object one can drop into a bucket of ice or water, as that operation is really only plausible as long as the hot object remains in a solid form, or at least a liquid form. Generally, if one aspires to give a measurement standard to the entire range of values that a quantity can take on, then one will have to fall back on patching. The best agreed-upon modern solution to the problem of thermometric extension is in fact a form of patching, represented in the International Practical Scale of Temperature. But there are more secure ways of patching than the Wedgwood patch.

Leapfrogging

The least ambitious and most cautious of all the strategies discussed so far, which I will call "leapfrogging," is exemplified very well in the development of metallic pyrometers. Initially the pattern of thermal expansion of a metallic substance was studied empirically in the lower temperature ranges, by means of the mercury thermometer. Then the phenomenological law established there was extrapolated into the domain of temperatures higher than the boiling point of mercury. Another exemplary case of leapfrogging was Cavendish's use of the alcohol thermometer for temperatures below the freezing point of mercury (see "Consolidating the Freezing Point of Mercury"). Extension by leapfrogging satisfies the overlap condition by design, since the initial establishment of the phenomenological law indicates exactly

---

[61] The search for a whole-range temperature standard continues to this day. A team at Yale University led by Lafe Spietz is currently working on creating a thermometer using an aluminum-based tunnel junction, which would be able to operate from very near absolute zero to room temperature. See Cho 2003.

how to ensure quantitative agreement in the overlapping domain. The conformity condition may or may not be satisfied depending on the particular case, but it seems to have been quite well satisfied in the cases we have discussed. The leapfrogging process could be continued indefinitely in principle. Once the new standard is established, a further extension can be made if a new phenomenological law can be established in the new domain by reference to the new standard, and that law is extrapolated into a further new domain.

Theoretical Unification

Instead of trying to find one material standard or a directly linked chain of material standards to cover the entirety of the desired domain, one could also try to establish an all-encompassing theoretical scheme that can provide justification for each proposed measurement standard. Using a common theory to validate various disparate standards would be a method of forging connections between them. If various new standards are linked to the original standard in this way, they could all be regarded as extensions of it. This is certainly a valid strategy in principle, but in the historical period I am presently discussing, there was no theory capable of unifying thermometric standards in such a way. Distinct progress using the strategy of theoretical unification was only made much later, on the basis of Kelvin's theoretical definition of absolute temperature, which I will discuss in detail in chapter 4.

Mutual Grounding as a Growth Strategy

The discussion in the last section makes it clear that we are faced with a problem of underdetermination, if that was not already clear in the narrative part. There are many possible strategies for the extension of a measurement standard, and each given strategy also allows many different possible extensions. So we are left with the task of choosing the best extension out of all the possible ones, or at least the best one among all the actual ones that have been tried. In each valid extension the original standard is respected, but the original standard cannot determine the manner of its own extension completely. And it can easily happen that in the new domains the existing meanings will not be precise enough to effect an unambiguous determination of the correct measurement standard. That is why the conformity condition, demanding coherence with pre-existing meanings, was often very easily satisfied.

In chapter 2 I discussed how Victor Regnault solved the problem of the choice of thermometric fluids by subjecting each candidate to a stringent test of physical consistency (comparability). At the end of "Minimalism against Duhemian Holism," I also alluded to the type of situations in which that strategy would not work. We have now come face to face with such a situation, in post-Wedgwood pyrometry. Wedgwood tested his instrument for comparability, and it passed. In other people's hands, too, the Wedgwood pyrometer seems to have passed the test of comparability, as long as standard clay pieces were used. Although that last qualification would be sufficient to discredit the Wedgwood pyrometer if we applied the same kind of rigor as seen in Regnault's work, every other pyrometer would also have failed such stringent tests of comparability, until the late nineteenth century. Even platinum pyrometers, the most controllable of all, yielded quite divergent results when employed by Guyton and by Daniell. On the whole, the quality and amount of available data were clearly not sufficient to allow each pyrometer to pass rigorous tests of comparability until much later.

This returns us to the question of how it was possible to reject the Wedgwood pyrometer as incorrect, when each of the alternative pyrometers was just about as poor as Wedgwood's instrument. In "Ganging Up on Wedgwood," we have seen how the Wedgwood pyrometer was rejected after it was shown that various other pyrometers disagreed with it and agreed much more with each other (the situation can be seen at a glance in table 3.2). To the systematic epistemologist, this will seem like a shoddy solution. First of all, if there is any justification at all involved in this process, it is entirely circular: the platinum pyrometer is good because it agrees with the ice calorimeter, and the ice calorimeter is good because it agrees with the platinum pyrometer; and so on. Second, relying on the convergence of various shaky methods seems like a poor solution that was accepted only because there was no single reliable method available. One good standard would have provided an operational definition of temperature in the pyrometric range, and there would have been no need to prop up poor standards against each other. These are cogent points, at least on the surface. However, I will argue that the state of post-Wedgwood pyrometry does embody an epistemic strategy of development that has positive virtues.

In basic epistemological terms, relying on the convergence of various standards amounts to the adoption of coherentism after a recognized failure of foundationalism. I will discuss the relative merits of foundationalist and coherentist theories of justification more carefully in chapter 5, but what I have in mind at this point is the use of coherence as a guide for a dynamic process of concept formation and knowledge building, rather than strict justification. A very suggestive metaphor was given by Otto Neurath, a leader of the Vienna Circle and the strongest advocate of the Unity of Science movement: "We are like sailors who have to rebuild their ship on the open sea, without ever being able to dismantle it in dry-dock and reconstruct it from the best components."[62]

As often noted, there is some affinity between Neurath's metaphor and W. V. O. Quine's later coherentist metaphor of the stretchable fabric: "The totality of our so-called knowledge or beliefs...is a man-made fabric which impinges on experience only along the edges. Or, to change the figure, total science is like a field of force whose boundary conditions are experience" (Quine [1953] 1961, 42). But there is one difference that is very important for our purposes. In Quine's metaphor, it does not really matter what shape the fabric takes; one presumes it will not rip. In contrast, when we are sailing in Neurath's leaky boat, we will drown if we do not actively do something about it, and do it right. In other words, Neurath's metaphor embodies a clear value judgment about the current state of knowledge, namely that it

---

[62]Neurath [1932/33] 1983, 92. For further discussions of Neurath's philosophy, and particularly the significance of "Neurath's boat," see Cartwright et al. 1996, 139 and 89ff.

is imperfect, and also a firm belief that it can be improved. Neurath's metaphor has a progressivist moral, which is not so central to Quine's.

Post-Wedgwood pyrometry was a very leaky boat. And if I may take some liberties with Neurath's metaphor, we must recognize that even such a leaky boat was already a considerable achievement, since there was no boat at all earlier. Investigators like Clément and Desormes were like shipwrecked sailors pulling together a few planks floating by, to form a makeshift lifeboat (however unrealistic that possibility might be in a real shipwreck). Guyton got on that boat bringing the plank of platinum pyrometry, which fitted well enough. They also picked up the plank of Wedgwood pyrometry, admired it for its various pleasing qualities, but reluctantly let it float away in the end, since it could not be made to fit. They did have the choice of floating around hanging on to the Wedgwood plank waiting for other planks that were compatible with it, but they decided to stay on the boat that they already had, leaky as it was. It is difficult to fault such prudence.

Metaphors aside, what exactly were the positive merits of this process, which I will call the "mutual grounding" of measurement standards? First of all, it is an excellent strategy of managing uncertainty. In the absence of a measurement standard that is demonstrably better than others, it is only sensible to give basically equal status to all initially plausible candidates. But a standard that departs excessively from most others needs to be excluded, just by way of pragmatics rather than by any absolute judgment of incorrectness. In metrological extension, we are apt to find just the sort of uncertainty that calls for mutual grounding. In the new domain the pre-existing meaning is not likely to be full and precise enough to dictate an unambiguous choice of a measurement standard: there is probably no sufficient basis of sensation; few general theories can cover unknown domains confidently; and any extensions of phenomenological laws from known domains face the problem of induction. All in all, there will probably be many alternative extensions with an underdetermined choice between them.[63]

Mutual grounding is not a defeatist compromise, but a dynamic strategy of development. First of all, it allows new standards to come into the convergent nest; the lack of absolute commitment to any of the standards involved also means that some of them can be ejected with relative ease if further study reveals them to be inconsistent with others. Throughout the process decisions are taken in the direction of increasing the degree of coherence within the whole set of mutually grounded standards. The aspect of the coherence that is immediately sought is a numerical convergence of measurement outcomes, but there are also other possible

---

[63] In Chang 1995a, I discussed the case of energy measurement in quantum physics in terms of mutual grounding. In that case, the uncertainty was brought on by theoretical upheaval. In nineteenth-century physics standards for energy measurement in various macroscopic domains had been firmly established, and their extension into microscopic domains was considered unproblematic on the basis of Newtonian mechanics and classical electromagnetic theory, which were assumed to have universal validity. With the advent of quantum mechanics, the validity of classical theories was denied in the microscopic domains, and suddenly the old measurement standards lost their unified theoretical justification. However, the ensuing uncertainty was dealt with quite smoothly by the mutual grounding of two major existing measurement methods and one new one.

aspects to consider, such as the relations between operational procedures, or shared theoretical justification.

The strategy of mutual grounding begins by accepting underdetermination, by not forcing a choice between equally valid options. In the context of choosing measurement standards, accepting plurality is bound to mean accepting imprecision, which we can actually afford to do in quite a few cases, with a promise of later tightening. If we let underdetermination be, multiple standards can be given the opportunity to develop and prove their virtues, theoretically or experimentally. Continuing with observations using multiple standards helps us collect a wide range of phenomena together under the rubric of one concept. That is the best way of increasing the possibility that we will notice previously unsuspected connections, some of which may serve as a basis of further development. A rich and loose concept can guide us effectively in starting up the inquiry, which can then double back on itself and define the concept more rigorously. This, too, is an iterative process of development, as I will discuss in more detail in chapter 5.

4

Theory, Measurement, and Absolute Temperature

Narrative: The Quest for the Theoretical Meaning of Temperature

> Although we have thus a strict principle for constructing a *definite* system for the estimation of temperature, yet as reference is essentially made to a specific body as the standard thermometric substance...we can only regard, in strictness, the scale actually adopted as *an arbitrary series of numbered points of reference sufficiently close for the requirements of practical thermometry*.
>
> William Thomson, "On an Absolute Thermometric Scale," 1848

A theoretically inclined reader may well be feeling disturbed by now to note that so much work on the measurement of temperature seems to have been carried out without any precise theoretical definition of temperature or heat. As seen in the last three chapters, by the middle of the nineteenth century temperature became measurable in a coherent and precise manner over a broad range, but all of that was achieved without much theoretical understanding. It is not that there was a complete lack of relevant theories—there had always been theories about the nature of heat, since ancient times. But until the late nineteenth century no theories of heat were successful in directing the practice of thermometry in a useful way. We have seen something of that theoretical failure in chapter 2. The discussion in this chapter will show why it was so difficult to make a productive and convincing connection between thermometry and the theory of heat, and how such a connection

---

The work on William Thomson's absolute temperature contained in this chapter was done in close collaboration with Sang Wook Yi. The results of our research are reported in full technical detail in Chang and Yi (forthcoming). The content of the section "Temperature, Heat, and Cold" has been adapted from Chang 2002.

was eventually made. In order to stay within the broad time period covered in this book, I will limit my discussion to theoretical developments leading up to classical thermodynamics. Statistical mechanics does not enter the narrative because it did not connect with thermometry in meaningful ways in the time period under consideration.

Temperature, Heat, and Cold

Practical thermometry achieved a good deal of reliability and precision before people could say with any confidence what it was that thermometers measured. A curious fact in the history of meteorology gives us a glimpse into that situation. The common attribution of the centigrade thermometer to the Swedish astronomer Anders Celsius (1701–1744) is correct enough, but his scale had the boiling point of water as 0° and the freezing point as 100°. In fact, Celsius was not alone in adopting such an "upside-down" thermometric scale. We have already come across the use of such a scale in "Can Mercury Be Frozen?" and "Can Mercury Tell Us Its Own Freezing Point?" in chapter 3, in the mercury thermometer designed by the French astronomer Joseph-Nicolas Delisle (1688–1768) in St. Petersburg (the Delisle scale can be seen in the Adams thermometer in fig. 1.1). In England the "Royal Society Thermometer" had its zero point at "extream heat" (around 90°F or 32°C) and increasing numbers going down the tube.[1] These "upside-down" scales were in serious scientific use up to the middle of the eighteenth century, as shown emblematically in figure 4.1.[2]

Why certain early pioneers of thermometry created and used such "upside-down" thermometers remains a matter for speculation. There seems to be no surviving record of the principles behind the calibration of the Royal Society thermometer, and Delisle's (1738) published account of his thermometers only concentrates on the concrete procedures of calibration. There is no clear agreement among historians about Celsius's motivations either. Olof Beckman's view is that "Celsius and many other scientists were used to both direct and reversed scales, and

---

[1] For more detail on the Celsius, Delisle, and Royal Society scales, see: Middleton 1966, 58–62, 87–89, 98–101; Van Swinden 1778, 102–106, 115–116, 221–238; and Beckman 1998. An original Delisle thermometer sent to Celsius by Delisle in 1737 is still preserved at Uppsala University; see Beckman 1998, 18–19, for a photo and description. The National Maritime Museum in Greenwich holds three Royal Society thermometers (ref. no. MT/Th.5, MT/BM.29, MT/BM.28), and also a late nineteenth-century thermometer graduated on the Delisle scale (ref. no. MT/Th.17(iv)).

[2] For instance, Celsius's original scale was adopted in the meteorological reports from Uppsala, made at the observatory that Celsius himself had founded, for some time in the late 1740s. From 1750 we find the scale inverted into the modern centigrade scale. The Royal Society thermometer provided the chief British standard in the early eighteenth century, and it was sent out to agents in various countries who reported their meteorological observations to the Royal Society, which were summarized for regular reports in the *Philosophical Transactions of the Royal Society of London*. The use of the Royal Society scale is in evidence at least from 1733 to 1738. Delisle's scale was recognized widely and remained quite popular for some time, especially in Russia.

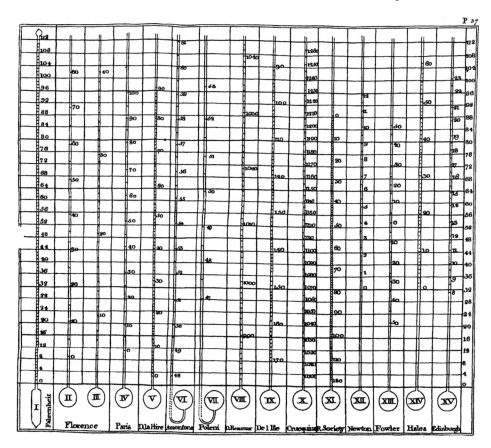

FIGURE 4.1. George Martine's comparison of fifteen thermometric scales. The ninth (Delisle) and eleventh (Royal Society) are of the "upside-down" type. This figure was attached at the beginning of Martine [1740] 1772, chart facing p. 37. Courtesy of the British Library.

simply did not care too much" about the direction.[3] My own hypothesis is that those who designed upside-down thermometers may have been thinking more in terms of measuring the degrees of cold than degrees of heat. If that sounds strange, that is only because we now have a metaphysical belief that cold is simply the absence of heat, not a real positive quality or entity in its own right. Although the existence of the upside-down temperature scales does not prove that their makers were trying to measure degrees of cold rather than heat, at least it reveals a lack of a sufficiently strong metaphysical commitment against the positive reality of cold.

---

[3] Private communication, 28 February 2001; I thank Professor Beckman for his advice.

Indeed, as we have seen in "Can Mercury be Frozen?" and "Can Mercury Tell Us Its Own Freezing Point?" in chapter 3, in seventeenth- and eighteenth-century discussions of low-temperature phenomena, people freely spoke of the "degrees of cold" as well as "degrees of heat."[4] The practical convenience of the Delisle scale in low-temperature work is obvious, as it gives higher numbers when more cooling is achieved. If we look back about a century before Celsius, we find that Father Marin Mersenne (1588–1648), that diplomat among scholars and master of "mitigated skepticism," had already devised a thermometer to accommodate all tastes, with one sequence of numbers going up and another sequence going down.[5] Similarly, the alcohol thermometer devised by the French physicist Guillaume Amontons (1663–1738) had a double scale, one series of numbers explicitly marked "degrees of cold" and the other "degrees of heat" (see fig. 4.2).

The history of cold is worth examining in some more detail, since it is apt to shake us out of theoretical complacency regarding the question of what thermometers measure. There have been a number of perfectly capable philosophers and scientists through the ages who regarded cold as real as heat—starting with Aristotle, who took cold and hot as opposite qualities on an equal footing, as two of the four fundamental qualities in the terrestrial world. The mechanical philosophers of the seventeenth century were not united in their reactions to this aspect of Aristotelianism. Although many of them subscribed to theories that understood heat as motion and cold as the lack of it, the mechanical philosophy did not rule out giving equal ontological status to heat and cold. In the carefully considered view of Francis Bacon (1561–1626), heat was a particular type of expansive motion and cold was a similar type of contractive motion; therefore, the two had equal ontological status. Robert Boyle (1627–1691) wanted to rule out the positive reality of cold, but had to admit his inability to do so in any conclusive way after honest and exhaustive considerations. The French atomist Pierre Gassendi (1592–1655) had a more complex mechanical theory, in which "calorific atoms" caused heat by agitating the particles of ordinary matter; Gassendi also postulated "frigorific atoms," whose angular shapes and sluggish motion made them suited for clogging the pores of bodies and damping down the motions of atoms.[6]

---

[4]For example, Bolton (1900, 42–43) quotes Boyle as saying in 1665: "The common instruments show us no more than the relative coldness of the air, but leave us in the dark as to the positive degree thereof...." Similarly, the records of the Royal Society (Birch [1756–57] 1968, 1:364–5) state that at the meeting of 30 December 1663 a motion was made "to make a standard of cold," upon which Hooke suggested "the degree of cold, which freezes distilled water." Even Henry Cavendish, who had quite a modern-sounding theory of heat, did not hesitate titling his article of 1783: "Observations on Mr. Hutchins's Experiments for Determining the Degree of Cold at which Quicksilver Freezes."

[5]See Bolton 1900, 30–31; Mersenne described this thermometer in 1644.

[6]For a summary of the mechanical philosophers' views on heat and cold, see Pyle 1995, 558–565. For further detail, see Bacon [1620] 2000, book II, esp. aphorism XX, 132–135; on Boyle, see Sargent 1995, 203–204.

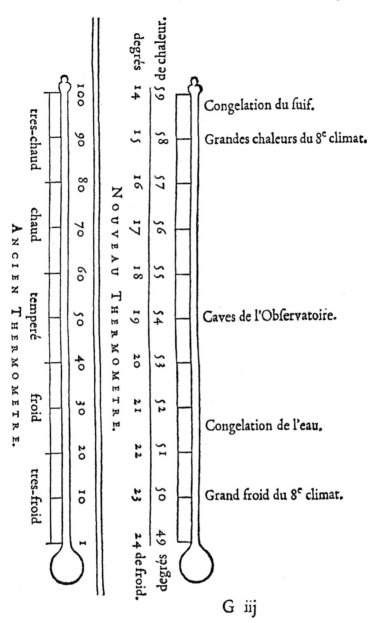

FIGURE 4.2. Amontons's thermometer (1703, 53), with a double scale. Courtesy of the British Library.

Gassendi's sort of theory seems to have enjoyed much popularity for some time, as reported in 1802 by Thomas Thomson (1773–1852), Scotland's premier "chemist breeder" and early historian of chemistry:[7]

> There have been philosophers... who maintained that cold is produced not by the abstraction of caloric merely, but by the addition of a positive something, of a peculiar body endowed with specific qualities. This was maintained by [Petrus van] Muschenbroek [1692–1761] and [Jean Jacques d'Ortous] De Mairan [1678–1771], and *seems to have been the general opinion of philosophers about the commencement of the eighteenth century*. According to them, cold is a substance of a saline nature, very much resembling nitre, constantly floating in the air, and wafted about by the wind in very minute corpuscles, to which they gave the name of frigorific particles. (T. Thomson 1802, 1:339; emphasis added)

Even by the late eighteenth century the question about the nature of cold had not been settled. The 2d edition of the *Encyclopaedia Britannica* in 1778 reported that there was no agreement on this question, but it came down on the side of supposing the independence of cold from heat, which led to the sort of discourse as the following: "if a body is heated, the cold ought to fly from it." This way of thinking persisted and even gathered strength by the third edition of *Britannica*, published at the end of the century. The author of the article on "heat" there admitted a good deal of uncertainty in current knowledge, and opined that the best way of proceeding was "to lay down certain principles established from the obvious phenomena of nature, and to reason from them fairly as far as we can." Ten such principles were offered, and the first one reads: "Heat and cold are found to expel one another. Hence we ought to conclude, that heat and cold are both *positives*."[8]

Into this confused field came a striking experiment that seemed like a direct confirmation of the reality of cold and generated the controversy that became the last and most crucial debate in the banishing of cold from the ontology of the universe. The experiment was originally the work of the Genevan physicist and statesman Marc-Auguste Pictet (1752–1825). Pictet (1791, 86–111) set up two concave metallic mirrors facing each other and placed a sensitive thermometer at the focus of one of the mirrors; then he brought a hot (but not glowing) object to the focus of the other mirror and observed that the thermometer started rising immediately. (Fig. 4.3 shows a later version of this experiment, by John Tyndall.[9]) After a trial with the mirrors separated by a distance of 69 feet that showed no time lag in the production of the effect, Pictet concluded he was observing the radiation of heat at an extremely high speed (like the passage of light) and certainly not its

---

[7] For more information about Thomson, who is an important source for the history of chemistry up to his time, see Morrell 1972 and Klickstein 1948.

[8] "Cold," in *Encyclopaedia Britannica*, 2d ed., vol. 3 (1778), 2065–2067; the quoted statement is on p. 1066. "Heat," in *Encyclopaedia Britannica*, 3d ed. (1797), 8:350–353; the quoted principle is on p. 351.

[9] Tyndall (1880, 289–292) found it wondrous to have the opportunity to use this apparatus at the Royal Institution. He recalled acquiring the yearning to become a natural philosopher in his youth from the excitement of reading an account of experiments performed by Humphry Davy with the very same apparatus, which had initially been commissioned by Count Rumford.

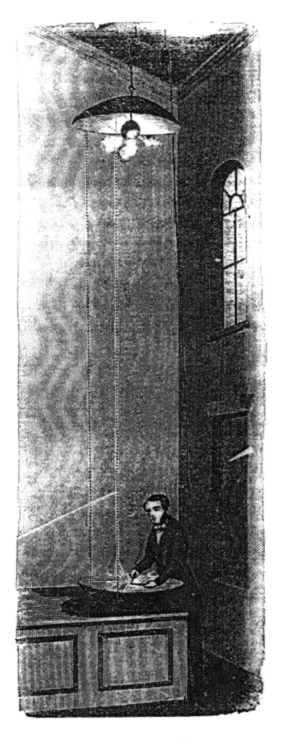

FIGURE 4.3. Illustration showing a version of Pictet's "double-reflection" experiment, from Tyndall 1880 (290). The spark ignited at the focus of the lower mirror causes the explosion of the hydrogen-chlorine balloon at the focus of the upper mirror. In an experiment more like Pictet's, a hot copper ball placed in the lower focus causes a blackened hydrogen-oxygen balloon at the upper focus to explode. Courtesy of the British Library.

slow conduction through the air. To appreciate how remarkable this result was, we need to remember that its publication was a full decade before the discovery of infrared heating effect in sunlight by William Herschel (1800b). Unmediated heat transfer through space was a very novel concept at that time.

This experiment was already stunning enough, but now Pictet described a variation on it that still inspires incredulity in those who read its description (1791, 116–118):

> I conversed on this subject with Mr. [Louis] Bertrand [1731–1812], a celebrated professor of mathematics in our academy, and pupil of the immortal Euler. He asked me if I believed cold susceptible of being reflected? I confidently replied no; that cold was only privation of heat, and that a negative could not be reflected. He requested me, however, to try the experiment, and he assisted me in it.

When Pictet introduced a flask filled with snow into the focus of one mirror, the thermometer at the focus of the other mirror immediately dropped "several degrees," as if the snow emitted rays of cold that were reflected and focused at the thermometer. When he made the snow colder by pouring some nitrous acid on it, the cooling effect was enhanced. But how could that be, any more than a dark object could emit rays of darkness that make a light dimmer at the receiving end? Pictet was initially "amazed" by the outcome of his own experiment, which he felt was "notorious." Bertrand's suggestion seemed a frivolous one at the outset, but now it had to be addressed seriously.

The situation here is reminiscent of the recent philosophical debates surrounding Ian Hacking's argument that we are entitled to believe in the reality of unobservable objects postulated in our theories if we can manipulate them successfully in the laboratory, for instance when we can micro-inject a fluid into a cell while monitoring the whole process with a microscope. Hacking (1983, 23) put forward a slogan that will be remembered for a long time: "[I]f you can spray them, then they're real." Hacking explains that this last insight came out of his own experience of overcoming his disbelief about the reality of the positron, the antiparticle of the electron. After learning how positrons can be "sprayed" onto a tiny niobium ball to change its electric charge (in a modern version of the Millikan oil-drop experiment), Hacking felt compelled to give up his previous notion that positrons were mere theoretical constructs. But what is Pictet's experiment, if not a successful *spraying* of cold onto the thermometer to lower its temperature, just as Hacking's physicists spray positrons onto a niobium ball to change its electric charge? Anyone wanting help in denying the reality of cold will have to look elsewhere, since the only answer based on Hacking's "experimental realism" will have to be that radiant cold is indeed real.

Pictet himself dispensed with the conundrum relatively quickly, by convincing himself that he was only really observing heat being radiated away from the thermometer and sinking into the ice; the thermometer loses heat in this way, so naturally its temperature goes down. But it is clear that someone with precisely the opposite picture of reality could give a perfectly good mirror-image explanation: "Heat doesn't really exist (being a mere absence of cold), yet certain phenomena could fool us into thinking that it did. When we observe a warmer object apparently heating a colder one by radiation, all that is happening is that the colder object is radiating cold to the warmer one, itself getting less cold in the process." The Edinburgh chemist John Murray (1778?–1820) summed up the quandary arising from Pictet's and some related experiments as follows:

> In these experiments, then, we have apparently the emanation from a cold body of a positively frigorific power, which moves in right lines, is capable of being intercepted, reflected and condensed, and of producing, in its condensed state, its accumulated cooling power; and they appear equally conclusive in establishing the existence of radiant cold, as the other experiments are in establishing the existence of radiant heat. (Murray 1819, 1:359–360)

A more generalized view was reported by the English polymath Thomas Young (1773–1829), best known now for his wave theory of light, in his Royal Institution lectures at the start of the nineteenth century: "Any considerable increase of heat gives us the idea of positive warmth or hotness, and its diminution excites the idea of positive cold. Both these ideas are simple, and each of them might be derived either from an increase or from a diminution of a positive quality."[10]

Elsewhere I have examined what I consider to be the last decisive battle in this long-standing dispute, which was instigated by Count Rumford (1753–1814), who took up Pictet's experiment and developed further variations designed to show that the observations could not be explained in terms of the radiation of heat.[11] Rumford is best known to historians and physicists for his advocacy of the idea that heat was a form of motion rather than a material substance, but in his own day he was celebrated even more as an inventor and a social reformer, who got rich by selling improved fireplaces and kitchens, invented the soup kitchen for the poor, reorganized the army and rounded up beggars into workhouses in Munich, and founded the Royal Institution in London.[12] Rumford's scientific work was always geared toward practical applications, and his interest in radiant heat and cold was no exception. He noted that a highly reflective surface would serve to retard the cooling of objects (and people) in cold weather, since such a surface would be effective in reflecting away the "frigorific radiation" impinging on the object from its colder surroundings. Rumford tested this idea by experiments with metallic cylinders "clothed" in different types of surfaces, and late in life he enhanced his reputation as an eccentric by defying Parisian fashion with his winter dress, which was "entirely white, even his hat."[13]

When the fundamental ontology of heat and cold was in such state of uncertainty, it was understandably difficult to reach a consensus about what it was

---

[10]Young 1807, 1:631. He continued: "[B]ut there are many reasons for supposing heat to be the positive quality, and cold the diminution or absence of that quality"; however, he did not care to state any of those "many reasons."

[11]See Chang 2002. Evans and Popp 1985 gives an informative account of the same episode from a different point of view.

[12]For brief further information, see Chang 2002, 141. The most detailed and authoritative modern account of his life and work is Brown 1979.

[13]About Rumford's winter dress, see "Memoirs" 1814, 397, and Brown 1979, 260.

that thermometers measured in theoretical terms. But, as we know, the question of the existence of positive cold did get settled in the negative in the end. In the previously mentioned article, I argue that the metaphysical consensus against Rumford's frigorific radiation was reached thanks to the clear and increasing prevalence of the caloric theory of heat, into which it was difficult to fit any notion of positive cold. Therefore, the caloric theory at least produced an agreement that what thermometers measured was the degree or intensity of heat, not cold. But did it enable a complete and detailed theoretical understanding of temperature? That is the subject of the next section.

Theoretical Temperature before Thermodynamics

The first detailed theoretical understanding of temperature arrived in the late eighteenth century, in the tradition of the caloric theory, which was introduced briefly in "Caloric Theories against the Method of Mixtures" in chapter 2. As there were many different theories about the matter of heat, most commonly referred to as "caloric" after Lavoisier, I will speak of the "caloric theories" in the plural. Within any material theory of heat, it was very natural to think of temperature as the density of heat-matter. But as we shall see shortly, there were many different ways of cashing out that basic idea, and there were many difficulties, eventually seen as insurmountable, in understanding the operation of thermometers in such terms.

Within the caloric tradition, the most attractive theoretical picture of temperature was given by the "Irvinists," namely followers of the Scottish physician William Irvine (1743–1787).[14] The Irvinists defined temperature as the total amount of heat contained in a body, divided by the body's capacity for heat. This "absolute" temperature was counted from the "absolute zero," which was the point of complete absence of caloric. One of the most compelling aspects of Irvinist theory was the explanation of latent heat phenomena, which were seen as consequences of changes in bodies' capacities for caloric. For instance, the change of state from ice to water was accompanied by an increase in the capacity for heat, which meant that an input of heat was needed just to maintain the same temperature while the ice was melting. Irvine confirmed by direct measurement that the specific heat (which he identified with heat capacity) of water was indeed higher than the specific heat of ice, in the ratio of 1 to 0.9. The Irvinist explanation of latent heat in state changes was illustrated by the analogy of a bucket that suddenly widens; the liquid level in the bucket (representing temperature) would go down, and more liquid (representing heat) would have to be put in to keep the level where it was before (see fig. 4.4). The heat involved in chemical reactions was explained similarly, by pointing to observed or presumed differences between the heat capacities of the reactants and the products.

Irvine's ideas were highly controversial and never commanded a broad consensus, but they were adopted and elaborated by some influential thinkers especially

---

[14]For further information about Irvine and his ideas, see Fox 1968, Fox 1971, 24ff., and Cardwell 1971, 55ff.

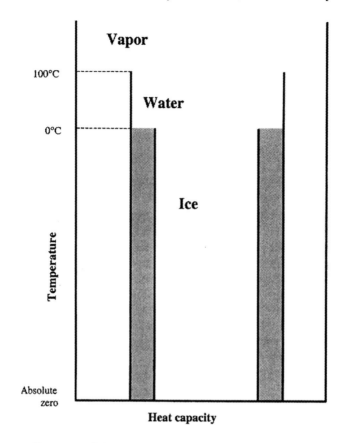

FIGURE 4.4. An illustration of the Irvinist theory of latent heat, adapted from Dalton 1808, figure facing p. 217.

in Britain. Perhaps the most effective advocate of Irvine's ideas was the Irish physician and chemist Adair Crawford (1748–1795), who was active in London in later life but had attended Irvine's lectures while he was a student in Glasgow. Crawford applied Irvine's ideas in order to arrive at a new solution to the long-standing puzzle on how warm-blooded animals generated heat in their bodies (basically, the air we breathe in has a higher heat capacity than the air we breathe out). Crawford's influential treatise on animal heat (1779, 1788) served as an important conduit of Irvine's ideas to other scholars, including John Dalton (1766–1844), the originator of the chemical atomic theory. Other important Irvinists included Sir John Leslie (1766–1832), professor of mathematics and then of natural philosophy in Edinburgh, and the chemist John Murray (1778?–1820), writer of influential textbooks.

The demise of Irvinism owed much to experimental failure. In fact what worked decisively against it was something that Karl Popper would have admired greatly: a high degree of testability. Irvine's hypothesis of heat capacity specified a precise quantitative relationship between heat capacity, temperature, and total heat, and the Irvinists were clearly agreed in how to render their theoretical notion of heat capacity measurable, by identifying it with the operational concept of specific heat. Therefore, measurements of specific heat, temperature, and latent heat in various chemical and physical processes yielded unequivocal tests of Irvinist theory. Most strikingly for thermometry, the Irvinists set out to locate the absolute zero by measuring how much latent heat was necessary to make up for known amounts of change in heat capacity while maintaining the temperature at the same level. For instance, in figure 4.4, the shaded area represents the latent heat involved in turning ice at 0°C into water at 0°C; the width of that shaded area is the difference in the specific heats of ice and water. The specific heat difference and the latent heat can be measured, and then dividing the latter by the former gives the height of the shaded rectangular area, which corresponds to the temperature difference between the melting point of ice and absolute zero. When Irvine performed this calculation, he obtained −900°F as the value of absolute zero (Fox 1971, 28). This was a great triumph for Irvinism: the "zeros" of all previous thermometric scales had been arbitrary points; with Irvinist theory one could calculate how far the conventional zeros were from the absolute zero.

However, most of the predictions made on the basis of Irvine's hypothesis were not confirmed. In the case of the measurement of absolute zero, the values obtained were so diverse that they rather indicated the nonexistence of absolute zero, in the opinion of the opponents of Irvinism. In their critique of Irvinism, Laplace and Lavoisier gave Irvinist calculations of absolute zero ranging from −600°C to −14,000°C. Even Dalton, who was doing the calculations in good Irvinist faith, admitted that the value he presented as the most probable, −6150°F, was derived from results that varied between −11,000°F and −4000°F (Cardwell 1971, 64, 127). Thus, Irvinist caloric theory stuck its neck out, and it was duly chopped off. Of course there were many other reasons why Irvinist theory fell out of favor, but the lack of convergence in the absolute-zero values was certainly one of its most important failures.

It is a pity that Irvinism did not work out. After its demise, theoretical notions of heat and temperature have never recovered such beautiful simplicity. In addition, Irvinism was the only theory up to that point that provided a clear, detailed, and quantitative link between thermometric measurements and the theoretical concepts of temperature and heat. The caloric theorists who rejected Irvinism could only engage in defensive and untestable moves when it came to the understanding of temperature.

The tradition of caloric theories described as "chemical" in chapter 2, led by Lavoisier, could avoid the experimental failure of Irvinism because it had postulated two different states of caloric (free/sensible vs. combined/latent). In the chemical caloric theory, temperature was conceived as the density (or pressure) of free caloric only, since latent caloric was usually conceived as caloric deprived of its characteristic repulsive force, hence unable to effect the expansion of thermometric substances. But it was impossible to refute the various chemical-calorist doctrines by direct experiment, since there was no direct way to measure how much caloric was going latent or becoming sensible again in any given physical or chemical process, not to mention how much total latent caloric was contained in a body. Rumford came up against this irrefutability when he tried to argue against the caloric theory on the basis of his famous cannon-boring experiment. In Rumford's own view, his experiment disproved the caloric theory because it showed that an indefinite amount of heat could be generated by friction (motion). For the leading calorists, Rumford's experiment only showed the disengagement of combined caloric as a consequence of the mechanical agitation, and no one knew just how much combined caloric was there to be released in such a manner.[15]

The chemical calorists also refused to accept the Irvinist identification of specific heat and heat capacity. For them the very concept of Irvinist heat capacity was meaningless, and specific heat was simply a measure of how much heat a body required to go from temperature $T°$ to temperature $(T + 1)°$, which was actually a variable function of T. We have seen in Haüy's critique of De Luc, in "Caloric Theories against the Method of Mixtures" in chapter 2, how indeterminate the relation between temperature and heat input became under the chemical caloric theory. Haüy's work was probably among the best contributions from that tradition to thermometry, but still it could only play a negative role. As discussed in "The Calorist Mirage of Gaseous Linearity" in chapter 2, the later theoretical work by Laplace further weakened the link between the theory and practice of thermometry, despite his aims to the contrary. Laplace redefined temperature as the density of the "free caloric of space," to which one had no independent observational access at all, and which was postulated to be so thinly spread that it would have been undetectable for that reason in any case. Laplace's derivations of the gas laws only amounted to a spurious rationalization of the existing practice of gas thermometry, rather than adding any real insight or concrete improvement to it. All in all, it is not surprising that the serious practitioners of thermometry, including Regnault, mostly disregarded the latter-day caloric theory for the purposes of their art.

If the caloric theories failed to link thermometry effectively to the theoretical concept of temperature, did other theories do any better? Here the modern commentator will jump immediately to the collection of theories that can be classified as "dynamic" or "mechanical," built on the basic idea that heat consists of motion, since that is the tradition that eventually triumphed, though after many decades of neglect during the heyday of caloric theories. Note, however, that the dynamic theories that were in competition with the caloric theories bore very little resemblance to the dynamic theories that came after thermodynamics. Despite a very long history including a phase of high popularity in the context of seventeenth-century mechanical philosophy, dynamic theories of heat were not developed in precise and quantitative ways until the nineteenth century.

---

[15]This famous experiment is described in Rumford [1798] 1968. For typical examples of calorist rebuttals, see Henry 1802, and Berthollet's discussion reproduced in Rumford [1804a] 1968, 470–474. Rumford's own conception of the caloric theory was clearly of the Irvinist variety, as shown in his attempt to argue that the heat could not have come from changes in heat capacity by confirming that the specific heat (by weight) of the metal chippings was the same as the specific heat of the bulk metal.

The first dynamic theorist that we should take note of is indeed Rumford, who has often been hailed as the neglected pioneer whose work should have toppled the caloric theory but did not. However, as intimated by his advocacy of positive cold radiation discussed briefly in the previous section, Rumford's theory of heat was very far from the modern kinetic theory. As was common during that time, Rumford envisaged molecules of matter as vibrating around fixed positions even in gases, not bouncing around in random motion. Rumford presumed temperature to be defined by the frequency of the vibrations, but he vacillated on this point and expressed a suspicion that it might really depend on the velocities that the molecules took on in the course of their oscillations.[16] It is important not to read modern conceptions too much into Rumford's work. He believed that the vibrating molecules would continually set off vibrations in the ether without themselves losing energy in that process. He also rejected the idea of absolute zero because he did not consider temperature as a proper quantity:

> Hot and cold, like fast and slow, are mere relative terms; and, as there is no relation or proportion between motion and a state of rest, so there can be no relation between any degree of heat and absolute cold, or a total privation of heat; hence it is evident that all attempts to determine the place of absolute cold, on the scale of a thermometer, must be nugatory. (Rumford [1804b] 1968, 323)

This opinion regarding the absolute zero was shared by Humphry Davy (1778–1829), Rumford's protégé in the early days of the Royal Institution, and fellow advocate of the dynamic theory of heat: "Having considered a good deal the subject of the supposed real zero, I have never been satisfied with any conclusions respecting it.... [T]emperature does not measure a quantity, but merely a property of heat."[17]

Instead of Rumford, we might look to the pioneers of the kinetic theory of gases, who believed in the random translational motion of molecules, because they are the ones who created the modern theoretical idea that temperature is a measure of the kinetic energy of molecules. But even there the situation is not so straightforward. (On the history of the kinetic theory, I will largely rely on Stephen Brush's [1976] authoritative account.) Very early on Daniel Bernoulli (1700–1782) did explain the pressure of gases as arising from the impact of molecular collisions and showed that it was proportional to the vis viva of individual molecules (mv^2 in modern notation), but he did not identify molecular vis viva as temperature. The first theoretical conception of temperature in the kinetic theory seems to have been articulated by John Herapath (1790–1868) in England, who composed an important early article on the subject in 1820. But Herapath's idea was that temperature was proportional to the velocity of the molecules, not to the square of the velocity as the modern theories would have it. In any case, Herapath's work was largely ignored by the scientific establishment in London, and for a time he was reduced to publishing some of his scientific articles in the Railway Magazine, which

---

[16]See Rumford [1804b] 1968, 427–428; see also the brief yet interesting commentary on this matter in Evans and Popp 1985, 749.

[17]Letter from Davy to John Herapath (6 March 1821), quoted in Brush 1976, 1:118.

he edited. The only real impact his work had was made through James Joule, who made use of some of Herapath's ideas in the late 1840s.[18]

Even worse reception than Herapath's awaited the work of the Scottish engineer John James Waterston (1811–1883), who is the first person on record to have identified temperature as proportional to mv^2, starting in 1843. Waterston's paper submitted to the Royal Society in 1845 was flatly rejected and did not see the light of day until Lord Rayleigh discovered it in the Society's archives in 1891 and had it published in the *Philosophical Transactions* finally. The same idea was put into real circulation only through the work of Rudolf Clausius (1822–1888), whose publications on the matter started in 1850.[19] By that time Thomson had already published his first definition of absolute temperature, and macroscopic thermodynamics was to be the most important vehicle of progress in the theory of heat for a time.

In short, up to the middle of the nineteenth century there was a good deal of disagreement about what temperature meant in a dynamic theory of heat. Besides, even if one accepted a certain theoretical definition, it was not easy to make a successful bridge between such a definition and actual measurement. In fact Herapath was probably the only one to build a clear operational bridge, by declaring that the pressure of a gas was proportional to the *square* of its "true temperature." But both his theoretical conception and his operationalization were seen as problematic, and there is no evidence that anyone took up Herapath's thermometry.[20] Meanwhile neither Waterston nor Clausius came up with a method of operationalizing their concept of temperature as proportional to the average kinetic energy of molecules. In fact, even in modern physics it is not entirely clear how easily that concept can be operationalized (except by sticking an ordinary thermometer in a gas).

William Thomson's Move to the Abstract

We can appreciate the true value of Thomson's work on thermometry when we consider the failure in both of the major traditions of heat theory, discussed in the last section. He crafted a new theoretical concept of temperature through an ingenious use of the new thermodynamic theory and also made that concept measurable. William Thomson (1824–1907), better known to posterity as Lord Kelvin, was one of the last great classical physicists, who made many key advances in thermodynamics and electromagnetism, and worked tenaciously to the end of his life to create a viable physical theory of the ether. Son of a Belfast-born Glasgow professor of mathematics, precocious Thomson received just about the best conceivable

---

[18] See Brush 1976, 1:20 on Bernouilli's work, 1:69, 110–111 on Herapath's ideas, and 1:115–130 on Herapath's general struggle to get his views heard (not to mention accepted). See Mendoza 1962–63, 26, for Joule's advocacy of Herapath's ideas.

[19] See Brush 1976, 1:70–71, for Waterston's basic idea; 1:156–157, for the circumstances of its publication; and 1:170–175, for Clausius's contribution.

[20] See Brush 1976, 1:111–112, for an explanation of Herapath's conception of temperature.

scientific education: early study at home was followed by attendance at Glasgow University, then a Cambridge degree in mathematics, rounded out by further study and apprenticeship in the heart of the scientific establishment in Paris. At the unlikely age of 22, Thomson was appointed to the chair of natural philosophy at Glasgow, which he held from 1846 to 1899. His early reputation was made in mathematical physics, especially through his able defense of Fourier's work and its application to electrodynamics. Later in life Thomson became absorbed in various practical applications of physics, including the laying of the much-celebrated Atlantic telegraph cable. It was characteristic of Thomson's work to insist on an abstract mathematical formulation of a problem, but then to find ways of linking the abstract formulation back to concrete empirical situations, and his work in thermometry was no exception.[21]

It is important to keep in mind just how much Thomson's thermometry arose out of an admiring opposition to Victor Regnault's work, which I examined in detail in chapter 2. Thomson's sojourn in Paris occurred when Regnault was beginning to establish himself as the undisputed master of precision experiment. Following an initial introduction by Biot, Thomson quickly became one of Regnault's most valued young associates. In the spring of 1845 he reported from Paris to his father: "It has been a very successful plan for me going to Regnault's laboratory. I always get plenty to do, and Regnault speaks a great deal to me about what he is doing.... I sometimes go to the lab as early as 8 in the morning, and seldom get away before 5, and sometimes not till 6." Thomson's duties in the lab evolved from menial duties, such as operating an air pump on command, to collaboration with Regnault on the mathematical analysis of data. By 1851 we find Thomson paying a visit back to Paris, and subsequently spending a good deal of time writing out an account of the new thermodynamic theory for Regnault. To the end of his life Thomson recalled Regnault's tutelage with great fondness and appreciation. At a Parisian banquet held in his honor in 1900, Thomson (now Lord Kelvin) recalled with emotion how Regnault had taught him "a faultless technique, a love of precision in all things, and the highest virtue of the experimenter—patience."[22]

Thomson's first contribution to thermometry, a paper of 1848 presented to the Cambridge Philosophical Society, began by recognizing that the problem of temperature measurement had received "as complete a practical solution...as can be desired" thanks to Regnault's "very elaborate and refined experimental researches." Still, he lamented, "the theory of thermometry is however as yet far from being in so satisfactory a state" (Thomson [1848] 1882, p.100). As I discussed in detail in "Regnault: Austerity and Comparability" in chapter 2, Regnault had just consolidated his precision air thermometry by shrinking from theoretical speculations

---

[21]The most extensive and detailed existing accounts of Thomson's life and work are Smith and Wise 1989 and Thompson 1910. A briefer yet equally insightful source is Sharlin 1979.

[22]For records of Thomson's work with Regnault, see Thompson 1910, 117, 121–122, 126, 128, 226, 947–948, 1154. Thomson's letter to his father (30 March 1845) is quoted from p. 128 and his 1900 speech from p. 1154.

about the nature of heat and temperature. Consonant with the disillusionment with abstract theory that was particularly widespread in France (see "Regnault and Post-Laplacian Empiricism" in chapter 2), Regnault's attitude was that basic measurements had to be justified in themselves. Those who wanted to found measurements on some theory which in itself was in need of empirical verification, had got the epistemic order of things exactly backwards. Much as the young Glaswegian professor admired Regnault's work, this austere antitheoretical manner of doing science apparently did not appeal to Thomson.

Thomson did appreciate quite well the powerful way in which Regnault had used the comparability criterion for rigorous practical thermometry, demonstrating that the air thermometer was a good instrument to use because it gave highly consistent readings even when its construction was made to vary widely. Even so, Thomson complained:

> Although we have thus a strict principle for constructing a *definite* system for the estimation of temperature, yet as reference is essentially made to a specific body as the standard thermometric substance, we cannot consider that we have arrived at an *absolute* scale, and we can only regard, in strictness, the scale actually adopted as *an arbitrary series of numbered points of reference sufficiently close for the requirements of practical thermometry.*[23]

So he set about looking for a general theoretical principle on which to found thermometry. He needed to find a theoretical relation expressing temperature in terms of other general concepts. The conceptual resource Thomson used for this purpose came from something that he had learned while working in Regnault's laboratory, namely the little-known theory of heat engines by the army engineer Sadi Carnot (1796–1832), which Thomson himself would help to raise into prominence later. It is useful here to recall that "Regnault's great work" was the result of a government commission to determine the empirical data relevant to the understanding of steam engines.[24] As Thomson was attempting to reduce temperature to a better established theoretical concept, the notion of mechanical effect (or, work) fitted the bill here. A theoretical relation between heat and mechanical effect is precisely what was provided by a theory of heat engines. Thomson explained:

> The relation between motive power and heat, as established by Carnot, is such that *quantities of heat*, and *intervals of temperature*, are involved as the sole elements in the expression for the amount of mechanical effect to be obtained through the agency of heat; and since we have, independently, a definite system for the measurement of quantities of heat, we are thus furnished with a measure for intervals according to which absolute differences of temperature may be estimated. (Thomson [1848] 1882, 102; emphases original)

---

[23]Thomson [1848] 1882, 102; emphases original; see also Joule and Thomson [1854] 1882, 393.
[24]See the introductory chapter in Regnault 1847 for the way he conceptualized the operation of a heat engine, which was actually not so different from Carnot's framework.

Carnot's theory provides a theoretical relation between three parameters pertaining to an abstractly conceived heat engine: heat, temperature, and work. If we can measure heat and work directly, we can infer temperature from the theory.

It is necessary to explain at this point some elements of Carnot's theory and its background. (This is a digression, but a necessary one for a true understanding of Thomson's work, which is the end point of our narrative. Readers who are already familiar with the history of heat engines and the technical aspects of Carnot's theory can skip this part and proceed directly to page 181.) The widespread interest in the theory of heat engines in the early nineteenth century owed much to the impressive role that they played in the Industrial Revolution. The phrase "heat engine" designates any device that generates mechanical work by means of heat. In practice, most of the heat engines in use in Carnot's lifetime were steam engines, which utilize the great increase in the volume and pressure of water when it is converted into steam (or their decrease when steam is condensed back into liquid water). Initially the most important use of steam engines was to pump water in order to drain deep mines, but soon steam engines were powering everything from textile mills to ships and trains. It is not the case that the steam engine was invented by the Scottish engineer James Watt (1736–1819), but he did certainly make decisive improvements in its design. I will note some aspects of his innovations here, because they are closely linked with certain features of Carnot's theory. Watt, like Wedgwood (see "Adventures of a Scientific Potter" in chapter 3), worked on a very interesting boundary between technology and science. Early in his career Watt was a "mathematical instrument-maker" for the University of Glasgow, where he worked with Joseph Black and John Robison. Later he established his firm with Matthew Boulton near Birmingham and became a key member of the Lunar Society, associating closely with Priestley and Wedgwood among others, and with De Luc, who was an occasional but important visitor to the Lunar circle.[25]

In a typical Watt engine, steam was generated in a boiler and led into a cylinder, pushing a piston and thereby generating mechanical work (see fig. 4.5). That is the easy part to understand. Watt's innovations concerned subtler points.[26] The steam, having filled the cylinder, needed to be condensed (back into the liquid state), so that the piston could return to its original position and the whole operation could be repeated. All steam engineers knew the necessity of this condensing phase,[27] but Watt was the first one to study the physics of that phase carefully. At that time, the condensation was normally achieved by spraying cold water into the cylinder. Watt found that the exiting engines used an excessive amount of cold water. That not only condensed the steam but also cooled down the cylinder, which resulted in a waste of heat because in the next cycle fresh steam was

---

[25] See Schofield 1963 on Watt's interaction with the Lunar Society.
[26] There are many expositions of Watt's innovations, but here I rely mostly on Cardwell 1971, 42–52.
[27] In fact, in earlier steam engines due to Savery and Newcomen, it was the contractive phase that performed the useful work, in the form of mechanical pull. Watt's work on the steam engine began when he was asked to fix a Newcomen engine used for lecture demonstrations at the University of Glasgow.

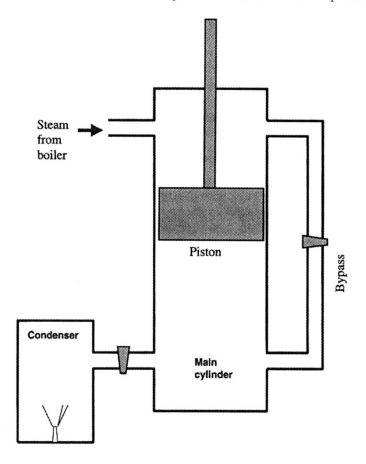

FIGURE 4.5. A diagram illustrating the workings of a Watt engine, adapted from Cardwell 1971, 49, and Sandfort 1964, 31. Hot steam enters the upper part of the main cylinder, pushing down the piston and performing work; then the piston is pulled back up; at the same time the bypass valve is opened to allow the steam to pass to the part of the cylinder below the piston; from there it gets sucked into the condenser, where a jet of cold water cools it and turns it into liquid water.

needed to heat up the cylinder all over again. Watt reduced the amount of cold water, putting in only as much cold water as required for condensing most of the steam. (He had an experimental knowledge of latent heat, which Black helped him understand theoretically.)

This innovation certainly cut down on the consumption of fuel; however, Watt found that it also reduced the power of the engine. The problem was that maintaining the cylinder at a relatively high temperature also meant that there still remained steam at a relatively high pressure, which made it difficult to push the piston back in when the contractive operation took place. In order to tackle this problem, Watt invented the "separate condenser," a vessel into which the used steam was evacuated in order to be condensed (see fig. 4.5). As cold water was sprayed into the separate condenser, the steam that had moved into it became condensed and created a relative vacuum, causing more steam to flow into the condenser from the main cylinder. All the while the main cylinder itself remained at a high temperature, but most of the hot steam got sucked away into the condenser. Watt's separate condenser allowed a great increase in efficiency, and it comes down as one of the landmarks in the history of technology.

However, Watt was not completely satisfied yet. He noticed that the steam *rushed* into the condenser from the steam vessel and reckoned that there must be useful work wasted in that rushing. This dissatisfaction led him to the "expansive principle" in 1769. Watt realized that steam introduced into the cylinder was capable of doing further work by expanding by its own force, until its pressure became equal to the external atmospheric pressure. Therefore he decided to cut off the supply of steam from the boiler well before the piston was pushed out all the way, in order to extract all the work that the steam was capable of doing. The significance of this innovation cannot be stressed too strongly. As D. S. L. Cardwell puts it (1971, 52): "By his invention of the expansive principle (1769) this meticulous Scotsman foreshadowed the progressive improvement of heat-engines and the postulation by Sadi Carnot of a general theory of motive power of heat. With astonishing insight Watt had laid one of the cornerstones of thermodynamics."

This, finally, brings us to Sadi Carnot. Carnot's theoretical investigation into the workings of heat engines, too, began with a concern about efficiency. In order to have a theoretically tidy measure, Carnot decided that efficiency should be reckoned in a cycle of operations. In the cycle originally conceived by Carnot, the engine receives a certain amount of caloric from a hot place, performs a certain amount of mechanical work by the agency of the caloric, and then releases the caloric into a cooler place (such as Watt's condenser); at the end of the process the engine returns to its original state. Carnot's metaphor for understanding the heat engine was the waterwheel, which produces mechanical work by harnessing the water falling from a higher place to a lower place; likewise, caloric performed work by falling from a place of higher temperature to a place of lower temperature. Efficiency in such a cycle is measured as the ratio of work performed to the amount of caloric passed through the engine. The task that Carnot set himself was to understand the factors that affect this efficiency.

The peculiar greatness of Carnot's work was to extract the essence of the functioning of all heat engines in a highly abstract form. The abstractness is also what makes Carnot's theory somewhat bewildering and difficult to grasp for most of those who come across it for the first time. As shown in figure 4.6, Carnot imagined a heat engine with an unspecified "working substance" enclosed in a cylinder throughout the cycle (rather than injected into the cylinder and then evacuated, as in a Watt engine). The working substance receives heat and performs work by expanding; then it has to be compressed and cooled in order to return to its original state if the cycle is to be complete; the latter process of condensation requires some work to be done to the substance, and some heat to be taken away from it at the same time. A steam engine operating in that way would be a cylinder closed in by a piston, containing in it a water-steam mixture in equilibrium, under a certain

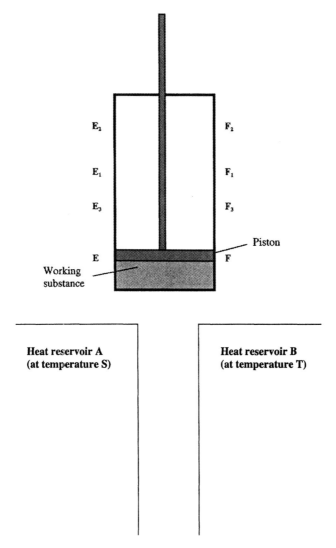

FIGURE 4.6. A schematic representation of a possible form of the Carnot engine, adapted from Thomson [1849] 1882, 121.

amount of pressure; heat is supplied to the system, generating more steam and pushing the piston out; afterwards heat is taken away and the piston pushed back into its original position, forcing some of the steam to condense back to a liquid state; hopefully the compression requires less work than the work generated by the expansion, so that we can have a positive net performance of work.

To finalize the shape of Carnot's cycle, we need to add one more element, which I can only imagine was inspired by Watt's "expansive principle" as Cardwell suggests, namely the insight that steam can do further work after the supply of heat is cut off. That is a process that modern physicists call "adiabatic expansion," namely expansion without any communication of heat from (or to) an external source. It was well known by Carnot's time that an adiabatic expansion of a gas resulted in the lowering of its temperature, though there were serious disputes about the underlying cause of this phenomenon.[28] Now we can see how Carnot's famous cycle is put together in its modern four-stroke form (refer to fig. 4.6):

> *First stroke.* The working substance, at initial temperature S, receives a quantity of heat (H) from the "heat reservoir" A, also kept at temperature S. The working substance maintains its temperature at S, and expands, pushing the piston out from position EF to position E_1F_1, doing a certain amount of work, which we denote by W_1.
>
> *Second stroke.* The heat reservoir is removed, and the working substance expands further on its own (adiabatically), cooling down from temperature S to temperature T. The piston is pushed out further, from E_1F_1 to E_2F_2, and further work, W_2, is done.
>
> *Third stroke.* The working substance is now compressed, staying at the lower temperature T and releasing some heat to the heat reservoir B, also kept at temperature T. In this process the piston is pushed in from E_2F_2 to E_3F_3, and some work, W_3, is done *to* the working substance. (In Carnot's original conception, this stroke is continued until the amount of heat released is same as H, the amount of heat absorbed in the first stroke.)
>
> *Fourth stroke.* The working substance is compressed further, adiabatically (without contact with the heat reservoir). More work, W_4, is done to it in the compression, in which the piston moves from E_3F_3 back to its original position EF. The temperature of the working substance goes up from T to S. (It is assumed that the cycle will "close" at the end of the fourth stroke, by returning the working substance exactly to its original state.)

The efficiency of this cycle of operations is defined as the ratio W/H, where W is the net work performed by the working substance ($W_1 + W_2 - W_3 - W_4$), and H is the amount of heat absorbed in the first stroke (which is also the amount of heat released in the third stroke, in Carnot's original theory). Through theoretical reasoning about such a cycle, Carnot derived the very important result that the engine efficiency only depended on the temperatures of the cold and hot reservoirs (S and T in the description of the four-stroke cycle).

Finally, it will be helpful to explain the abstract graphic representation of the Carnot cycle that is almost universally encountered in textbooks. That representation originates from the 1834 exposition of Carnot's theory by Émile Clapeyron (1799–1864), an engineer trained at the École Polytechnique. Thomson initially learned Carnot's theory from Clapeyron's article and adopted the latter's graphic presentation. Most likely, Clapeyron's representation was inspired by another practice of Watt's, namely his use of the "indicator diagram." Watt monitored the performance of his steam engines by a mechanical device that automatically plotted the changing values of pressure and volume inside the steam vessel. According to

---

[28]See, for example, Dalton 1802a, which contains some of the pioneering experimental results and a nice Irvinist explanation.

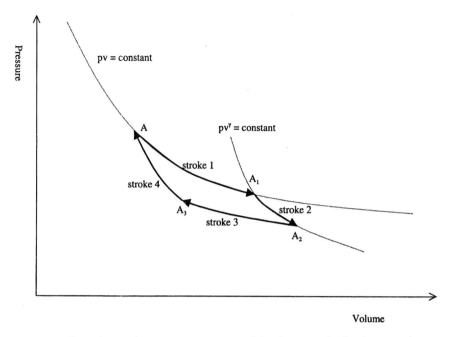

FIGURE 4.7. The indicator-diagram representation of the Carnot cycle, for the case of an air engine, adapted from Clapeyron [1834] 1837, 350.

Cardwell (1971, 80–81), the indicator was invented in 1796 by Watt's assistant John Southern, and it was a closely guarded trade secret. It is not clear how Clapeyron would have got access to this secret, but he plotted the Carnot cycle as a closed curve in this pressure-volume representation, which helped a great deal in creating a visual image of what was going on in the abstract engine. One very nice feature of the pressure-volume diagram is that the amount of mechanical work involved in a process is straightforwardly represented as the area under the curve representing the process (the work being the integral $\int p dv$). For a closed cycle of operations, the net work produced is represented as the area enclosed in the closed curve. Figure 4.7 is an example of the Carnot cycle plotted this way, representing an engine filled with a simple gas rather than the steam-water mix. The isothermal line representing the first stroke (AA_1) is a curve following Boyle's law (pv = constant); the adiabatic line representing the second stroke (A_1A_2) is a curve following the adiabatic gas law (pv^γ = constant, where γ is the ratio between the specific heat of the gas at constant pressure and the specific heat at constant volume).

With that background, we can now make full sense of Thomson's conception of absolute temperature. (For a complete understanding of the rest of the narrative part of this chapter, some familiarity with elementary physics and calculus is required; however, readers without such technical background will still be able to understand the gist of the arguments by following the verbal parts of the exposition. The analysis part refrains from technical discussions.) Initially in 1848, Thomson's basic idea was to define the interval of one degree of temperature as the amount that would result in the production of unit amount of mechanical work in a Carnot engine operating in that temperature interval. More precisely, in Thomson's own words:

> The characteristic property of the scale which I now propose is, that all degrees have the same value; that is, that a unit of heat descending from a body A at the temperature $T°$ of this scale, to a body B at the temperature $(T - 1)°$, would give out the same mechanical effect, whatever be the number T. *This may justly be termed an absolute scale, since its characteristic is quite independent of the physical properties of any specific substance.* (Thomson [1848] 1882, p.104; emphasis added)

This definition is what I will refer to as Thomson's "first absolute temperature." It is very important to note that Thomson's sense of "absolute" here has nothing to do with counting temperature from the absolute zero. In fact, as Thomson clarified later, his 1848 temperature did not have a zero point at all. The common notion of "absolute zero" that survives into modern conceptions is in fact much older than Thomson's work. It can be traced back to Guillaume Amontons's idea that an objective scale of temperature could be obtained if the zero point were found by extrapolating the observed pressure-temperature relation of air until the pressure became zero, since zero pressure would indicate a complete absence of heat. I will refer to this notion of temperature as the Amontons air temperature, or simply *Amontons temperature*. It is quite close to what people commonly mean by "absolute temperature" nowadays if they have not studied carefully what is meant by the notion in thermodynamic theory. As we shall see shortly, Thomson later modified his absolute temperature concept to bring it more into line with Amontons temperature, and from that point on the two different senses of "absolute" (not being related to particular materials, and having an absolute zero) became forever conflated.

Thomson's Second Absolute Temperature

Interestingly, almost as soon as Thomson advanced his initial concept of absolute temperature, he began to abandon the entire theoretical framework in which that concept was couched. This was in large part a consequence of his encounter with James Prescott Joule (1818–1889), the gentleman scientist from a Manchester family of brewers, who is credited with a crucial role in establishing the principle of conservation of energy.[29] (The course of Joule and Thomson's collaboration is well known to historians of science, so I will not repeat many details here.[30]) When Thomson heard Joule present his idea about the interconvertibility of heat and mechanical work at the 1847 meeting of the British Association for the Advancement of Science in Oxford, he was interested but skeptical. After reading Thomson's 1848 article on absolute temperature, Joule wrote urging him to reformulate his idea on the basis of the interconvertibility of heat and work, rather than holding on to Carnot's assumption that heat passed through the heat engine intact: "I dare say

---

[29]For further details about Joule's life and work, see Cardwell 1989 and Smith 1998.
[30]See, for example, Cardwell 1989, chs. 5 and 8.

they [your ideas] will lose none of their interest or value even if Carnot's theory be ultimately found incorrect." Thomson sent a congenial reply, but he was not quite ready to renounce Carnot's theory, which he proceeded to elaborate in an article published in 1849.[31]

By early 1851, however, Thomson had committed himself to a serious modification of Carnot's theory in light of Joule's ideas about the interconversion of heat and work. The few years after Thomson's conversion to interconversion were both highly unsettled and highly productive. The entire basis on which he had defined absolute temperature in 1848 had to be changed, because the understanding of the Carnot engine had to be revised fundamentally if heat was no longer considered to be a conserved quantity, and the generation of mechanical effect was seen as the conversion of a part of the heat input into work, rather than a by-product of the movement of heat.[32]

It would be fascinating to follow all the twists and turns that Thomson took in reshaping his concept of absolute temperature, but in the interest of clarity and accessibility I will only present a streamlined account here. There were three major steps, through which Thomson arrived at a very simple and pleasing definition of temperature (expressed as $T_1/T_2 = Q_1/Q_2$, as I will explain shortly), which was very different from the original definition but related to it.

The first step was to reformulate Carnot's theory itself so that it was compatible with energy conservation; this was achieved in a series of articles on the "dynamical theory of heat," starting in March 1851 (Thomson [1851a, b, c] 1882). The most important part of this reformulation, for the purpose of thermometry, was the concept of "Carnot's function." Recall that the original definition of temperature was based on the amount of mechanical effect produced in a Carnot cycle, for a given amount of heat passing through the engine. A crucial factor in such consideration of engine efficiency was what Thomson called "Carnot's coefficient" or "Carnot's multiplier," the parameter μ in the following relation:

$$W = H\mu(T_1 - T_2), \qquad (1)$$

where W is the mechanical work produced in the cycle, H the amount of heat passing through the engine, and T_1 and T_2 are the absolute temperatures of the hot and cold reservoirs.[33] When Thomson revised Carnot's theory, he preserved a very

---

[31]Joule to Thomson, 6 October 1848, and Thomson to Joule, 27 October 1848, Kelvin Papers, Add. 7342, J61 and J62, University Library, Cambridge.

[32]Although Thomson clearly preserved as much as he could from the old analyses in the formal sense, the following claim made many years later seems to me like unhelpful bravado: "This paper [of 1848] was wholly founded on Carnot's uncorrected theory.... [T]he consequently required corrections... however do not in any way affect the absolute scale for thermometry which forms the subject of the present article." What Thomson did demonstrate was that there was a simple numerical conversion formula linking his two definitions of absolute temperature: $T_1 = 100(\log T_2 - \log 273)/(\log 373 - \log 273)$. Note that $T_2 = 0$ puts T_1 at negative infinity, and T_1 is set at 0 when T_2 is 273. See the retrospective note attached to Thomson [1848] 1882, 106.

[33]I have deduced this relation from formula (7) given in Thomson [1849] 1882, 134, §31, by assuming that μ is a constant, which is a consequence of Thomson's first definition of absolute temperature.

similar factor, still denoted by μ and called "Carnot's function."[34] This was similar to the old Carnot coefficient, but there were two important differences. First, because heat was no longer a conserved quantity, H in equation (1) became meaningless; therefore Thomson substituted for it the heat *input*, namely the amount of heat absorbed in the first stroke (isothermal expansion). Second, this time μ was a function of temperature defined for an *infinitesimal* Carnot cycle, with an infinitesimal difference between the temperatures of the two heat reservoirs. With those adjustments, Thomson defined μ through the following work–heat relation parallel to equation (1):

$$W = Q\mu dT, \qquad (2)$$

where dT is the infinitesimal temperature difference and Q is the heat input.

Thomson's second step, simple yet crucial, was to liberalize the theoretical concept of temperature. Carnot's function was related to engine efficiency, to which Thomson still wanted to tie the temperature concept. But he now realized that nothing theoretical actually dictated the exact relation that Carnot's function should bear to temperature; his initial notion of 1848, defining temperature so that μ became a constant, was too restrictive for no compelling reason. Much more freely, Thomson wrote in an article of 1854 co-authored with Joule: "Carnot's function (derivable from the properties of any substance whatever, but the same for all bodies at the same temperature), *or any arbitrary function of Carnot's function*, may be defined as temperature."[35]

The third step, made possible by the second one, was to find a function of μ that matched existing practical temperature scales reasonably well. In a long footnote attached to an article of 1854 on thermo-electric currents, Thomson admitted a shortcoming of his first definition of absolute temperature, namely that the comparison with air-thermometer temperature showed "very wide discrepancies, even inconveniently wide between the fixed points of agreement" (as shown in table 4.1 in the next section). The most important clue in improving that shortcoming came from Joule, quite unsolicited: "A more convenient assumption has since been pointed to by Mr Joule's conjecture, that Carnot's function is equal to the mechanical equivalent of the thermal unit divided by the temperature by the air thermometer from its zero of expansion" (Thomson [1854] 1882, 233, footnote). What Thomson called "the temperature by the air thermometer from its zero of expansion" is Amontons temperature, introduced at the end of the last section. What he called "Mr. Joule's conjecture" can be expressed as follows:

$$\mu = J/(273.7 + t_c), \text{ or } \mu = J/t_a, \qquad (3)$$

---

[34]The following derivation of the form of Carnot's function is taken from Thomson [1851a] 1882, 187–188, §21. Here and elsewhere, I have changed Thomson's notation slightly for improved clarity in relation with other formulae.

[35]Joule and Thomson [1854] 1882, 393; emphasis added. It is stated there that this liberal conception was already expressed in Thomson's 1848 article, but that is not quite correct.

where t_c is temperature on the centigrade scale, t_a is Amontons temperature, and J is the constant giving the mechanical equivalent of heat.[36] Thomson called this proposition Joule's "conjecture" because he had serious doubts about its rigorous truth. In fact, Thomson considered it a chief objective of the "Joule-Thomson experiment" to determine the value of μ empirically, which in itself would imply that he considered any empirical proposition regarding the value of μ as an as-yet unverified hypothesis. However, he thought Joule's conjecture was probably approximately true, and therefore capable of serving as a point of departure in finding a concept of absolute temperature closely aligned with practical temperature scales. Thus, Thomson used Joule's unverified conjecture as a heuristic device "pointing to" a new theoretical definition of temperature. In a joint article with Joule, Thomson wrote:

> Carnot's function varies very nearly in the inverse ratio of what has been called "temperature from the zero of the air-thermometer" [Amontons temperature]... and we may *define* temperature simply as the reciprocal of Carnot's function. (Joule and Thomson [1854] 1882, 393–394; emphasis added)

This new idea can be expressed mathematically as follows:

$$\mu = J/T, \qquad (4)$$

where T denotes the theoretical absolute temperature (compare with equation (3), which has the same form but involves Amontons temperature (t_a), which is defined operationally by the air thermometer).

After giving this definition, the Joule-Thomson essay added another formulation, which would prove to be much more usable and fruitful:

> If any substance whatever, subjected to a perfectly reversible cycle of operations, takes in heat only in a locality kept at a uniform temperature, and emits heat only in another locality kept at a uniform temperature, the temperatures of these localities are proportional to the quantities of heat taken in or emitted at them in a complete cycle of operations.[37]

In mathematical form, we may write this as follows:

$$T_1/T_2 = Q_1/Q_2, \qquad (5)$$

---

[36] For this expression, see Thomson [1851a] 1882, 199; here Thomson cited Joule's letter to him of 9 December 1848. I have taken the value 273.7 from Joule and Thomson [1854] 1882, 394.

[37] Joule and Thomson [1854] 1882, 394. Essentially the same definition was also attached to Thomson's article on thermo-electricity published in the same year: "Definition of temperature and general thermometric assumption.—If two bodies be put in contact, and neither gives heat to the other, their temperatures are said to be the same; but if one gives heat to the other, its temperature is said to be higher. The temperatures of two bodies are proportional to the quantities of heat respectively taken in and given out in localities at one temperature and at the other, respectively, by a material system subjected to a complete cycle of perfectly reversible thermodynamic operations, and not allowed to part with or take in heat at any other temperature: or, the absolute values of two temperatures are to one another in the proportion of the heat taken in to the heat rejected in a perfect thermo-dynamic engine working with a source and refrigerator at the higher and lower of the temperatures respectively" (Thomson [1854] 1882, 235).

where the T's indicate the absolute temperatures of the isothermal processes (strokes 1 and 3 of the Carnot cycle) and the Q's indicate the amounts of heat absorbed or emitted in the respective processes.

How is this alternate formulation justified? The Joule-Thomson article itself is not very clear on that point, but it is possible to show, as follows, that definition (4) follows as a consequence of definition (5), which means that we can take (5) as the primary definition. Take a Carnot cycle operating between absolute temperatures T and T' (where $T > T'$), in which the engine absorbs heat Q in the first stroke and releases Q' in the third stroke ($Q > Q'$). Energy conservation dictates that the net mechanical work produced in that cycle is $J(Q - Q')$, where J is the constant giving the mechanical equivalent of heat, and ($Q - Q'$) gives the amount of heat destroyed (converted into mechanical work). Now, using the definition of absolute temperature given in (5), we can express the work as follows:

$$W = J(Q - Q') = JQ(1 - T'/T) = JQ(T - T')/T. \quad (6)$$

If we consider a cycle in which the temperature difference is infinitesimal, we may write equation (6) as follows:

$$W = JQ(dT/T). \quad (7)$$

Now recall the definition of Carnot's function given in equation (2), $W = Q\mu dT$. Equating that with (7) gives $\mu = J/T$, which is the definition expressed in equation (4), so we have the desired result.

Definition (5) marked a point of closure in Thomson's theoretical work on thermometry, although he would return to the subject many years later. In subsequent discussions I will refer to definitions (4) and (5) together as Thomson's "second absolute temperature." This closes my discussion of the theoretical development of the temperature concept. But how was this abstract concept to be made susceptible to measurement? That is the subject of the next two sections.

Semi-Concrete Models of the Carnot Cycle

In a way, creating a theoretical definition of temperature was the easy part of Thomson's task. Anyone can make up a theoretical definition, but the definition will not be useful for empirical science, unless it can be connected to the realm of physical operations. Linking up an abstract theoretical concept with concrete physical operations is a challenging task in general, as I will discuss more carefully in the analysis part of this chapter, but it was made starkly difficult in Thomson's case, since he had deliberately fashioned the absolute temperature concept to make sure that any connections whatsoever to any particular objects or materials were severed. How was Thomson going to turn around later and say, excuse me, but now I would like to have those connections back? The operationalization of Thomson's absolute temperature was a problem that remained nontrivial well into the twentieth century. Herbert Arthur Klein reports:

> The situation is well summarized by R. D. Huntoon, former director of the Institute for Basic Standards of the U.S. National Bureau of Standards. Surveying the status

of standards for physical measurement in the mid-1960s, he noted that the actual relationships between the universally used IPTS [International Practical Temperature Scale] scales and the corresponding thermodynamic scale are "not precisely known." (Klein 1988, 333)

In order to have a true appreciation of the problem that Thomson faced, we need to pause for a moment to reflect on the nature of Carnot's theory. As noted earlier, Carnot wanted a completely general theory of heat engines, which meant that the "working substance" in his theoretical engine was conceived as an abstract body possessing only the properties of pressure, volume, temperature, and heat content. How was Carnot able to deduce anything useful at all about the behavior of a substance of such skeletal description? He made use of some general assumptions, such as the conservation of heat and some propositions found in the latest physics of gases, but they were still not sufficient to allow the deduction of anything definite about engine efficiency. Quite plainly, in actual situations the efficiency will depend on the particular design of the engine and the nature of the particular working substance.

Carnot made headway by treating only a very restricted class of heat engines, though he still avoided invoking particular properties of the working substance in the general theory. The following were the most important restrictions. (1) Carnot only treated engines that worked through a cyclical process, in which the working substance returned, at the end of a cycle, exactly to the state in which it started. (2) The Carnot engine was not merely cyclical, but cyclical in a very specific way, with definite strokes constituting the cycle. (3) Finally, and most crucially, the Carnot cycle was also perfectly reversible; reversibility implied not only the absence of friction and other forms of dissipation of heat and work within the engine but also no transfer of heat across any temperature differences, as I will discuss further shortly. Those restrictions allowed Carnot to prove some important results about his abstract heat engine.

Still, Thomson faced a great difficulty as long as the object of Carnot's theory remained so removed from any actually constructable heat engine. Let us consider how Thomson endeavored to deal with that difficulty, first going back to his original definition of absolute temperature in 1848. The conceptually straightforward scheme for measuring Thomson's *first* absolute temperature would have been the following: take an object whose temperature we would like to measure; use it as a Carnot heat reservoir and run a Carnot engine between that and another reservoir whose temperature is previously known; and measure the amount of mechanical work that is produced, which gives the difference between the two temperatures. The difficulty of realizing that procedure can only be imagined, because there is no record of anyone ever who was crazy enough to attempt it. In order to meet the standard of precision in thermometry established by Regnault, the instrument used would have needed to be frighteningly close to the theoretical Carnot engine. That route to the operationalization of absolute temperature was a nonstarter.

So Thomson took a conceptual detour. Instead of attempting to measure temperature directly with a thermometer constructed out of an actual, fully concrete Carnot engine, he theorized about versions of the Carnot engine that were concrete enough to allow the use of certain empirical data in the description of its workings. The key to making a reliable connection between the abstract definition of absolute temperature and actual empirical data was to use a model that was sufficiently concrete, but still ideal in the sense of satisfying Carnot's propositions about engine efficiency. Thomson worked out two such quasi-concrete models, following some moves made by Carnot himself: a system made of water and steam, and a system with only air in it. Here I will only give the details of Thomson's water-steam system.[38] The important advantage of this system is that the pressure of "saturated" steam is a function of temperature only (see "The Understanding of Boiling" in chapter 1), which simplifies the reasoning a great deal as we shall see later. This quasi-concrete model allowed Thomson to compute the heat–work relation from empirical data. As we shall see, the relevant empirical data were certain parameters measured as functions of temperature measured by an air thermometer. Putting such data into the definition of absolute temperature yielded a relation between absolute temperature and air-thermometer temperature, with which he could convert air-thermometer temperature into absolute temperature. Let us now see how this calculation was made.

Thomson was still working with the original version of the Carnot theory before energy conservation when he was trying to measure his first absolute temperature. To recap the relevant parts of that theory briefly: a Carnot engine produces a certain amount of work, W, when a certain amount of heat, H, is passed through it; we need to evaluate W, which is visually represented by the area enclosed by the four-sided figure $AA_1A_2A_3$ in figure 4.8, representing the amount of work done by the steam-water mixture in strokes 1 and 2, minus the amount of work done to the steam-water mixture in strokes 3 and 4. Thomson estimated the area in question actually by performing the integration along the pressure axis, as follows:

$$W = \int_{p_1}^{p_2} \xi \, dp, \tag{8}$$

where p_1 and p_2 are the pressures in strokes 3 and 1, which are constant because the temperature is constant in each stroke; ξ is the length of the line spanning the curved sides of the figure (AA_3 and A_1A_2).

Now, what does ξ represent physically? That is the crucial question. Thomson answered it as follows:

> We see that ξ is the difference of the volumes below the piston at corresponding instants of the second and fourth operations, or instants at which the saturated steam and the water in the cylinder have the same pressure p, and, consequently, the same temperature which we may denote by t. Again, throughout the second operation [curve A_1A_2 in the figure] the entire contents of the cylinder possess a greater amount of heat by H units than during the fourth [curve A_3A]; and, therefore, at any instant of the second operation there is as much more steam as

---

[38]For the treatment of the air engine, see Thomson [1849] 1882, 127–133.

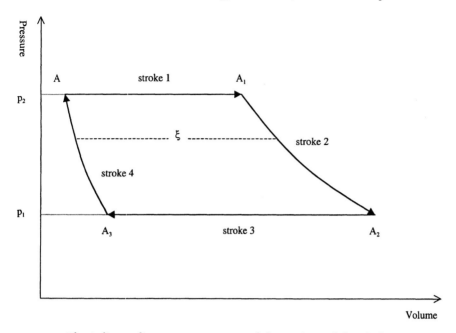

FIGURE 4.8. The indicator-diagram representation of the working of the ideal steam–water cycle, adapted from Thomson [1849] 1882, 124.

contains H units of latent heat, than at the corresponding instants of the fourth operation. (Thomson [1849] 1882, 125–126)

The crucial assumption here is that there is a strict correlation between the temperature and pressure of saturated steam; by the time Thomson was writing, this was generally accepted as an empirical law (see "The Understanding of Boiling" in chapter 1). Now we must ask how much increase of volume results from the production of the amount of steam embodying latent heat H. That volume increment is given as follows:

$$\xi = (1 - \sigma)H/k, \tag{9}$$

where k denotes the latent heat per unit volume of steam at a given temperature, and σ is the ratio of the density of steam to the density of water. The formula makes sense as follows: the input of heat H produces H/k liters of steam, for which $\sigma H/k$ liters of water needs to be vaporized; the *net* increase of volume is given by subtracting that original water volume from the volume of steam produced.

Substituting that expression into equation (8), we have:

$$W = \int_{p_1}^{p_2} (1 - \sigma) \frac{H}{k} dp. \tag{10}$$

Now, because all of the parameters in this equation except H are functions of air-thermometer temperature t, we can rewrite the integral in terms of t (taking H out as a constant), as follows:

$$W = H \int_T^S (1-\sigma)\frac{dp}{kdt} dt, \qquad (11)$$

where S and T are the temperatures of the working substance in the first and the third strokes. According to Thomson's first definition of absolute temperature, the difference between those two temperatures on the absolute scale is proportional to W/H, which can be evaluated by performing the integration in equation (11) after putting in the relevant empirical data. Comparing the absolute temperature difference estimated that way with the temperature difference $(S-T)$ as measured on the air-thermometer scale gives the conversion factor expressing how many air-thermometer degrees correspond to one degree of absolute temperature interval, at that point in the scale.

Therefore the measurement of absolute temperature by means of the steam–water cycle came down to the measurement of the parameters occurring in the integral in (11), namely the pressure, density, and latent heat of saturated steam as functions of air-thermometer temperature. Fortunately, detailed measurements of those quantities had been made, by none other than Regnault. Using Regnault's data, Thomson constructed a table with "a comparison of the proposed scale with that of the air-thermometer, between the limits 0° and 230° of the latter." Table 4.1 gives some of Thomson's results, converted into a more convenient form. Note that the relationship between the air temperature and the absolute temperature is not linear; the size of one absolute temperature degree becomes smaller and smaller in comparison to the size of one air-temperature degree as temperature goes up. (As noted before, this absolute scale in fact had no zero point but stretched to negative infinity; that was no longer the case with Thomson's second definition of absolute temperature.)

Let us now consider whether Thomson at this stage really succeeded in his self-imposed task of measuring absolute temperature. There are three major difficulties. The first one was clearly noted by Thomson himself: the formulae given above require the values of k, the latent heat of steam by volume, but Regnault had only measured the latent heat of steam by weight. Lacking the facility to make the required measurements himself, Thomson converted Regnault's data into what he needed by assuming that steam obeyed the laws of Boyle and Gay-Lussac. He knew that this was at best an approximation, but thought there was reason to believe that it was a sufficiently good approximation for his purposes.[39]

Second, in calculating the amount of mechanical work, the entire analysis was premised on the assumption that the pressure of saturated steam depended only on the temperature. As noted earlier, that pressure–temperature relation was not something deducible a priori, but an empirically obtained generalization. The rigorous reliability of this empirical law was not beyond doubt. Besides, does not the

---

[39]See Thomson [1848] 1882, 104–105. Thomson also had to use Boyle's and Gay-Lussac's laws to reason out the air-engine case; see Thomson [1849] 1882, 129, 131.

TABLE 4.1. Thomson's comparison of air-thermometer temperature and his first absolute temperature

Air-thermometer temperature	Absolute temperature (first definition)
0°C	0[a]
5	5.660
10	11.243
15	16.751
20	22.184
25	27.545
30	32.834
35	38.053
40	43.201
45	48.280
50	53.291
55	58.234
60	63.112
65	67.925
70	72.676
75	77.367
80	82.000
85	86.579
90	91.104
95	95.577
100	100
150	141.875
200	180.442
231	203.125

Source: This table, announced in Thomson [1848] 1882, 105, was published in Thomson [1849] 1882, 139 and 141, in slightly different forms than originally envisaged.

[a] This scale had no "absolute zero," and it was calibrated so that it would agree with the centigrade scale at 0° and 100°.

use of the pressure–temperature relation of steam amount to a reliance on an empirical property of a particular substance, just the thing Thomson wanted to avoid in his definition of temperature? In Thomson's defense, however, we could argue that the strict correlation between pressure and temperature was probably presumed to hold for all liquid–vapor systems, not just for the water–steam system. We should also keep in mind that his use of the pressure–temperature relation was not in the definition of absolute temperature, but only in its operationalization. Since Carnot's theory gave the assurance that all ideal engines operating at the same temperatures had the same efficiency, calculating the efficiency in any particular system was sufficient to provide a general answer.

Finally, in the theoretical definition itself, absolute temperature is expressed in terms of heat and mechanical work. I have quoted Thomson earlier as taking comfort in that "we have, independently, a definite system for the measurement of quantities of heat," but it is not clear what he had in mind there. The standard laboratory method for measuring quantities of heat was through calorimetry based on the measurement of temperature changes induced in a standard substance (e.g. water), but of course that had to rely on a thermometer. Recall that Thomson's scheme for operationalizing absolute temperature was to express W/H as a function of air-thermometer temperature. A good deal of complexity would have arisen if the measure of H itself depended on the use of the air thermometer (that is, if it had to be kept inside the integral in equation (11)). In one place Thomson ([1848] 1882, 106) mentions using the melting of ice for the purpose of calorimetry, but there were significant difficulties in any actual use of the ice calorimeter (see "Ganging Up on Wedgwood" in chapter 3). Still, we could say that in principle heat could be measured by ice calorimetry (or any other method using latent heat), in which case the measure of heat would be reduced to the measure of weight and the latent heat of the particular change-of-state involved. But the last step would end up involving an empirical property of a particular substance, again contrary to Thomson's original intention!

Using Gas Thermometers to Approximate Absolute Temperature

How Thomson himself might have proposed to deal with the difficulties mentioned in the last section is an interesting question. However, it is also a hypothetical question, because Thomson revised the definition of absolute temperature before he had a chance to consider carefully the problems of operationalizing the original concept. Let us therefore proceed to a consideration of how he attempted to measure his second absolute temperature, which is expressed in equations (4) and (5). Thomson, now in full collaboration with Joule, faced the same basic challenge as before: a credible Carnot engine (or, more to the point, a reversible cycle of operations) could not be constructed in reality. The operationalization of Thomson's second absolute temperature was a long and gradual process, in which a variety of analytical and material methods were tried out by Joule and Thomson, and by later physicists. All of those methods were based on the assumption, explicit or implicit, that an ideal gas thermometer would give the absolute temperature exactly. If so, any possible measure of how much actual gases deviate from the ideal might also give us an indication of how much the temperatures indicated by actual gas thermometers deviate from the absolute temperatures.

The first thing we need to get clear about is why an ideal gas thermometer would indicate Thomson's absolute temperature. The contention we want to support is that an ideal gas expands uniformly with absolute temperature, under fixed pressure (or, its pressure increases uniformly with temperature when the volume is fixed). Now, if we attempt to verify that contention by direct experimental test, we will get into a circularity: how can we tell whether a given sample of gas is ideal or not, unless we already know how to measure absolute temperature so that we can monitor the gas's behavior as a function of absolute temperature? Any successful argument that an ideal gas indicates absolute temperature has to be made in the realm of theory, rather than practical measurement. It is not clear to me whether

Thomson himself made any such argument directly, but at least a clear enough reconstruction can be made on the basis of the discussions he did give.[40]

The argument is based on the consideration of the isothermal expansion of an ideal gas. In that process the gas absorbs an amount of heat H, while its temperature remains the same.[41] The added heat causes the gas to expand, from initial volume v_0 to final volume v_1, while its pressure decreases from p_0 to p_1; in this expansion the gas also does some mechanical work because it pushes against an external pressure. The amount of the mechanical work performed, by definition of work, is expressed by the integral $\int pdv$. If we indicate Amontons temperature by t_a as before, then a constant-pressure gas thermometer dictates (by operational definition) that t_a is proportional to volume; likewise, a constant-volume gas thermometer dictates that t_a is proportional to pressure. With an ideal gas the constant-volume and the constant-pressure instruments would be comparable, so we can summarize the two relations of proportionality in the following formula:

$$pv = ct_a, \qquad (12)$$

where c is a constant specific to the given sample of gas. That is just an expression of the commonly recognized gas law. Putting that into the expression for mechanical work, and writing W_i to indicate the work performed in the isothermal process, we have:

$$W_i = \int_{v_0}^{v_1} pdv = \int_{v_0}^{v_1} \frac{ct_a}{v} dv. \qquad (13)$$

Since we are concerned with an isothermal process, the ct_a term can be treated as a constant. So the integration gives:

$$W_i = ct_a \log(v_1/v_0). \qquad (14)$$

If we ask how W_i varies with respect to t_a, we get:

$$\partial W_i / \partial t_a = c \log(v_1/v_0) = W_i/t_a. \qquad (15)$$

Now, the variation of W_i with temperature also has a simple relation to Carnot's function (and therefore to Thomson's second absolute temperature). Equation (2), $W = Q\mu dT$, expresses the net work done in a Carnot cycle in which the temperatures of the first and third strokes (the isothermal processes) differ by an infinitesimal amount, dT. The net work in that infinitesimal cycle is (to the first

---

[40] I am helped here by the exposition in Gray 1908, esp. 125. Gray was Thomson's successor in the Glasgow chair, and his work provides the clearest available account of Thomson's scientific work. The argument reconstructed here is not so far from the modern ones, for example, the one given in Zemansky and Dittman 1981, 175–177.

[41] The same by what standard of temperature? That is actually not very clear. But it is probably an innocuous enough assumption that a phenomenon occurring at a constant temperature by any good thermometer will also be occurring at a constant absolute temperature.

order) the infinitesimal difference between the work produced in the first stroke and the work consumed in the third stroke. So we may write:

$$W = \frac{\partial W_i}{\partial T} dT. \quad (16)$$

From (2) and (16), we have:

$$\partial W_i / \partial T = Q\mu = JQ/T. \quad (17)$$

In getting the second equality in (17) I have invoked Thomson's second definition of absolute temperature expressed in equation (4).

Now compare (15) and (17). The two equations, one in t_a and the other in T, would have exactly the same form, *if it were the case that* $JQ = W_i$. In other words, the Amontons temperature given by an ideal gas would behave like absolute temperature, if the condition $JQ = W_i$ were satisfied in an isothermal expansion. But the satisfaction of that condition, according to thermodynamic theory, is the mark of an ideal gas: all of the heat absorbed in an isothermal expansion gets converted into mechanical energy, with nothing going into changes in the internal energy (and similarly, in adiabatic heating by compression, all the work spent on the gas being converted into heat). In fact, this condition is none other than what is often called "Mayer's hypothesis," because the German physician Julius Robert Mayer (1814–1878) was the first one to articulate it. Joule and Clausius also assumed that Mayer's hypothesis was empirically true, and it was the basis on which Joule produced his conjecture discussed in "Thomson's Second Absolute Temperature."[42]

Pondering Mayer's hypothesis led Thomson into the Joule-Thomson experiment. Thomson, who was more cautious than Joule or Clausius about accepting the truth of Mayer's hypothesis, insisted that it needed to be tested by experiment and persuaded Joule to collaborate with him on such an experiment. In an actual gas Mayer's hypothesis is probably not strictly true, which means that Amontons temperature defined by an actual gas is not going to be exactly equal to Thomson's absolute temperature. Then, the extent and manner in which an actual gas deviated from Mayer's hypothesis could be used to indicate the extent and manner in which actual-gas Amontons temperature deviated from absolute temperature.

In testing the empirical truth of Mayer's hypothesis, Joule and Thomson investigated the passage of a gas through a narrow opening, which is a process that ought to be isothermal for an ideal gas but not for an actual gas. They reasoned that an ideal gas would not change its temperature when it expands freely. But an actual gas has some cohesion, so it would probably require some energy to expand it (as it takes work to stretch a spring). If that is the case, a free expansion of an actual gas would cause it to cool down because the energy necessary for the expansion would have to be taken from the thermal energy possessed by the gas itself, unless there is

---

[42] For a history of Mayer's hypothesis and its various formulations, see Hutchison 1976a. For an account of how Joule arrived at his conjecture, which the correspondence between Joule and Thomson reveals, see Chang and Yi (forthcoming).

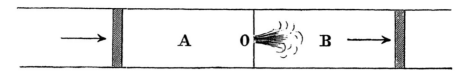

FIGURE 4.9. A schematic representation of the Joule-Thomson experiment (Preston 1904, 801, fig. 231).

an appropriate external source of energy. The amount of any observed cooling would give a measure of how much the actual gas deviates from the ideal.

The basic scheme for Joule and Thomson's experiment designed in order to get at this effect, often called the "porous plug experiment," was laid out by Thomson in 1851, inspired by an earlier article by Joule (1845). The procedure, shown schematically in figure 4.9, consisted in forcing a continuous stream of gas through two (spiral) pipes connected to each other through a very small orifice. Because it was difficult to measure the temperature of the gas exiting from the orifice precisely, Joule and Thomson instead measured the amount of heat that was required in bringing the gas back to its original temperature after its passage. From this amount of heat and the specific heat of the gas, the temperature at which the gas had exited from the orifice was inferred.

How were the results of the Joule-Thomson experiment used to get a measure of how much the actual gas-thermometer temperatures deviated from absolute temperature? Unfortunately the details of the reasoning are much too complicated (and murky) for me to summarize effectively here.[43] What is more important for my present purpose, in any case, is to analyze the character of Joule and Thomson's results. They used the experimental data on cooling in order to derive a formula for the behavior of actual gases as a function of absolute temperature, showing how they deviated from the ideal gas law. Joule and Thomson's "complete solution" was the following:[44]

$$v = \frac{CT}{p} - \frac{1}{3}AJK\left(\frac{273.7}{T}\right)^2. \tag{18}$$

This equation expresses T, "the temperature according to the absolute thermodynamic system of thermometry," in terms of other parameters, all of which are presumably measurable: v is the volume of a given body of gas; p is its pressure; C is a parameter "independent of both pressure and temperature"; A is a constant that is characteristic of each type of gas; J is the mechanical equivalent of heat; and K is specific heat (per unit mass) under constant pressure. So equation (18) in principle indicates a straightforward way of measuring absolute temperature T. The second term on the right-hand side gives the measure of deviation from the ideal; without

---

[43]Interested readers can review the Joule-Thomson reasoning in detail in Hutchison 1976a and Chang and Yi (forthcoming).

[44]See Joule and Thomson [1862] 1882, 427–431; the equation reproduced here is (a) from p. 430.

that term, the equation would simply reduce to the ideal gas law, which would mean that the gas thermometer correctly indicated the "temperature according to the absolute thermodynamic system of thermometry."

This marks a closing point of the problem of measuring absolute temperature. Thomson and Joule were confident that they had finally succeeded in reducing absolute temperature to measurable quantities, and they in fact proceeded to compute some numerical values for the deviation of the air thermometer from the absolute scale. The results were quite reassuring for the air thermometer (see table 4.2): although the discrepancy increased steadily as the temperature increased, it was estimated to be only about 0.4°C even at around 300°C for Regnault's standard air thermometer. One can imagine Joule and Thomson's pleasure at obtaining such a close agreement. Although their achievement received further refinement toward the end of the nineteenth century, it was never seriously challenged, to the best of my knowledge. However, it is difficult to regard Thomson and Joule's work on the measurement of absolute temperature as complete, in terms of epistemic justification.

TABLE 4.2. Joule and Thomson's comparison of absolute temperature (second definition) and air-thermometer temperature

Absolute temperature, minus 273.7°[a]	Air-thermometer temperature[b]
0	0
20	20 + 0.0298
40	40 + 0.0403
60	60 + 0.0366
80	80 + 0.0223
100	100
120	120 − 0.0284
140	140 − 0.0615
160	160 − 0.0983
180	180 − 0.1382
200	200 − 0.1796
220	220 − 0.2232
240	240 − 0.2663
260	260 − 0.3141
280	280 − 0.3610
300	300 − 0.4085

Source: Joule and Thomson [1854] 1882, 395–396.

[a]Joule and Thomson designed the size of one absolute degree to be the same as one centigrade degree as much as possible, by making the two scales agree exactly at the freezing and the boiling (steam) points by definition. But this new absolute temperature had a zero point, which Joule and Thomson estimated at −273.7°C. Hence the absolute temperature of the freezing point is 273.7° (we would now say 273.7 Kelvin), though it is displayed here as 0 (and all other absolute temperature values shifted by 273.7) in order to facilitate the comparison with ordinary air-thermometer temperatures.
[b]Given by a constant-volume air thermometer filled with a body of air at atmospheric pressure at the freezing point of water.

The clearest problem is that they used the uncorrected mercury thermometer in the Joule-Thomson experiment, so that it is far from clear that the corrections of the air thermometer on the basis of their data can be trusted. In the analysis I will enter into detailed philosophical and historical discussions of that matter.

Analysis: Operationalization—Making Contact between Thinking and Doing

> We have no right to measure these [Joule-Thomson] heating and cooling effects on any scale of temperature, as we have not yet formed a thermometric scale.
>
> William Thomson, "Heat," 1880

In the narrative part of this chapter, I traced the long and tortuous process by which temperature became established as a concept that is both theoretically cogent and empirically measurable. Now I will attempt to reach a more precise understanding of the nature of that achievement. To simplify our progress so far, the stories in the first three chapters of this book were about how to construct a coherent quantitative concept based on concrete physical operations. But such operational concept building is not sufficient in a genuinely theoretical science. For many important reasons scientists have desired abstract theoretical structures that are not purely and directly constructed from observable processes. However, if we also want to maintain our commitment to having an empirical science by giving empirical significance and testability to our abstract theories, we must face up to the challenge of operationalization: to link abstract theoretical structures to concrete physical operations. To put it in the terminology developed in "Beyond Bridgman" in chapter 3, operationalization is the act of creating operational meaning where there was none before.

Operationalizing an abstract theory involves operationalizing certain individual concepts occurring in it, so that they can serve as clear and convenient bridges between the abstract and the concrete. And one sure way of operationalizing a concept is to make it physically measurable, although the category of the operational includes much more than what can be considered measurements in the narrow sense of the term. Therefore measurement is the most obvious place where the challenges of operationalization can manifest themselves, though it is not the only place. In the following sections I will revisit the history of the attempts to measure Thomson's absolute temperature, in relation to the two main tasks involved in operationalization: to do it and to know whether it has been done well.

The Hidden Difficulties of Reduction

Most authors who have given accounts of the physics and the history of absolute temperature measurement seem to take comfort in the fact that absolute temperature

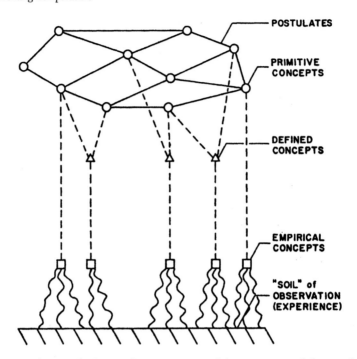

FIGURE 4.10. Herbert Feigl's (1970, 6) representation of the connection of abstract theoretical concepts to empirical observations. Reprinted by permission of the University of Minnesota Press.

can be expressed as a function of other quantities, which are measurable. Thomas Preston (1860–1900), the Irish physicist who discovered the anomalous Zeeman effect, expressed the situation perspicaciously in his textbook on heat:

> If we possess any thermodynamic relation, or any equation involving τ [absolute temperature] and other quantities which can be expressed in terms of p [pressure] and v [volume], then each such relation furnishes a means of estimating τ when the other quantities are known. (Preston 1904, 797)

In other words, absolute temperature can be measured through reduction to pressure and volume, which are presumably easily measurable. The structure of that reductive scheme is in fact quite similar to what one had in Irvinism: Irvinist absolute temperature could be expressed in terms of heat capacity and latent heat, both of which were regarded as straightforwardly measurable. Both Preston's and Irvine's views both fit nicely into a rather traditional philosophical idea, that the operationalization of a theoretical concept can be achieved by a chain of relations that ultimately link it to directly observable properties. Figure 4.10 gives a graphic representation of that reductive conception, taken from Herbert Feigl; figure 4.11 is a similar picture, due to Henry Margenau. In figure 4.12 I have made a graphic representation of Preston's idea about absolute temperature, which fits perfectly with the other pictures.

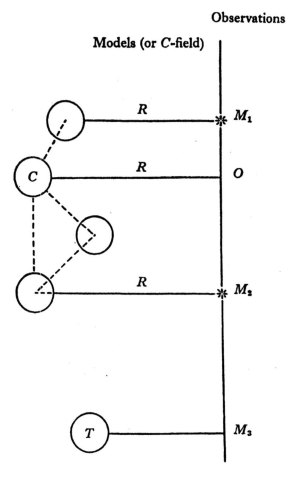

FIGURE 4.11. Henry Margenau's (1963, 2) representation of the connection of abstract theoretical concepts to empirical observations. Reprinted by permission of the University of Chicago Press.

The reductive view of operationalization is comforting, but we must not be lulled into epistemological sleep so easily. Seemingly operational concepts that allegedly serve as the reductive base of operationalization are often unoperationalized themselves, and in fact very difficult to operationalize. In Preston's case, the apparent comfort in his statement is provided by an ambiguity in the meaning of the terms "pressure" and "volume." The p and v that occur in thermodynamic theory are abstract concepts; they do not automatically become operationalized simply because we designate them by the same words "pressure" and "volume" as we use in everyday circumstances, in which those concepts are indeed full of operational meaning. In other words, figure 4.12 is an embodiment of an unwarranted equivocation. A more accurate picture is given in figure 4.13, where the thermodynamic relations linking p, v, and τ are placed squarely in the realm of the abstract, and the

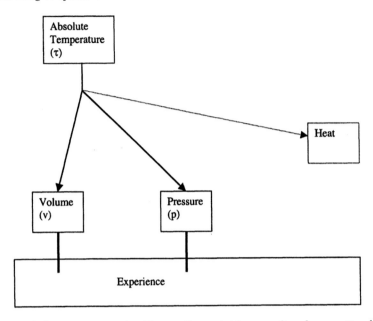

FIGURE 4.12. A diagram representing Thomas Preston's idea regarding the operationalization of absolute temperature.

necessity to operationalize p and v becomes apparent. Preston's equivocation is just the kind of thing that we have learned to be wary of in "Travel Advisory from Percy Bridgman" in chapter 3. There Bridgman warned that "our verbal machinery has no built-in cutoff": it is easy to be misled by the use of the same word or mathematical symbol in various situations into thinking that it means the same thing in all of those situations. (Bridgman focused most strongly on the unwarranted jump from one domain of phenomena to another, but his warning applies with equal force to the jump between the abstract and the concrete.)

If pressure and volume in thermodynamics were themselves very easy to operationalize, my disagreement with Preston would be a mere quibble, and the redrawing of the picture as given in figure 4.13 would be the end of the matter. But the discussions of temperature in earlier chapters of this book should be a sufficient reminder that concepts that seem easy to operationalize may actually be very difficult to operationalize. Consider pressure. Even if we take a primitive and macroscopic theoretical notion of pressure, there is no reason to assume that the mercury manometer was any more straightforward to establish than the mercury thermometer.[45] When we get into thermodynamic theory, the pressure concept is meant to apply universally, even in situations to which no actual manometer can be applied; among those situations would be microscopic processes and perfectly

---

[45]See the extensive history of the barometer in Middleton 1964b.

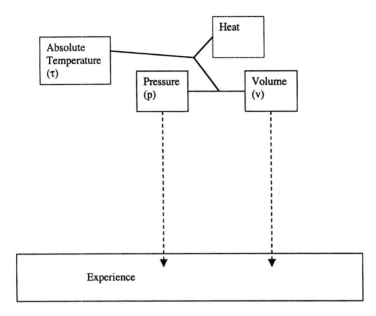

FIGURE 4.13. A revised version of Preston's picture of the operationalization of absolute temperature.

reversible processes. And as soon as we engage in thermodynamic theorizing of any power, we are dealing with differential equations involving pressure as a continuous variable, well defined down to infinitesimal precision. If we take one step further and also pick up the kinetic theory of gases, then pressure is theoretically defined in terms of the aggregate impact of countless molecules bouncing off a barrier. By that point there is simply no hope that the operationalization of the pressure concept will be trivial in general; most likely, it will be of the same order of difficulty as the operationalization of temperature.

A slightly different kind of problem emerges if we look back at Thomson's own definitions of absolute temperature. In Thomson's second definition, absolute temperature was expressed in terms of *heat*, not pressure and volume. With the first definition, Thomson's attempt at operationalization had led to its expression in terms of pressure, density, air-thermometer temperature, and latent heat (see equation (11) in "Semi-Concrete Models of the Carnot Cycle"). Even Preston gives an operationalization that involves latent heat in addition to volume and pressure, right after the statement quoted above. Generally speaking, any definition of temperature referring to the efficiency of the Carnot cycle (via Carnot's function μ) will have to involve the measure of heat in some way. But defining temperature in terms of quantities of heat raises an intriguing problem of circularity, since the quantitative concept of heat is usually operationalized by reduction to temperature. Consider Thomson's own experimental practice, in which the unit of heat was: "the quantity necessary to elevate the temperature of a kilogram of water from 0° to 1° of the air-thermometer" (Thomson [1848] 1882, 105). In the Joule-Thomson experiment, heat measurement was made by means of Joule's mercury thermometer. One way to avoid this circularity would be to measure heat in some other way. But as I have noted already, at least in the period covered in this chapter there were no viable calorimetric methods that did not rely on thermometers. Given that state of calorimetry, any attempts to operationalize temperature by reducing it to heat would not have got very far.

All in all, no simple reductive scheme has been adequate for the operationalization of absolute temperature. More generally speaking, if operationalization by reduction seems to work, that is only because the necessary work has already been done elsewhere. Reduction only expresses the concept to be operationalized in terms of other concepts, so it achieves nothing unless those other concepts have been operationalized. Operationalization has to come before reduction. Somewhere, somehow, some abstract concepts have to become operationalized without reference to other concepts that are already operationalized; after that, those operationalized concepts can serve as reductive bases of reduction for other concepts. But how can that initial operationalization be done?

Dealing with Abstractions

In order to reach a deeper understanding of the process of operationalization, we need to be clearer about the distinction between the abstract and the concrete. At this point it will be useful to be more explicit about what I mean by "abstraction."[46] I conceive abstraction as the act of removing certain properties from the description of an entity; the result is a conception that can correspond to actual entities but cannot be a full description of them. To take a standard example, the geometrical triangle is an abstract representation of actual triangular objects, deprived of all qualities except for their form. (A triangle is also an *idealization* by virtue of having perfectly straight lines, lines with no width, etc., but that is a different matter.) In principle the deletion of any property would constitute an abstraction, but most pertinent for our purposes is the abstraction that takes away the properties that individuate each entity physically, such as spatio-temporal location, and particular sensory qualities.

Thomson should be given credit for highlighting the issue of abstraction, by insisting that the theoretical concept of temperature should make no reference to the properties of any particular substances. Thus, he pointed to the gap between the concreteness of an operational concept and the abstractness that many scientists considered proper for a theoretical concept. It is instructive to make a comparison with earlier heat theorists discussed in "Theoretical Temperature before Thermodynamics." The Irvinists made an unreflective identification between the abstract concept of heat capacity and the operational concept of specific heat; some of them may even have failed to see the distinction very clearly. The chemical calorists, on the other hand, did not do very much to operationalize their concept of combined

---

[46]For a more nuanced view on the nature of abstraction in scientific theories, see Cartwright 1999, 39–41 and passim.

caloric. What truly set Thomson apart was his dual insistence: that theoretical concepts should be defined abstractly, and that they should be operationalized.

In one of his less phenomenalistic moments, Regnault stated the desired relationship between a concept and its measurement operation: "In the establishment of the fundamental data of physics...it is necessary that the adopted procedure should be, so to speak, the *material realization* of the definition of the element that is being sought."[47] (Whether or not Regnault himself really ever put this maxim into practice, it is a very suggestive idea.) In order to follow that idea straightforwardly in the case of Thomson's absolute temperature, we would need to build a physical Carnot engine. But, as discussed in "Semi-Concrete Models of the Carnot Cycle," it is impossible to make an actual heat engine that qualifies as a Carnot cycle. One might imagine that this impossibility is only a practical problem; the Carnot cycle is an idealized system, but surely we could try to approximate it, and see how close our approximation is?[48] That line of thinking misses the most essential feature of the situation: the Carnot cycle of post-Thomson thermodynamics is abstract, and real heat engines are concrete. It is possible that Carnot himself conceived his cycle simply as an idealized version of actual heat engines, but in later thermodynamics the Carnot cycle is an abstraction.

At the heart of the abstract character of the Carnot cycle is the fact that the concept of temperature that occurs in its description is not temperature measured by a concrete thermometer, but Thomson's absolute temperature. Therefore no such thing as an "actual Carnot cycle" (or even approximations to it) can exist at all, unless and until the absolute temperature concept has been operationalized. But the operationalization of absolute temperature requires that we either create an actual Carnot cycle or at least have the ability to judge how much actual systems deviate from it. So we end up with a nasty circularity: the operationalization of absolute temperature is impossible unless the Carnot cycle has sufficient operational meaning, and the Carnot cycle cannot have sufficient operational meaning unless absolute temperature is operationalized.

A similar problem plagues the use of the Joule-Thomson effect for the operationalization of absolute temperature. The Joule-Thomson experiment was intended to give us a comparison between actual gases and ideal gases, but such a comparison is not so straightforward as it may sound. That is because the concept of the "ideal gas" is an abstract one. It would be a mistake to think that the ideal gas of modern physics is one that obeys the ideal gas law, $PV = nRT$, where T

---

[47]Regnault, quoted in Langevin 1911, 49. This statement suggests that Regnault, despite his hostility to theories, still recognized very clearly the fact that measurement must answer to a pre-existing concept. In other words, perhaps in contrast to Bridgman's variety of empiricism, Regnault's empiricism only sought to eliminate the use of hypotheses in measurements, not to deny non-operational components to a concept's meaning.

[48]Shouldn't it be easy enough to tell how much the efficiency of an actual heat engine departs from that of the ideal Carnot engine? Carnot himself did make such estimates, and Thomson carried on with the same kind of calculation; see, for example, Thomson [1849] 1882, sec. 5 of the appendix (pp. 150–155). But it is an epistemic illusion that a direct comparison can be made between the Carnot cycle and a physical heat engine, for reasons that will become clearer in the next section.

is temperature measured by an ordinary thermometer. No, T in that equation is Thomson's absolute temperature (or some other abstract theoretical temperature concept). If we ever found a gas in which the product of pressure and volume actually were proportional to temperature as measured by an ordinary thermometer (in degrees Kelvin), it would certainly *not* be an ideal gas, unless our ordinary thermometer actually indicated absolute temperature, which is highly unlikely. Not only is it highly unlikely but it is impossible for us to know whether it is the case or not, without having done the Joule-Thomson experiment and interpreted it correctly.

So we are faced with the following puzzle: how were Joule and Thomson able to judge how much actual gases deviated from the ideal, when they only had ordinary thermometers and would not have known how an ideal gas should behave when monitored by ordinary thermometers? Their experiments employed Joule's mercury thermometers for the estimation of three crucial parameters: not only the size of the cooling effect but also the specific heat of the gases (K) and the mechanical equivalent of heat (J), all of which entered into the correction factor in equation (18) in "Using Gas Thermometers to Approximate Absolute Temperature." In that situation, how can we be assured that the correction of thermometers on the basis of the Joule-Thomson measurements was correct? (Note that the situation here is fundamentally different from that of Thomson's scheme for operationalizing his first absolute temperature. In that case, the use of empirical data taken with the air thermometer did not pose a problem because Thomson was seeking an explicit correlation of absolute temperature and air-thermometer temperature.)

Thomson recognized this problem clearly in his entry on "heat" in the ninth edition of the *Encyclopaedia Britannica* (1880, 49, §55), where he confessed with great insight: "[W]e have no right to measure these [Joule-Thomson] heating and cooling effects on any scale of temperature, as we have not yet formed a thermometric scale." Thomson indicated how the problem could be avoided, in principle: "Now, instead of reckoning on any thermometric scale the cooling effect or the heating effect of passage through the plug, we have to measure the quantity of work (δw) required to annul it." But he admitted that "the experiments as actually made by Joule and Thomson simply gave the cooling effects and heating effects shown by mercury thermometers." The justification that Thomson produced at the end of this remarkable discourse is disappointing:

> The very thermometers that were used [in the Joule-Thomson experiment] had been used by Joule in his original experiments determining the dynamical equivalent of heat [J], and again in his later experiments by which for the first time the specific heat of air at constant pressure [K] was measured with sufficient accuracy for our present purpose. Hence by putting together different experiments which had actually been made with those thermometers of Joule's, the operation of measuring δw, at all events for the case of air, was virtually completed. Thus according to our present view the mercury thermometers are merely used as a step in aid of the measurement of δw, and their scales may be utterly arbitrary.

What Thomson claims here is that the temperature measurements are merely ways of getting at the value of the quantity δw, and the final result is independent of the particular method by which it is obtained. Thomson's claim is hollow, unless it happens to be the case that the resulting empirical formula for δw is not a function of mercury temperature at all. But δw is a function of mercury temperature (t) in general, according to Joule and Thomson's own results. The empirical formula derived from their experiments is the following:

$$-d\vartheta/dp = A(273.7/t)^2, \qquad (19)$$

where ϑ is the temperature change in the gas, and A is a constant whose value depends on the nature of the gas (A was determined to be 0.92 air, and 4.64 for carbon dioxide); δw is given by multiplying ϑ by K (the specific heat of the gas) and J (the mechanical equivalent of heat).[49] I do not see how it can be argued that δw would in general have no dependence on t. The same point can be seen even more clearly if we take the view that certain errors are introduced into the measured values of ϑ, K, and J, if those values are obtained on the basis of the assumption that the mercury thermometer readings indicate the absolute temperature. Thomson's claim amounts to insisting a priori that all such errors cancel each other out when the three quantities are worked into a formula to produce the final result. That is possible in particular cases, but by no means guaranteed.

This, however, seems to be where Thomson left the problem. In the corpus of his work after the *Britannica* article and a couple of related articles published around the same time, I have not found any further contributions to the measurement of absolute temperature. Thomson was apparently quite satisfied with the theoretical understanding of absolute temperature that he had been able to secure in the framework of a fully developed theory of thermodynamics, and in practical terms he was happy with the old Joule-Thomson empirical results that seemed to give a sufficient indication that the deviations of gas-thermometer temperature from his second absolute temperature were quite small.

Let us take stock. I have noted a circularity: the operationalization of temperature seems to require the use of an already operationalized concept of temperature. Thomson and Joule broke this circularity by using an existing operationalization of temperature that was fully admitted to be unjustified—namely, a mercury thermometer, which was by no means guaranteed to indicate absolute temperature. Was it possible to guard against the errors that might have been introduced in using a thermometer that was admittedly open to corrections? Thomson does not seem to have given a very convincing answer to that question. Can we go beyond Thomson here?

Operationalization and Its Validity

The discussion in the last section brought us to a dead end of justification. I believe that the problem arises from applying an inappropriate notion of justification to operationalization, as I will try to show in this section and the next. It is helpful

---

[49] Joule and Thomson [1862] 1882, 428–429; for further exposition on how this result was obtained, see Chang and Yi (forthcoming). In the case of hydrogen, $d\phi/dp$ was apparently not a function of temperature; see Thomson 1880, 49.

here to start with Nancy Cartwright's (1999, 180) unequivocal statement of the relation between theories and the world:[50] "theories in physics do not generally represent what happens in the world; only models represent in this way, and the models that do so are not already part of any theory." As Cartwright explains, this amounts to an important revision of her earlier view that was famously expressed in *How the Laws of Physics Lie*. I take her current view to mean that the laws of physics are not even capable of lying when it comes to states of affairs that are amenable to actual physical operations, because the theoretical laws cannot say anything about the actual situations—unless and until the concepts occurring in the laws have been operationalized.

To get a clearer view on operationalization, it is helpful to see it as a two-step process, as represented in figure 4.14. A theoretical concept receives its abstract characterization through its various relations with other theoretical concepts; those relations constitute an abstract system. The first step in operationalization is *imaging*: to find a concrete image of the abstract system that defines the abstract concept. I call this an "image" rather than a "model," in order to avoid any possible suggestion that it might be an actual physical system; the concrete image is still an imagined system, a conception, consisting of conceivable physical entities and operations, but not actual ones. The concrete image is not a physical embodiment of the abstract system.[51] Finding the concrete image is a creative process, since the abstract system itself will not dictate how it should be concretized. When we take the Carnot cycle and say it could be realized by a frictionless cylinder-and-piston system filled with a water-steam mixture that is heated and cooled by heat reservoirs of infinite capacities, we are proposing a concrete image of an abstract system. And there are many other possible concrete images of the Carnot cycle.

After finding a concrete image of the abstract system, the second step in operationalization is matching: to find an actual physical system of entities and operations that matches up with the image. Here it cannot be taken for granted that we will be able to find such a matching system. If it seems clear that there cannot be any matching actual systems, then we pronounce the image to be idealized.[52] We can expect that the concrete images we need to deal with in many cases will be idealized ones. In the context of thermodynamics and statistical mechanics, perfect reversibility is an idealization in the macroscopic realm. But it is possible to estimate how closely actual systems approximate idealized ones.

With the help of this two-step view of operationalization, let us return to the question of the justification or validation of operationalization. The first thing we need to do is lose the habit of thinking in terms of simple correctness. It is very tempting to think that the ultimate basis on which to judge the validity of

---

[50]She attributes the inspiration for this view partly to Margaret Morrison (1998).

[51]Although I do not use the term "model," some of the recent literature on the nature and function of models is clearly relevant to the discussion here. See especially Morrison and Morgan's view (1999, 21–23) that theoretical models can function as measuring instruments.

[52]Strictly speaking, this notion of idealization does not apply to abstract systems; however, if an abstract system only seems to have idealized concrete images, then we could say that the abstract system itself is idealized.

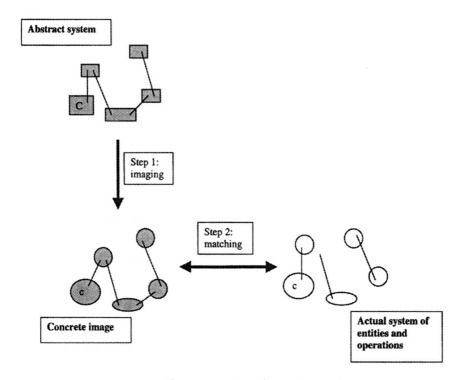

FIGURE 4.14. The two-step view of operationalization.

an operationalization should be whether measurements made on its basis yield values that correspond to the real values. But what are "the real values"? Why do we assume that unoperationalized abstract concepts, in themselves, possess any concrete values at all? For instance, we are apt to think that each phenomenon must possess a definite value of absolute temperature, which we can in principle find out by making the correct measurement. It is very difficult to get away from this sort of intuition, but that is what we must to do in order to avoid behaving like the fly that keeps flying into the windowpane in trying to get out of the room. An unoperationalized abstract concept does not correspond to anything definite in the realm of physical operations, which is where values of physical quantities belong. To put it somewhat metaphorically: equations in the most abstract theories of physics contain symbols (and universal constants), but no numerical values of any concrete physical property; the latter only appear when the equations are applied to concrete physical situations, be they actual or imagined.[53] Once an operationalization is made, the abstract concept possesses values in concrete

---

[53] I am not advocating an ontological nihilism regarding the reality of possessed values of properties, only a caution about how meaningful our concepts are at various levels. A concept that is fully meaningful

situations. But we need to keep in mind that those values are products of the operationalization in question, not independent standards against which we can judge the correctness of the operationalization itself. That is the root of the circularity that we have encountered time and again in the attempt to justify measurement methods.

If we come away from the idea of correspondence to real values, how can the question of validity enter the process of operationalization at all? That is a difficult question, to which I am not able to offer a complete answer. But it seems to me that the answer has to lie in looking at the correspondence between systems, not between the real and measured values of a particular quantity. A valid operationalization consists in a good correspondence between the abstract system and its concrete image, and between the concrete image and some system of actual objects and operations. The correspondence between systems is not something that can be evaluated on a one-dimensional scale, so the judgment on the validity of a proposed operationalization is bound to be a complex one, not a yes–no verdict.

Let us now return to the case of Thomson's absolute temperature and review the strategies of operationalization that he attempted, which were initially described in "Semi-Concrete Models of the Carnot Cycle" and "Using Gas Thermometers to Approximate Absolute Temperature" in the narrative. Logically the simplest method would have been to make a concrete image of the Carnot cycle and to find actual operations to approximate it; however, any concrete images of the Carnot cycle were bound to be so highly idealized that Thomson never made a serious attempt in that direction. The most formidable problem is the requirement of perfect reversibility, which any concrete image of the Carnot cycle would have to satisfy. As noted earlier, this is not merely a matter of requiring frictionless pistons and perfect insulation. Reversibility also means that all heat transfer should occur between objects with equal temperatures (since there would be an increase of entropy in any heat transfer across a temperature difference). But Thomson himself stated as a matter of principle, or even definition, that no net movement of heat can occur between objects that are at the same temperature.[54] The only way to make any sense of the reversibility requirement, then, is to see it as a limiting case, a clearly unattainable ideal that can only be approximated if the temperature differences approach zero, which would mean that the amount of time required for the transfer of any finite amount of heat will approach infinity. The fact that the Carnot cycle is a cycle exacerbates the difficulty because it requires that not just one, but a few different physical processes satisfying the reversibility requirement need to be found and put together.

Therefore Thomson's preference was not to deal with a Carnot cycle directly, but to deduce the relevant features of the abstract Carnot cycle from simpler abstract

---

at the theoretical level may be devoid of meaning at the operational level or vice versa. Differentiating the levels of meaning should also make it clear that I am not advocating an extreme operationalism of the kind that I rejected in "Beyond Bridgman" in chapter 3. An abstract concept within a coherent theory is not meaningless even if it is not operationalized, but it must be operationalized in order to have empirical significance.

[54] See the definition from 1854 quoted in note 38.

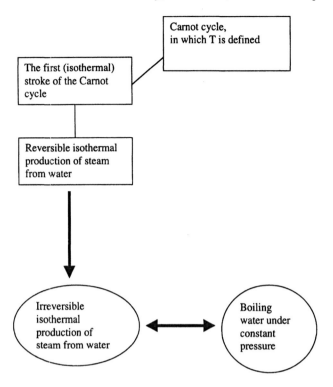

FIGURE 4.15. Thomson's operationalization of absolute temperature, by means of the water–steam system.

processes and to operationalize the simpler processes. Thomson's most sophisticated attempt in this direction came in 1880, in a renewed attempt to operationalize his second absolute temperature. He was assisted in this task by the principle of energy conservation, which allowed him to deduce theoretically the work-heat ratio for the cycle (and hence the heat input–output ratio needed for the definition of temperature) merely by considering isothermal expansion.[55] Then he only had to find the concrete image of the first stroke of the cycle rather than the whole cycle. That stroke was still an ideal reversible process, but Thomson thought that its concrete image could be an irreversible process, if it could be demonstrated to be equivalent in the relevant sense to the reversible process. This scheme of operationalization is summarized graphically in figure 4.15.

Thomson's favorite concrete image of the first stroke of the Carnot cycle was the production of steam from water. Thomson was able to allow the image to be a nonreversible process because the amount of mechanical work generated in this process, for a given amount of heat input, was not thought to be affected by

[55] For further details on this maneuver, see Chang and Yi (forthcoming).

whether the process was reversible or not. (That would be on the assumption that the internal energy of the steam–water system is a state function, so it has fixed values for the same initial and final states of the system regardless of how the system gets from one to the other.) Now, as described earlier, the particularly nice thing about the steam–water system is that the generation of steam from liquid water is a process that takes place isothermally under constant pressure, which makes it easy to compute the mechanical work generated (simply the constant pressure multiplied by the volume increment). That is not the case for gas-only systems, for instance, because an isothermal expansion of a gas cannot take place under constant pressure, so the computation of mechanical work produced in that expansion requires an exact knowledge of the pressure-volume relation. In addition, if the pressure varies during the heat intake process, computing the amount of heat entering the gas requires an exact knowledge of the variation of specific heat with pressure. All in all, the work-heat ratio pertaining to the isothermal expansion of a simple gas is surrounded with uncertainty, and it is not obvious whether and how much the work–heat ratio would depend on whether the expansion is carried out reversibly or not. There is no such problem in the steam–water case.

At last, Thomson had identified clearly a simple concrete image of an abstract thermodynamic system embodying the absolute temperature concept, an image that also had obvious potential for correspondence to actual systems. This explains his elation about the steam–water system, expressed in his 1880 article in the *Encyclopaedia Britannica*:

> We have given the steam thermometer as our first example of thermodynamic thermometry because intelligence in thermodynamics has been hitherto much retarded, and the student unnecessarily perplexed, and *a mere quicksand has been given as a foundation for thermometry*, by building from the beginning on an ideal substance called perfect gas, with none of its properties realized rigorously by any real substance, and with some of them unknown, and *utterly unassignable*, even by guess. (Thomson 1880, 47, §46; emphases added)

Unfortunately, Thomson was never able to bring his proposed steam-water thermometer into credible practical use. For one thing, no one, not even Regnault, had produced the necessary data giving the density of steam (both in itself and in relation to the density of water) as a function of temperature in the whole range of temperatures. Lacking those data, Thomson could not graduate a steam–water thermometer on the absolute scale without relying on unwarranted theoretical assumptions. Despite his attraction to the virtues of the steam–water system, Thomson was forced to calibrate his steam–water thermometers against gas thermometers.[56]

The actual operationalization of absolute temperature was effected by a slightly different strategy, in which the concrete image used was not the steam–water system, but the gas-only system. As explained in "Using Gas Thermometers to Approximate Absolute Temperature," the new thermodynamic theory yielded the

---

[56]Thomson 1880, 46, §44. See also Thomson [1879–80a] 1911 and Thomson [1879–80b] 1911.

proposition that an ideal gas would expand uniformly with increasing absolute temperature. By taking the expansion of the ideal gas as the theoretical framework for the definition of absolute temperature, Thomson freed himself from the need to look for a plausible concrete image of the Carnot cycle. The expansion of an ideal gas had a straightforward concrete image (an actual body of gas expanding against some external pressure), and it was certainly possible to perform actual operations matching that image: heat a gas in a cylinder with a piston pressing down on it; the temperature is indicated by the product of the pressure and the volume of the gas. (We have seen in "The Achievement of Observability, by Stages" in chapter 2 that high-quality gas thermometers were much more complicated, but we can ignore such complications for the moment.)

As there were no insurmountable obstacles in making and using gas thermometers, it was easy to proceed with the second step of this operationalization. The interesting problems in this case emerged in the process of examining the exact match between the image and the actual operations. Regnault had demonstrated very clearly that thermometers made with different gases disagreed with each other except at the fixed points. Therefore it was impossible to get an exact match between the image of the ideal gas thermometer and the collection of all actual gas thermometers, because the image stipulated only one temperature value in each situation, and the collection of the gas thermometers returned multiple values. There are various options one could take in order to get an exact match. (1) One option, which I have never seen advocated, is to modify the image by saying that temperature can have multiple values in a given situation, to negate what I have called the principle of single value in "Comparability and the Ontological Principle of Single Value" in chapter 2. (2) Another is to declare that one of the actual thermometers is the correct measurer of absolute temperature, and the rest must be corrected to conform to it. There is no logical problem in making such a conventionalist declaration, but it would tie the absolute temperature concept to the properties of one particular substance, for no convincing reason. As I have already emphasized, that was something that Thomson explicitly set out to avoid. (3) That leaves us with the option that Thomson (and Joule) actually took, which is to say that most likely none of the actual gas thermometers are exactly accurate indicators of absolute temperature, and they all have to be corrected so that they agree with each other. This scheme for operationalizing absolute temperature is summarized graphically in figure 4.16.

The challenge in that last option was to come up with some unified rationale according to which all the thermometers could be corrected so that they gave converging values of absolute temperature. The reasoning behind the Joule-Thomson experiment provided such a rationale, as explained in "Using Gas Thermometers to Approximate Absolute Temperature." It makes sense within energy-based thermodynamics that a gas which expands regularly with increasing temperature would also remain at the same temperature when it expands without doing work. Both types of behavior indicate that the internal energy of such a gas does not change merely by virtue of a change in its volume, and both can be considered different manifestations of Mayer's hypothesis (see Hutchison 1976a, 279–280). When thermodynamic theory is supplemented by the kinetic theory of gases, this condition can be

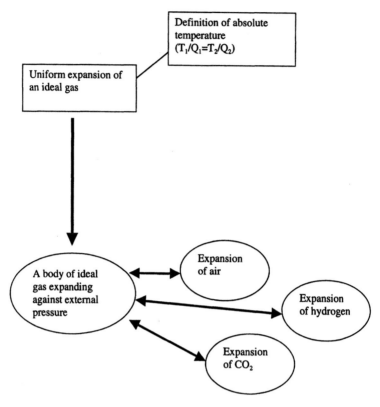

FIGURE 4.16. Joule and Thomson's operationalization of absolute temperature, by means of the gas-only system.

understood as the stipulated absence of intermolecular forces in an ideal gas. But in actual gases one would expect some degree of intermolecular interactions, therefore some irregularity in thermal expansion as well as some cooling (or heating) in Joule-Thomson expansion. As all actual gas thermometers could be corrected on this same basis, the temperature concept itself would maintain its generality (unlike in option 2). If these coordinated corrections did result in bringing the measured values closer together, then such convergence would be a step toward perfecting the match between the actual gas thermometers and the image of the ideal gas thermometer. The corrections carried out by Joule and Thomson, and by later investigators, seem to have produced just such a move toward convergence. (However, it would be worthwhile to compile the data more carefully in order to show the exact degree and shape of the convergence achieved.)

Accuracy through Iteration

One major puzzle still remains. I began my discussion in the last section by saying that we should avoid thinking about operationalization in terms of correctness, but ended by discussing corrections made to actual measuring instruments. If we are not concerned with correctness, what can the "corrections" possibly be about? The logic of this situation needs a more careful examination. The point of rejecting the question of correctness was to acknowledge that there are no pre-existing values of abstract concepts that we can try to approach; as mentioned earlier, the "real" values only come into being as a result of a successful operationalization. But I need to clarify and articulate the sense of validity that is hiding behind the word "successful" there.

Thomson himself did not quite give a clear explanation of the epistemic character of his maneuvers even in his 1880 article, as discussed in "Dealing with Abstractions." Fortunately, a much more satisfying understanding of the problem of operationalizing absolute temperature was to emerge within a decade, apparently starting with the work of Hugh Longbourne Callendar (1863–1930), English physicist and engineer who made important observations on the properties of steam and crucial contributions to electric-resistance thermometry. My discussion of Callendar's work will rely on the exposition given by Henri Louis Le Chatelier (who was introduced in the discussion of pyrometry in chapter 3), which is much more helpful than Callendar's own presentation.[57] The Callendar–Le Chatelier operationalization of absolute temperature can be understood as an instance of the process of epistemic iteration, first introduced in "The Iterative Improvement of Standards" in chapter 1.

The starting point of epistemic iteration is the affirmation of a certain system of knowledge, which does not have an ultimate justification and may need to be changed later for various reasons. The initial assumption for Callendar was that air-thermometer temperature and absolute temperature values were very close to each other. We start by writing the law governing the thermal behavior of actual gases as follows:

$$pv = (1 - \phi)RT, \qquad (20)$$

where R is a constant, T is absolute temperature, and ϕ is an as-yet unknown function of T and p. The factor ϕ is what makes equation (20) different from the ideal gas law, and it is a different function for each type of gas; it is *presumed* to be small in magnitude, which amounts to an assumption that actual gases roughly obey the ideal gas law. Such an assumption is not testable (or even fully meaningful) at that stage, since T is not operationalized yet; however, it may be vindicated if the correction process is in the end successful or discarded as implausible if the correction process cannot be made to work.

The next step is to estimate ϕ, which is done by means of the results of the Joule-Thomson experiment, following Thomson's method. Le Chatelier gives the following empirical result, calculated from the data obtained in experiments with atmospheric air:

$$\phi = 0.001173 \frac{p}{p_0} \left(\frac{T_0}{T} \right)^3, \qquad (21)$$

---

[57] Callendar 1887, 179, gives the starting point of his analysis and the final results, but does not give the details of the reasoning; my discussion follows Le Chatelier and Boudouard 1901, 23–26.

where p_0 is the standard atmospheric pressure and T_0 is the absolute temperature of melting ice. I have presented the derivation of this result in some detail elsewhere, but one important point can be gathered from merely inspecting the final outcome. Equation (21) is supposed to be an empirical result, but it expresses φ as a function of absolute temperature T, not as a function of t_a, (Amontons) temperature measured by an ordinary thermometer in the Joule-Thomson experiment. What happens in this derivation is a deliberate conflation of absolute temperature and air temperature (or mercury temperature), as Callendar and Le Chatelier take the empirical Joule-Thomson formula expressed in t_a and simply substitutes it into theoretical formulas expressed in T, letting t_a stand in for T. This is allowed, as an approximation, on the assumption that T and t_a are roughly equal because φ is very small.[58]

Unlike Thomson, Le Chatelier was very clear that equation (21) did not give the final correction (Le Chatelier and Boudouard 1901, 25): "This is still an approximate result, for we have depended upon the experiments of Joule and Thomson and on the law of adiabatic expansion." Here Le Chatelier was also acknowledging the fact that in the derivation of (21) he had helped himself to the adiabatic gas law, knowing that it was not known to be exactly true but assuming that it was approximately true. A further round of corrections could be made with the help of the correction indicated in (21). This would involve recalibrating the air thermometer, according to the law of expansion that is obtained by inserting (21) into (20); recall that the air thermometer was initially calibrated on the basis of the assumption that the expansion of air was exactly regular (φ = 0). With the recalibration of the air thermometer, one could either do the Joule-Thomson experiments again or just reanalyze the old data. Either way, the refined version of the Joule-Thomson experiment would yield a more refined estimate of φ, giving an updated version of (21). This process could be repeated as often as desired. A similar assessment of the situation was given twenty years later by A. L. Day and R. B. Sosman, with the most succinct conceptual clarity on the matter that I have seen:

> It is important at this point to recall that our initial measurements with the gas-thermometer tell us nothing about whether the gas in question obeys the law $pv = k\theta$ or not. Only measurements of the energy-relations of the gas can give us that information. But since such measurements involve the measurement of temperature, *it is evident that the realisation of the temperature scale is logically a process of successive approximations.* (Day and Sosman 1922, 837; emphasis added)

However, it seems that in practice no one was worried enough to enter into second-round corrections or beyond. Callendar calculated the first-round corrections on air up to 1000°C; although the corrections got larger with increasing temperature, they turned out to be only 0.62° at 1000°C for the constant-volume air thermometer, and 1.19° for the constant-pressure air thermometer. It was seen that the corrections would grow rapidly beyond that point, but that was not so

---

[58] See Chang and Yi (forthcoming) for details.

much of a practical concern since 1000°C was about the very limit at which any gas thermometers could be made to function at all in any case.[59] Le Chatelier was happy to declare:

> The deviations of the air-thermometer at high temperatures are thus very slight if concordance is established at 0° and 100°; we shall not have to occupy ourselves further with the differences between the indications of the thermodynamic thermometer and those of the gas-thermometer. (Le Chatelier and Boudouard 1901, 26)

One only needed to avoid gases like carbon dioxide, for which the corrections were significantly larger. Day and Sosman gave a similar view (1922, 837): "Practically, the first approximation is sufficient, so nearly do the gases commonly used in gas-thermometers conform to the 'ideal' behaviour expressed in the law $pv = k\theta$."

This is a pleasing result, but we must also keep in mind that the smallness of the first-round correction is hardly the end of the story, for two reasons. First of all, for each gas we would need to see whether the corrections actually continue to get smaller in such a way as to result in convergence. In mathematics conditions of iterative convergence can be discerned easily enough because the true function we are trying to approximate is already known or at least knowable. In epistemic iteration the true function is not known, and in cases like the present one, not even defined. So the only thing we can do is to carry on with the iteration until we are pragmatically satisfied that a convergence seems destined to happen. In the case of absolute temperature one iterative correction seemed to be enough for all practical purposes, for several gases usable in thermometers. However, to the best of my knowledge no one gave a conclusive demonstration that there would be convergence if further rounds of corrections were to be carried out.

The second point of caution emerges when we recall the discussion at the end of the last section. If we are to respect Thomson's original aim of taking the definition of temperature away from particular substances, it is not good enough to have convergence in the corrections of one type of gas thermometer. Various gas thermometers need to converge not only each in itself but all of them with each other. Only then could we have a perfect match between the single-valued image of absolute temperature and the operational absolute temperature measured by a collection of gas thermometers. It is perhaps plausible to reject some particular gases as legitimate thermometric fluids if there are particular reasons that should disqualify them, but at least some degree of generality would need to be preserved if we are to remain faithful to the spirit of Thomson's enterprise.

Seeing the "correction" of actual thermometers as an iterative process clarifies some issues that have been left obscure in my analysis so far. The clarification stems from the realization that in an iterative process, point-by-point justification of each and every step is neither possible nor necessary; what matters is that each stage leads on to the next one with some improvement. The point here is not just that

---

[59] See Callendar 1887, 179. According to Day and Sosman (1922, 859), up to that time only four attempts had been made to reach 1000°C with gas thermometers. See also the discussion about the high-temperature limits of the air thermometer in "Ganging Up on Wedgwood" in chapter 3.

slightly incorrect information fed into an iterative process may well be corrected (as Peirce pointed out in the example shown in "The Iterative Improvement of Standards" in chapter 1). As already noted, the question of correctness does not even apply, unless and until the iterative process produces a successful outcome. Therefore it makes sense to relax the sort of demand for justification that can lead us to seek illusory rigor.

There are several aspects of this relaxation. (1) First of all, one's exact starting point may not be important. In the case of absolute temperature, assuming that the concrete image of the ideal gas law was approximately true of actual gases happened to hit the nail nearly on the head, but the iterative correction process could also have started from other initial approximations and reached similar final results. (2) Some looseness can also be allowed in the process of reasoning adopted beyond the initial starting point. Thomson was able to make certain shortcuts and apparently unwarranted approximations in his various derivations without much of a tangible consequence.[60] Similarly, Le Chatelier helped himself to the adiabatic gas law, knowing full well that it was not guaranteed to be exactly right. (3) Empirical data that may not be exactly right can also be used legitimately. Therefore Thomson's defense of the use of Joule's mercury thermometer in the Joule-Thomson experiment was not only invalid (as I have argued in "Dealing with Abstractions") but also unnecessary. A recognition of the nature of the iterative process would have spared Thomson from an illusory problem and a pseudo-solution to it. (4) Just as different starting points may lead toward the same conclusion, different paths of reasoning may do so as well. Thomson himself proposed various methods of operationalizing absolute temperature, though only one was pursued sufficiently so it is difficult to know whether the same outcome would have been reached through his other strategies. But I think it is safe to say that the Joule-Thomson experiment was not the only possible way to obtain the desired results. In fact Joule and Thomson themselves noted in 1862, with evident pleasure, that Rankine had used Regnault's data to obtain a formula for the law of expansion of actual gases that was basically the same as their own result based on the Joule-Thomson experiment (equation (18)).[61]

One more important issue remains to be clarified. In the process of operationalizing an abstract concept, what exactly do we aim for, and what exactly do we get? The hoped-for outcome is an agreement between the concrete image of the abstract concept and the actual operations that we adopt for an empirical engagement with the concept (including its measurement). That is the correspondence that makes the most sense to consider, not the complacently imagined correspondence between theory and experience, or theory and "reality." With an iterative process we do not expect ever to have an exact agreement between the

---

[60]For a more detailed sense of this aspect of Thomson's work, see Chang and Yi (forthcoming).

[61]Joule and Thomson [1862] 1882, 430. They refer to Rankine's article in the *Philosophical Transactions*, 1854, part II, p. 336. In an earlier installment of their article Joule and Thomson ([1854] 1882, 375–377) had already reproduced a private communication from Rankine on the subject, with commentary.

operational image and the actual operations, but we hope for a gradual convergence between them. Such convergence would be a considerable achievement, especially if it could be achieved with a high degree of quantitative precision.

This convergence provides a basis for a workable notion of accuracy. We can say that we have an accurate method of measurement, if we have good convergence. How about truth? Can we ever say whether we have obtained the true values of an abstract concept like absolute temperature? The question of truth only makes sense if there is an objectively determinate value of the concept in each physical situation.[62] In case we have a convergent operationalization, we could consider the limits of convergence as the "real" values; then we can use these values as the criteria by which we judge whether other proposed operationalizations produce true values. But we must keep firmly in mind that the existence of such "real values" hinges on the success of the iterative procedure, and the successful operationalization is constitutive of the "reality." If we want to please ourselves by saying that we can approach true values by iterative operationalization, we also have to remember that this truth is a destination that is only created by the approach itself.

Theoretical Temperature without Thermodynamics?

The foregoing analysis points to an interesting way in which the understanding and measurement of temperature could be tidied up and possibly refined further. The suggestion arises most of all from acknowledging a clear defect in my own analysis. That defect was the postulation of a sharp dichotomy between the abstract and the concrete. The discomfort in my discussion created by that oversimplification may have already been evident to the reader. The dichotomy between the abstract and the concrete has been enormously helpful in clarifying my thinking at the earlier stages, but I can now afford to be more sophisticated. What we really have is a continuum, or at least a stepwise sequence, between the most abstract and the most concrete. This means that the operationalization of a very abstract concept can proceed step by step, and so can the building-up of a concept from concrete operations. And it may be positively beneficial to move only a little bit at a time up and down the ladder of abstraction. Thomson was much too ambitious, in trying to connect up a resolutely abstract concept of temperature with the most concrete measurement operations all at once. In the end he had to rely on smaller steps, and a brief review of his steps may suggest some ways of progressing in small steps more deliberately.

The Joule-Thomson work on gases and gas thermometers suggests that it would be possible to postulate a *slightly* abstract concept of temperature that is defined by the behavior of the ideal gas, with no reference to heat engines or any general thermodynamic theory. As was well known long before Thomson, most gases not too near their temperature of liquefaction have similar patterns of thermal

---

[62] It may seem odd to speak of the reality of temperature, which is a property and not an object. However, there are good arguments to the effect that it makes more sense to address the realism question to properties than to entities; see Humphreys 2004, sec. 15.

expansion. It would be a modest leap from that empirical observation to postulate a theoretical concept of temperature (let's call it "ideal-gas temperature") that fits into the following story: "the behavior a gas is governed by temperature according to a simple law; of course, there are some complications because we do not have a completely accurate way of measuring real temperature, and also the different gases are all slightly imperfect in their own various ways; but the ideal gas would expand with real temperature in a perfectly linear fashion." If we want to believe this pleasing theoretical story, we have to fill in some blanks successfully. We have to come up with some independent way of detecting and measuring these various imperfections in actual gases, in the hope that this measure will fit in with their deviations from linear expansion with temperature. This is not so far from what Thomson, Joule, and their followers actually did, picking up on the concept of Amontons temperature. The Joule-Thomson cooling effect is just such a property that we can conceive as being tightly correlated with deviations from regular expansion. To get that far, we do not need a highly theoretical story of the kind that Thomson sought to provide about why the Joule-Thomson cooling should be related to peculiarities in the expansion of gases. Such a highly theoretical story belongs in later steps going up the ladder of abstraction and does not need to constrain the construction of the concept of ideal-gas temperature.

In fact, in historical retrospect, it is not at all clear that the theory of heat engines was the best place to anchor the theoretical temperature concept. The use of the Carnot cycle in the definition of temperature was most unnecessary. It was certainly helpful initially for the purpose of getting Thomson to see how an abstract concept of temperature might be constructed, but he could have kicked that ladder away. The definition of 1854 was still overly restrictive, since it remained tied to heat input and output in isothermal processes occurring within a reversible cycle. Even the most liberalized concept of temperature based on the Carnot cycle would be restrictive, because such a concept cannot be adapted to nonreversible or noncyclical processes easily. Bypassing the Carnot cycle, and even all of classical thermodynamics, ideal-gas temperature can be linked up with the kinetic theory of gases very naturally. After that, it is of course plausible to think about incorporating the kinetic conception of temperature into an even more general and abstract theoretical framework, such as classical or quantum statistical mechanics. I am not saying that Thomson around 1850 should have or could have foreseen the development of the modern kinetic theory and statistical mechanics. What I think he could have perceived is that the notion of temperature applicable to the expansion of gases did not have to be housed in a theory of heat engines.

In "The Defense of Fixity" in chapter 1, I have noted the robustness of low-level laws; for those laws to be meaningful and useful, the concepts occurring in them need to be well defined at that semi-concrete level and well connected to adjacent levels of abstraction. Something like the concept of ideal-gas temperature is at the right level of abstraction to support the phenomenological physics of gases. Other theories at that same phenomenological level (e.g. those governing conduction and radiation) may well require different semi-concrete concepts of temperature. The various semi-concrete concepts may become united by more abstract theories. Such concept-building processes moving toward greater abstraction should be harmonized with operationalization processes that start with the most abstract concepts and concretize them step-by-step. Such a well-ordered two-way process of abstraction and concretization would enable us to build a conceptual-operational system that can evolve with maximum flexibility and minimum disruption in the face of change and new discoveries.

5

Measurement, Justification, and Scientific Progress

> Washing dishes and language can in some respects be compared. We have dirty dishwater and dirty towels and nevertheless finally succeed in getting the plates and glasses clean. Likewise, we have unclear terms and a logic limited in an unknown way in its field of application—but nevertheless we succeed in using it to bring clearness to our understanding of nature.
>
> Niels Bohr (1933), quoted by Werner Heisenberg,
in Physics and Beyond (1971)

The preceding chapters have examined the development of thermometry, concentrating on the justification of standards and assumptions. These stories of measurement bring out a particular mode of scientific progress, which I will try to articulate briefly and systematically now, building on various insights expressed earlier. I did not set out to advance any overarching epistemological doctrines in this book. However, it would be dishonest for me to hide the image of a particular mode of scientific progress that has emerged in the course of considering the concrete episodes. As I will elaborate further in "The Abstract and the Concrete," the content of this chapter should not be taken as a generalization from the preceding chapters, but as the articulation of an abstract framework that has been necessitated for the construction of the concrete narratives. Nor should this chapter be regarded as the summary of all of the epistemologically significant points made in earlier chapters. Here I will pull together only the ideas and arguments that can be strengthened and deepened through the synthesis; the rest can stand as they were developed in earlier chapters.

The overall argument of this chapter can be summarized as follows. In making attempts to justify measurement methods, we discover the circularity inherent in empiricist foundationalism. The only productive way of dealing with that circularity is to accept it and admit that justification in empirical science has to be coherentist. Within such coherentism, epistemic iteration provides an effective method of scientific progress, resulting in the enrichment and self-correction of the initially affirmed system. This mode of scientific progress embraces both conservatism and pluralism at once.

Measurement, Circularity, and Coherentism

In his famous discussion of the difficulties of the empirical testing of scientific theories, Pierre Duhem made the curious statement that "the experimental testing of a theory does not have the same logical simplicity in physics as in physiology" ([1906] 1962, sec. 2.6.1, 180–183). The physiologists can make their observations by means of laboratory instruments that are based on the theories of physics, which they take for granted. However, in testing the theories of physics, "it is impossible to leave outside the laboratory door the theory we wish to test." The physicists are forced to test the theories of physics on the basis of the theories of physics. Among physicists, those who are involved in the testing of complicated and advanced theories by means of elementary observations would be in a relatively straightforward epistemic position, much like Duhem's physiologists. But for those who try to justify the reasoning that justifies the elementary observations themselves, it is very difficult to escape circularity. The basic problem is clear: empirical science requires observations based on theories, but empiricist philosophy demands that those theories should be justified by observations. And it is in the context of quantitative measurement, where the justification needs to be made most precisely, that the problem of circularity emerges with utmost and unequivocal clarity.

In each of the preceding chapters I examined how this circularity of justification manifested itself in a particular episode in the development of thermometry, and how it was dealt with. Chapter 1 asked how certain phenomena could have been judged to be constant in temperature, when no standards of constancy had been established previously. The answer was found within the self-improving spiral of quantification—starting with sensations, going through ordinal thermoscopes, and finally arriving at numerical thermometers. Chapter 2 asked how thermometers relying on certain empirical regularities could be tested for correctness, when those regularities themselves would have needed to be tested with the help of thermometer readings. The answer was that thermometers could be tested by the criterion of comparability, even if we could not verify their theoretical justification. Chapter 3 asked how extensions of the established thermometric scale could be evaluated, when there were no pre-existing standards to be used in the new domains. The answer was that the temperature concept in a new domain was partly built through the establishment of a convergence among various proposed measurement methods applying there. Chapter 4 asked how methods of measuring abstract concepts of temperature could be tested, when establishing the correspondence between the abstract concept and the physical operations relied on some theory that would itself have required empirical verification using results of temperature measurement. The answer was found in the iterative investigation based on the provisional assumption of an unjustified hypothesis, leading to a correction of that initial hypothesis.

In each of my episodes, justification was only found in the coherence of elements that lack ultimate justification in themselves. Each episode is an embodiment of the basic limitation of empiricist foundationalism. I take as the definition of *foundationalism* the following statement by Richard Foley (1998, 158–159): "According to foundationalists, epistemic justification has a hierarchical structure. Some beliefs are self-justifying and as such constitute one's evidence base. Others are justified only if they are appropriately supported by these basic beliefs." The main difficulty in the foundationalist project is actually finding such self-justifying beliefs. There have been great debates on that matter, but I think most commentators would agree that any set of propositions that seem self-justifying tend not to be informative enough to teach us much about nature. Formal logic and mathematics are cases in point. In the realm of experience, the theory-ladenness of language and observation forces us to acknowledge that only unarticulated immediate experience can be self-justifying. And as Moritz Schlick conceded, such immediate experience (which he called "affirmations") cannot be used as a basis on which to build systems of scientific knowledge ([1930] 1979, 382): "Upon affirmations no logically tenable structure can be erected, for they are already gone at the moment building begins."[1]

Faced with this difficulty of foundationalist justification, we could try to escape by giving up on the business of empirical justification altogether. However, I do not think that is a palatable option. In the context of physical measurements, I can see only two ways to avoid the issue of justification altogether. First, we could adopt a simplistic type of conventionalism, in which we just decide to measure quantity Q by method M. Then M is the correct method, by fiat. One might think that something like the meter stick (or the standard kilogram, etc.), chosen by a committee, embodies such a conventionalist strategy. That would be ignoring all the considerations that went into selecting the particular platinum alloy for the meter stick, which led to the conclusion that it was the most robust and least changeable of all the available materials. Simplistic conventionalism comes with an unacceptable degree of arbitrariness. But going over to a more sophisticated form of conventionalism will bring back the necessity for justification; for example, in Henri Poincaré's conventionalism, one must justify one's judgment about which definitions lead to the simplest system of laws.

The second method of eliminating the question of justification is the kind of extreme operationalism that I examined critically in "Beyond Bridgman" in chapter 3, according to which every measurement method is automatically correct because it defines its own concept. The problem with that solution is that it becomes mired in endless specifications that would actually prevent measurement altogether, because any variation whatsoever in the operation would define a new concept. To get a flavor of that problem, review the following passage from Bridgman:

> So much for the length of a stationary object, which is complicated enough. Now suppose we have to measure a moving street car. The simplest, and what we may call the 'naïve' procedure, is to board the car with our meter stick and repeat the operations we would apply to a stationary body.... But here there may be new questions of detail. *How shall we jump on to the car with our stick in hand?* Shall we run and jump on from behind, or shall we let it pick us up in front? Or perhaps

---

[1] Nonetheless, in the "protocol sentence debate" within the Vienna Circle, Schlick remained a foundationalist in opposition to Neurath, arguing that affirmations still served as the foundation of the empirical testing of knowledge.

does now the material of which the stick is composed make a difference, although previously it did not? (Bridgman 1927, 11; emphasis added)

This kind of worrying is very effective as a warning against complacency, but not conducive to plausible practice. As it is impossible to specify all of the potentially relevant circumstances of a measurement, it is necessary to adopt and justify a generally characterized procedure.

Therefore, we have no choice but to continue seeking justification, despite the absence of any obvious self-justifying foundations. Such perseverance can only lead us to *coherentism*. I use the word in the following sense, once again adopting the formulation by Foley (1998, 157): "Coherentists deny that any beliefs are self-justifying and propose instead that beliefs are justified in so far as they belong to a system of beliefs that are mutually supportive." We have seen coherentism in positive action in each of the preceding chapters. The simplest cases were seen in chapter 2, where Regnault employed the criterion of "comparability" to rule out certain thermometers as candidates for indicators of real temperature, and in chapter 3, where the "mutual grounding" of various measurement methods served as a strategy for extending the concept of temperature into far-out domains. In the remainder of this chapter I wish to articulate a particular version of coherentism that can serve as a productive framework for understanding scientific progress. (However, that articulation in itself does not constitute a vindication of coherentism over foundationalism.)

Before moving on to the articulation of a progressive coherentism, it is interesting to note that the chief foundationalist metaphor of erecting a building on the firm ground actually points to coherentism, if the physical situation in the metaphor is understood correctly. The usual foundationalist understanding of the building metaphor is as outdated as flat-earth cosmology. There was allegedly a certain ancient mythological picture of the universe in which a flat earth rested on the back of very large elephants and the elephants stood on the back of a gigantic turtle. But what does the turtle stand on? We can see that the question of what stands at the very foundation is misplaced, if we think about the actual shape of the earth on which we rest. We build structures outward on the round earth, not upward on a flat earth. We build on the earth because we happen to live on it, not because the earth is fundamental or secure in some ultimate sense, nor because the earth itself rests on anything else that is ultimately firm. The ground itself is not grounded. The earth serves as a foundation simply because it is a large, solid, and dense body that coheres within itself and attracts other objects to it. In science, too, we build structures around what we are first given, and that does not require the starting points to be absolutely secure. On reflection, the irony is obvious: foundationalists have been sitting on the perfect metaphor for coherentism!

In fact, this metaphor of building on a round earth has a great potential to help us make coherentism more sophisticated. It is more useful than Quine's membrane or even Neurath's boat (both mentioned in "Mutual Grounding as a Growth Strategy" in chapter 3), because those established metaphors of coherentism do not convey any sense of hierarchical structure. Although our round earth is not an immovable foundation, the gravitational physics of the earth still gives us a sense of direction—"up" and "down" at a given location, and "inward" and "outward" overall. That direction is a powerful constraint on our building activities, and the constraint can be a very useful one as well (consider all the difficulties of building a space station, where that constraint is absent). This constraint provides a clear overall direction of progress, which is to build up (or outward) on the basis of what is already put down (or within). There is a real sense in which elements of the inner layers support the elements of the outer layers, and not vice versa.[2] Round-earth coherentism can incorporate the most valid aspects of foundationalism. It allows us to make perfect sense of hierarchical justification, without insisting that such justification should end in an unshakeable foundation or fearing that it is doomed to an infinite regress.[3]

Making Coherentism Progressive: Epistemic Iteration

So far I have argued that the quest for justification is bound to lead to coherentism. But the real potential of coherentism can be seen only when we take it as a philosophy of *progress*, rather than *justification*. Of course there are inherent links between progress and justification, so what I am advocating is only a change of emphasis or viewpoint, but it will have some real implications. This reorientation of coherentism was already hinted in the contrast between the two treatments of fixed points given in "The Validation of Standards" and "The Iterative Improvement of Standards" in chapter 1. The question I would like to concentrate on is how to go on in the development of scientific knowledge, not how to justify what we already have.

In the framework of coherentism, inquiry must proceed on the basis of an affirmation of some existing system of knowledge. That point has been emphasized by a wide variety of major philosophers, including Wittgenstein (1969), Husserl (1970), Polanyi (1958), and Kuhn (1970c). (As Popper conjectured, the historical beginning of this process was probably inborn expectations; the question of ultimate origin is not very important for my current purposes.) Starting from an existing system of knowledge means building on the achievements of some actual past group of intelligent beings. As Lawrence Sklar (1975, 398–400) suggested tentatively (in a curiously foundationalist metaphor), a "principle of conservatism" may

---

[2]But it is possible occasionally to drill underneath to alter the structure below, as long as we do not do it in such a way as to make the platform above collapse altogether. See Hempel 1966, 96.

[3]One interesting question that remains is how often scientific inquiry might be able to get away with pretending that certain assumptions are indubitable truths. Recall Duhem's view (discussed in "Measurement, Circularity, and Coherentism") that the physiologist does not need to worry about the correctness of the principles of physics that underwrite the correct functioning of his measuring instruments. Although foundationalism may not work as a general epistemology, there are scientific situations that are in effect foundationalist. To return to the metaphor of buildings, it is quite true that most ordinary building work is done as if the earth were flat and firmly fixed. It is important to discern the types of situations in which the breakdown of foundationalism does or does not affect scientific research in a significant way.

be "a foundation stone upon which all justification is built." This gives knowledge an indelibly historical character. The following analogy, used by Harold Sharlin (1979, 1) to frame his discussion of William Thomson's work, has general significance:

> The father–son relationship has an element that is analogous to the historical basis of scientific research. The son has his father to contend with, and he rejects him at his peril. The scientific tradition may obstruct modern science, but to deny that tradition entirely is to undermine the basis for scientific investigations. For the son and a new generation of scientists, there are two courses open: submit to the past and be a duplicate hemmed in by the lessons of someone else's experience, or escape. Those who seek to escape the past without doing violence to the historical relationship between the present and the past are able to maintain their independence and make original contributions.

I summarized a similar insight in the "principle of respect" in "The Validation of Standards" in chapter 1. It is stronger than what William G. Lycan (1988, 165–167, 175–176) calls the "principle of credulity," which only says that a belief one holds initially should not be rejected without a reason, but it should be rejected whenever there is a reason, however insignificant. The principle of respect does not let the innovator off so easily. Those who respect the affirmed system may have quite strong reasons for rejecting it, but will continue to work with it because they recognize that it embodies considerable achievement that may be very difficult to match if one starts from another basis.

The initial affirmation of an existing system of knowledge may be made uncritically, but it can also be made while entertaining a reasonable suspicion that the affirmed system of knowledge is imperfect. The affirmation of a known system is the only option when there is no alternative that is clearly superior. A simple example illustrates this point. Fahrenheit made some important early experimental contributions to the study of specific heats, by mixing measured-out amounts of fluids at different initial temperatures and observing the temperature of the resulting mixture. In these experiments he was clearly aware of an important source of error: the initial temperature of the mixing vessel (and the thermometer itself) would have an effect on the outcome. The only way to eliminate this source of error was to make sure that the mixing vessel started out at the temperature of the resulting mixture, but that temperature was just what the experiment was trying to find out. The solution adopted by Fahrenheit was both pragmatic and profound at once. In a letter of 12 December 1718 to Boerhaave, he wrote:

> (1) I used wide vessels which were made of the thinnest glass I could get. (2) I saw to it that these vessels were heated to approximately the same temperature as that which the liquids assumed when they were poured into them. (3) I had learned this approximate temperature from some tests performed in advance, and found that, if the vessel were not so approximately heated, it communicated some of its own temperature (warmer or colder) to the mixture. (van der Star 1983, 80–81)

I have not been able to find a record of the exact procedure of approximation that Fahrenheit used. However, the following reconstruction would be a possibility, and would be quite usable independently of whether Fahrenheit used it himself. Start with the vessel at the halfway temperature between the initial temperatures of the hot and the cold liquids. Measure the temperature of the mixture in that experiment, and then set the vessel at that temperature for the next experiment, whose outcome will be slightly different from the first. This procedure could be repeated as many times as desired, to reduce the error arising from the initial vessel temperature as much as we want. In the end the initial vessel temperature we set will be nearly identical to the temperature of the mixture. In this series of experiments, we knowingly start with an ill-founded guess for the outcome, but that guess serves as a starting point from which a very accurate result can be reached.

This is an instance of what I have named "epistemic iteration" in "The Iterative Improvement of Standards" in chapter 1, which I characterized as follows: "Epistemic iteration is a process in which successive stages of knowledge, each building on the preceding one, are created in order to enhance the achievement of certain epistemic goals.... In each step, the later stage is based on the earlier stage, but cannot be deduced from it in any straightforward sense. Each link is based on the principle of respect and the imperative of progress, and the whole chain exhibits innovative progress within a continuous tradition." Iteration provides a key to understanding how knowledge can improve without the aid of an indubitable foundation. What we have is a process in which we throw very imperfect ingredients together and manufacture something just a bit less imperfect. Various scientists and philosophers have noted the wonderful, almost-too-good-to-be-true nature of this process, and tried to understand how it works. I have already mentioned Peirce's idea about the self-correcting character of knowledge in "The Iterative Improvement of Standards" in chapter 1. George Smith (2002, 46) argues convincingly that an iterative engagement with empirical complexities was what made Newton's system superior to his competitors': "In contrast to the rational mechanics of Galileo and Huygens, the science coming out of the *Principia* tries to come to grips with actual motions in all their complexity—not through a single exact solution, however, but through a sequence of successive approximations." Among contemporary philosophers, it is perhaps Deborah Mayo (1996) who has made the most extensive efforts to explain the nature of self-correction and "ampliative inference."[4]

Of course, there is no guarantee that the method of epistemic iteration will always succeed. A danger inherent in the iterative process is the risk of self-destruction (cf. Smith 2002, 52). Since the initially affirmed system is subject to modification, there is a possibility that the validity of the inquiry itself will be jeopardized. How can it be justifiable to change elements of the initially affirmed system, which is the very basis of our attempt at making progress? It is not that there is any problem about changing one's mind in science. The concern is that the whole process might become a morass of self-contradiction. What we need to ensure is that the changes in the initially affirmed system do not invalidate the very outcomes that prompted the changes. Whether that is possible is a contingent empirical

---

[4] For a quick introduction to Mayo's ideas, see my review of her major work (Chang 1997).

question for each case. If all attempted iterative improvements to a system result in self-contradiction, that may be taken as a failure of the system itself. Such repeated self-destruction is as close as we can get to an empirical falsification of the initially affirmed system. In earlier chapters we have had various glimpses of such rejection of initially affirmed beliefs: for instance, the essential fluidity of mercury ("Can Mercury be Frozen?" and "Can Mercury Tell Us Its Own Freezing Point?" in chapter 3), Irvine's doctrine of heat capacity ("Theoretical Temperature before Thermodynamics" in chapter 4), and the linearity of the expansion of alcohol (end of "Regnault: Austerity and Comparability" in chapter 2). If a system of knowledge is judged to be unable to support any progressive inquiry, that is a damning verdict against it.

When iteration is successful, how do we judge the degree of progress achieved by it? Here one might feel a hankering back to foundationalism, in which self-justifying propositions can serve as sure arbiters of truth; then we may have a clear sense in which scientific progress can be evaluated, according to how closely we have approached the truth (or at least how well we have avoided falsity). Without an indubitable foundation, how will we be able to judge whether we have got any closer to the truth? What we need to do here is look away from truth. As even the strictest foundationalists would admit, there are a variety of criteria that we can and should use in judging the merits of systems of knowledge. These criteria are less than absolute, and their application is historically contingent to a degree, but they have considerable force in directing our judgments.

In Carl Hempel's (1966, 33–46) discussion of the criteria of empirical confirmation and the acceptability of hypotheses in the framework of hypothetico-deductivism, many aspects that go beyond simple agreement between theory and observation are recognized as important factors. First of all Hempel stresses that the quality of theory-observation agreement has to be judged on three different criteria: the quantity, variety, and precision of evidence. In addition, he gives the following as criteria of plausibility: simplicity, support by more general theories, ability to predict previously unknown phenomena, and credibility relative to background knowledge. Thomas Kuhn (1977, 322) lists accuracy, consistency, scope, simplicity, and fruitfulness as the "values" or "standard criteria for evaluating the adequacy of a theory," which allow the comparative judgment between competing paradigms despite incommensurability. Bas van Fraassen (1980, 87) mentions elegance, simplicity, completeness, unifying power, and explanatory power only to downgrade these desirables as mere "pragmatic virtues," but even he would not suggest that they are without value, and others argue that these pragmatic virtues can be justificatory. William Lycan (1988; 1998, 341) gives the following as examples of epistemic or "theoretical" virtues: simplicity, testability, fertility, neatness, conservativeness, and generality (or explanatory power). I will refer to all of these various criteria of judgment as "epistemic values" or "epistemic virtues," using the terms somewhat interchangeably. (The same list of things will be referred to as epistemic *values* when they are taken as criteria by which we make judgments on systems of knowledge, and as epistemic *virtues* when they are taken as good qualities possessed by the systems of knowledge.)

Whether and to what extent an iterative procedure has resulted in progress can be judged by seeing whether the system of knowledge has improved in any of its epistemic virtues. Here I am defining progress in a pluralistic way: the enhancement of any feature that is generally recognized as an epistemic virtue constitutes progress. This notion of progress might be regarded as inadequate and overly permissive, but I think it is actually specific enough to serve most purposes in philosophy of science, or in science itself. I will quickly dispense with a few major worries. (1) A common normative discourse will not be possible if people do not agree on the list of recognized epistemic values, but there is actually a surprising degree of consensus about the desirability of those epistemic values mentioned earlier. (2) Even with an agreement on the values there will be situations in which no unequivocal judgment of progress can be reached, in which some epistemic values are enhanced and others are diminished. However, in such situations there is no reason why we should expect to have an unequivocal verdict. Some have attempted to raise one epistemic virtue (or one set of epistemic virtues) as the most important one to override all others (for example, van Fraassen's empirical adequacy, Kuhn's problem-solving ability, and Lakatos's novel predictions), but no consensus has been reached on such an unambiguous hierarchies of virtues.[5] (3) One might argue that the only virtue we can and should regard as supreme over all others is truth, and that I have gone astray by setting truth aside in the first place when I entered into the discussion of epistemic virtues. I follow Lycan (1988, 154–156) in insisting that the epistemic virtues are valuable in their own right, regardless of whether they lead us to the truth. Even if truth is the ultimate aim of scientific activity, it cannot serve as a usable criterion of judgment. If scientific progress is something we actually want to be able to assess, it cannot mean closer approach to the truth. None of the various attempts to find usable watered-down truthlike notions (approximate truth, verisimilitude, etc.) have been able to command a consensus.[6]

Fruits of Iteration: Enrichment and Self-Correction

Let us now take a closer look at the character of scientific progress that can be achieved by the method of epistemic iteration. There are two modes of progress enabled by iteration: *enrichment*, in which the initially affirmed system is not negated but refined, resulting in the enhancement of some of its epistemic virtues; and *self-correction*, in which the initially affirmed system is actually altered in its content as a result of inquiry based on itself. Enrichment and self-correction often occur simultaneously in one iterative process, but it is useful to consider them separately to begin with.

---

[5]See Kuhn 1970c, 169–170 and 205; Van Fraassen 1980, 12 and ch. 3; Lakatos 1968–69 and Lakatos 1970, or Lakatos [1973] 1977 for a very quick introduction.
[6]Psillos (1999, ch. 11) gives a useful summary and critique of the major attempts in this direction, including those by Popper, Oddie, Niiniluoto, Aronson, Harré and Way, and Giere. See also Psillos's own notion of "truth-likeness" (pp. 276–279), and Boyd 1990, on the notion of approximate truth and its difficulties.

Enrichment

How enrichment works can be illustrated in the following true story. Driving along the scenic Route 2 in Western Massachusetts on the way from Boston to my alma mater, Northfield Mt. Hermon School, I used to spot a wonderfully puzzling road sign that announced: "Bridge St. Bridge." It seemed to designate a bridge named after the street that was named after the bridge itself. For a few years I passed this sign every so often and wondered how that name ever came to be. It finally occurred to me that I could do some research in local history to find out the story, but from that point on I could not find that sign again, nor remember which town it was in. Nevertheless, I did arrive at the following plausible historical hypothesis. Initially the town was so small that there was only one bridge (referred to as "The Bridge") and no named streets. Then came enough streets so they had to be named, and the street leading to the bridge was named "Bridge Street," naturally. Then came other bridges, necessitating the naming of bridges. One of the easiest ways of naming bridges is by the streets they connect to (as in the "59th Street Bridge" celebrated in the Simon and Garfunkel song also known as "Feelin' Groovy"). When that was done our original bridge was christened: Bridge Street Bridge! If my hypothesis is correct, the apparent circular nonsense in that name is only a record of a very sensible history of iterative town development.

We have seen iterative enrichment in action first of all in the process of quantifying the operational temperature concept, analyzed in chapter 1. Initially the judgment of temperature was only qualitative, based on the sensation of hot and cold. Then came thermoscopes, certified by their broad agreement with sensations; thermoscopes allowed a decisive and consistent comparison and ordering of various phenomena possessing different temperatures. Afterwards numerical thermometers, arising iteratively from thermoscopes, went further by attaching meaningful numbers to the degrees of hot and cold. In this developmental process temperature evolved from an unquantified property to an ordinal quantity, then to a cardinal quantity. Each stage built on the previous one, but added a new dimension to it. A very similar type of development was seen in the process of creating temperature standards in the realm of the very hot and the very cold in chapter 3. The chief epistemic virtue enhanced in these processes can be broadly termed "precision." However, this story also reveals the complexity of what we casually think of as precision. Going from purely qualitative to the quantitative is certainly enhancement of precision, but so is the move from the ordinal to the cardinal, as is the increase in numerical precision after the move into the cardinal. All three of these enhancements occurred in the invention of the thermometer as we know it.

A different aspect of iterative enrichment was also seen in the extension of the temperature scale discussed in chapter 3. After the establishment of the numerical temperature concept by means of thermometers operating in a narrow band of temperatures, that concept was extended to previously inaccessible domains by means of thermometers capable of withstanding extreme conditions. The establishment of a temperature standard in a new domain shows the familiar process of going from the qualitative (this time based on simple instrumental operations, rather than pure sensation) to the ordinal to the cardinal, in that new domain. An overview of the whole temperature scale gives us a picture of iterative extension, in which the concept in the initial domain is preserved but augmented into new domains. The chief epistemic virtue enhanced in this process is scope. (An extension of scope also occurred already in the replacement of sensation by thermoscopes, discussed in "The Validation of Standards" and "The Iterative Improvement of Standards" in chapter 1.) These are just some illustrative examples of iterative enrichment. I predict that many others will be seen if we examine other areas of scientific progress in the same vein.

Self-Correction

The other major aspect of iterative progress, self-correction, can also be illustrated at first with a tale taken from everyday life (with a slight exaggeration). Without wearing my glasses, I cannot focus very well on small or faint things. Therefore if I pick up my glasses to examine them, I am unable to see the fine scratches and smudges on them. But if I *put on* those same glasses and look at myself in the mirror, I can see the details of the lenses quite well. In short, my glasses can show me their own defects. This is a marvelous image of self-correction. But how can I trust the image of defective glasses that is obtained through the very same defective glasses? In the first instance, my confidence comes from the sensible clarity and acuity of the image itself, regardless of how it was obtained. That gives me some reason to accept, provisionally, that certain defects in the glasses somehow do not affect the quality of the image seen (even when the image is of those defects themselves). But there is also a deeper layer in this mechanism of self-correction. Although at first I delight in the fact that my glasses can give me clear, detailed pictures despite its defects, on further observation I realize that some defects do distort the picture, sometimes recognizably. Once that is realized, I can attempt to correct the distortions. For example, a large enough smudge in the central part of a lens will blur the whole picture seen through it, including the image of the smudge itself. So when I see a blurry smudge on my left lens in the mirror, I can infer that the boundaries of that smudge must actually be sharper than they seem. I could go on in that way, gradually correcting the image on the basis of certain features of the image itself that I observe. In that case my glasses would not only tell me their own defects but also allow me to get an increasingly accurate understanding of those defects.

In chapter 2 we saw how an initial assumption of the correctness of a certain type of thermometer could defeat itself, by producing observations that show the lack of comparability between different individual thermometers of that same type. In that episode all that was possible was falsification rather than positive correction, but in "The Validation of Standards" and "The Iterative Improvement of Standards" in chapter 1 we saw how a later standard initially based on a prior standard could proceed to overrule and correct that prior standard. This can be regarded as a self-correction of the prior standard, if we take the later standard as an evolved version of the prior standard. In chapter 4 we saw various instances of self-correction, most clearly exhibited in the Callendar–Le Chatelier method of operationalizing the absolute temperature concept, in which the initial assumption that actual gases obeyed the ideal gas law was used in the calculation of their deviation from it (see "Accuracy through Iteration" for further details). In his critique of Heinrich Hertz's formulation of mechanics, Simon Saunders has elucidated a very similar process for the improvement of time measurement; in short, "the accuracy of the time-piece must itself be revisable in accordance with dynamical principles." Saunders points out that:

> we have *only* an iterative process, and it is only guaranteed to yield determinate results given the assumptions of the theory, including estimates on interplanetary matter densities and the rapid fall-off of tidal effects of stellar objects.... [T]he notion of 'test' in mechanics is in large part a question of consistency in application under systematically improving standards of precision.[7]

I think many more examples will be found if we make a careful survey of the exact sciences.

Tradition, Progress, and Pluralism

I have articulated in this chapter a particular mode of scientific progress: epistemic iteration in the framework of coherentism. Before closing the discussion of coherentist iteration, I would like to add a few brief observations pertaining to the politics of science. Almost by definition epistemic iteration is a *conservative* process because it is based on the principle of respect, which demands the affirmation of an existing system of knowledge. However, the conservatism of iteration is tempered by a pervasive *pluralism*. There are several aspects to this pluralism.

First of all, the principle of respect does not dictate which system of knowledge one should affirm initially. One can certainly start by affirming the currently orthodox system, as the modern climate in science encourages. However, even orthodoxy is a choice one makes. Nothing ultimately forces us to stay entirely with the system in which we have been brought up. In Kuhn's description of "normal science" it is assumed that there is only one paradigm given to the scientists within a given scientific discipline, but I think it is important to recognize that orthodoxy can be rejected without incurring nihilism if we can find an alternative pre-existing system to affirm as a basis of our work. That alternative system may be an earlier version of the current orthodoxy, or a long-forgotten framework dug up from the history of science, or something imported from an entirely different tradition. Kuhn has argued persuasively that paradigm shifts occur as a consequence of adhering to the orthodox paradigm and pushing it until it breaks, but he has not argued that sticking faithfully to a paradigm is the only legitimate way, or even the most effective way, of moving on to a different paradigm.

The affirmation of an existing system does not have to be wholesale, either. Scientists do often adopt something like a Kuhnian paradigm that consists of an entire "disciplinary matrix," and such an inclination may also be a pervasive trend

---

[7] See Saunders 1998, 136–142; the quoted passages are from 137 and 140 (emphasis original). I thank an anonymous referee for Oxford University Press for pointing out this work to me.

in professionalized modern science. However, affirmation does not have to be so complete, as we can in fact see by going back to Kuhn's own original meaning of "paradigm": an exemplary piece of work that is widely emulated, an achievement that "some particular scientific community acknowledges for a time as supplying the foundation for its further practice" (Kuhn 1970c, 10; Kuhn 1970b, 271–272). The emulation of an exemplar will not necessarily generate a comprehensive disciplinary matrix. Also the emulation of the same exemplar by different people can lead to different outcomes, in the absence of strict educational and professional enforcement. Even when there are communities that adhere to paradigms in the more complete sense, anyone not belonging fully in such a community has no obligation to adopt all elements of its dominant paradigm. And if there are competing communities that study the same subject, then the individual has the option of creating a coherent hybrid system to affirm.

It is also possible to choose the *depth* of affirmation. For example, if there is a sufficient degree of despair or disillusionment about all existing systems of knowledge, scientists may decide to reject all systems that are overly developed and start again with the affirmation of something that seems more basic and secure. This is how we should understand most cases of phenomenalism or positivism when it appears in the actions of practicing scientists, as it often did in nineteenth-century physics and chemistry. Disillusioned with overly intricate and seemingly fruitless theories about the microphysical constitution and behavior of matter, a string of able scientists (e.g. Wollaston, Fourier, Dulong, Petit, Carnot, Regnault, Mach, and Duhem) retreated to more observable phenomena and concepts closely tied to those phenomena (cf. "Regnault and Post-Laplacian Empiricism" in chapter 2). But even the positivists, to the extent they were practicing scientists, started by affirming existing systems, just not such metaphysically elaborate ones.

Finally, the affirmation of an existing system does not fix the direction of its development completely. The point is not merely that we do not know which direction of development is right, but that there may be no such thing as the correct or even the best direction of development. As noted in "Making Coherentism Progressive," the desire to enhance different epistemic virtues may lead us in different directions, since enhancing one virtue can come at the price of sacrificing another. Even when we only consider the enhancement of one given epistemic virtue, there may be different ways of achieving it, and more than one way of achieving it equally well. We are often led away from this pluralistic recognition by an obsession with truth because mutually incompatible systems of knowledge cannot all be true. The achievement of other virtues is not so exclusive. There can be different ways of enhancing a certain epistemic virtue (e.g., explanatory power or quantitative precision in measurement) that involve belief in mutually incompatible propositions. Generally speaking, if we see the development of existing knowledge as a creative achievement, it is not so offensive that the direction of such an achievement is open to some choice.

All in all, the coherentist method of epistemic iteration points to a *pluralistic traditionalism*: each line of inquiry needs to take place within a tradition, but the researcher is ultimately not confined to the choice of one tradition, and each tradition can give rise to many competing lines of development. The methodology of epistemic iteration allows the flourishing of competing traditions, each of which can progress on its own basis without always needing to be judged in relation to others. This pluralism should be distinguished clearly from any reckless relativism. In an amorphous type of coherentism, any self-consistent system of knowledge is deemed equally valid; in contrast, the coherentism I advocate is driven by the imperative of progress, so each tradition is continually judged by its record of enhancing various epistemic virtues. Even Feyerabend (1975, 27), with his "anarchism" in epistemology, was concerned with assessing systems of knowledge in terms of progress made, though he declined to define "progress" explicitly. In the coherentist framework of iterative progress, pluralism and traditionalism can coexist happily. It is doubtful whether the intellectual and social constraints governing specialist communities of professional scientists can allow a full functioning of the freedom inherent in epistemic iteration. However, not all scientific activity is subject to those constraints. This last insight provides the basis of my kind of work in history and philosophy of science, as I will elaborate further in chapter 6.

The Abstract and the Concrete

The abstract insights summarized and developed in this chapter arose from the concrete studies contained in earlier chapters, but they are not simple generalizations from the few episodes examined there. It would be foolish to infer how science in general does or should progress from what I have seen in a small number of particular episodes, all from the same area of science. The problem of generalization continually plagues the attempts to integrate the history of science and the philosophy of science, so I think at least a few brief remarks are in order here outlining my view on how that problem can be dealt with.

I agree with Lakatos (1976) as far as saying that all historiography of science is philosophical. Abstract ideas emerge from concrete studies because they are necessary ingredients of narratives. Abstract ideas are needed for the full understanding of even just one concrete episode; it is a mistake to think that they can be eliminated by a conscientious avoidance of generalizations. We cannot understand actions, not to mention judge them, without considering them in abstract terms (such as "justified," "coherent," "observation," "measurement," "simple," "explanation," "novel," etc., etc.). An instructive concrete narrative cannot be told at all without using abstract notions in the characterization of the events, characters, circumstances, and decisions occurring in it. Therefore what we do when we extract abstract insights from a particular episode is not so much generalization as an articulation of what is already present. It may even be an act of self-analysis, in case our episode was initially narrated without much of an awareness of the abstractions that guided its construction.

In each episode studied in this book I have been asking abstract questions. Most generally put, the recurring questions have concerned the building and justification of human knowledge: How can we say we really know what we know? How can we come to know more and better than we did before? The asking of these questions requires an abstract conception of "justification" and "progress," even if the intended answers are only about particular episodes. The questions retain their abstract character even when they are narrowed down to particular epistemic values, because those values are also abstract.

In this chapter I have advanced one central abstract idea: epistemic iteration is a valid and effective method of building scientific knowledge in the absence of infallible foundations. But I have not proposed anything resembling "the scientific method." The ideas I advanced in this chapter are abstract, but they are not presumed to have universal applicability. What I have attempted is to identify one particular method by which science can progress, and discern the circumstances under which that method can fruitfully be applied. That does not rule out other methods that may be used as alternatives or complements. It is important not to confuse the *abstract* and the *universal*. An abstraction is not general until it is applied widely.

An abstract idea needs to show its worth in two different ways. First, its *cogency* needs to be demonstrated through abstract considerations and arguments; that is what I have started to do in this chapter for the idea that epistemic iteration is a good method of progress when there are no firm foundations to rely on. Second, the *applicability* of the idea has to be demonstrated by showing that it can be employed in the telling of various concrete episodes in instructive ways. The idea of epistemic iteration helped me in understanding the possibility of scientific progress in each of the historical episodes explored in this book. Much more concrete work is needed, of course, for getting a good sense of the general extent of its applicability.

6

Complementary Science—History and Philosophy of Science as a Continuation of Science by Other Means

> Criticism is the lifeblood of all rational thought.
>
> Karl Popper, "Replies to My Critics," 1974

> To turn Sir Karl's view on its head, it is precisely the abandonment of critical discourse that marks the transition to a science.
>
> Thomas S. Kuhn, "Logic of Discovery or Psychology of Research?" 1970

This book has been an attempt to open up a new way of improving our knowledge of nature. If I have been successful in my aim, the studies contained in the preceding chapters of this book will defy classification along traditional disciplinary lines: they are at once historical, philosophical, and scientific. In the introduction I gave a very brief characterization of this mode of study as *complementary science*. Having engaged in several concrete studies, I am now ready to attempt a more extensive and in-depth general discussion of the aims and methods of complementary science. The focus here will be to present complementary science as a productive direction in which the field of history and philosophy of science can advance, without denying the importance of other directions. Such a programmatic statement has a threefold aim. First, it will state explicitly some goals that have already been motivating much work in history and philosophy of science, including my own. Second, a strong statement of these goals will hopefully stimulate further work directed toward them. Finally, a clear definition of the mode of study I am

---

Some of the ideas elaborated here were initially published in Chang 1999.

advocating may encourage other related modes of study to be defined more clearly in opposition or comparison.[1]

The Complementary Function of History and Philosophy of Science

My position can be summarized as follows: history and philosophy of science can seek to generate scientific knowledge in places where science itself fails to do so; I will call this the *complementary* function of history and philosophy of science, as opposed to its *descriptive* and *prescriptive* functions. Lest the reader should reach an immediate verdict of absurdity, I hasten to add: by the time I have finished explaining the sense of the above statement, some peculiar light will have been thrown on the sense of the expressions "generate," "scientific knowledge," "science," "fails," and "history and philosophy of science" itself. (In the following discussion I will use the common informal abbreviation "HPS" for history and philosophy of science, not only for brevity but also in order to emphasize that what I envisage is one integrated mode of study, rather than history of science and philosophy of science simply juxtaposed to each other. HPS practiced with the aim of fulfilling its complementary function will be called *HPS in its complementary mode* or, synonymously, *complementary science* as I have already done in the introduction.)

In tackling the question of purpose, one could do much worse than start by looking at the actual motivations that move people: why does anyone want to study such a thing as HPS, even devote an entire lifetime to it? Here the only obvious starting point I have is myself, with a recognition that different people approach the field with different motivations. What drove me initially into this field and still drives me on is a curious combination of delight and frustration, of enthusiasm and skepticism, about science. What keeps me going is the marvel of learning the logic and beauty of conceptual systems that had initially seemed alien and nonsensical. It is the admiration in looking at routine experimental setups and realizing that they are actually masterpieces in which errors annihilate each other and information is squeezed out of nature like water from rocks. It is also the frustration and anger at the neglect and suppression of alternative conceptual schemes, at the interminable calculations in which the meanings of basic terms are never made clear, and at the necessity of accepting and trusting laboratory instruments whose mechanisms I have neither time nor expertise to learn and understand.

Can there be a common thread running through all of these various emotions? I think there is, and Thomas Kuhn's work gives me a starting point in articulating it. I am one of those who believe that Kuhn's ideas about normal science were at least as important as his ideas about scientific revolutions. And I feel an acute dilemma about normal science. I think Kuhn was right to emphasize that science as we know it can only function if certain fundamentals and conventions are taken for granted

---

[1]The expository models I wish to emulate for these purposes are the "Vienna Circle Manifesto" of the logical positivists (Neurath et al. [1929] 1973), and David Bloor's statement of the strong program in the sociology of scientific knowledge (1991, ch. 1).

and shielded from criticism, and that even revolutionary innovations arise most effectively out of such tradition-bound research (see Kuhn 1970a, Kuhn 1970b, etc.). But I also think Karl Popper was right to maintain that the encouragement of such closed-mindedness in science was "a danger to science and, indeed, to our civilization," a civilization that often looks to science as the ideal form of knowledge and even a guide for managing social affairs (Popper 1970, 53). The practice of HPS as a complement to specialist normal science offers a way out of this dilemma between destroying science and fostering dogmatism. I believe that this is one of the main functions that HPS could serve, at once intellectual and political.

In other words, a need for HPS arises from the fact that specialist science[2] cannot afford to be completely open. There are two aspects to this necessary lack of openness. First, in specialist science many elements of knowledge must be taken for granted, since they are used as foundations or tools for studying other things. This also means that certain ideas and questions must be suppressed if they are heterodox enough to contradict or destabilize those items of knowledge that need to be taken for granted. Such are the necessities of specialist science, quite different from a gratuitous suppression of dissent. Second, not all worthwhile questions can be addressed in specialist science, simply because there are limits to the number of questions that a given community can afford to deal with at a given time. Each specialist scientific community will have some degree of consensus about which problems are most urgent, and also which problems can most plausibly be solved. Those problems that are considered either unimportant or unsolvable will be neglected. This is not malicious or misguided neglect, but a reasonable act of prioritization necessitated by limitations of material and intellectual resources.

All the same, we must face up to the fact that suppressed and neglected questions represent a loss of knowledge, actual and potential. The complementary function of HPS is to recover and even create such questions anew and, hopefully, some answers to them as well. Therefore the desired result of research in HPS in this mode is an enhancement of our knowledge and understanding of nature. HPS can recover useful ideas and facts lost in the record of past science, address foundational questions concerning present science, and explore alternative conceptual systems and lines of experimental inquiry for future science. If these investigations are successful, they will complement and enrich current specialist science. HPS can enlarge and deepen the pool of our knowledge about nature; in other words, HPS can generate scientific knowledge.

The following analogy may be helpful in illustrating my ideas about this complementary function of HPS, though it is rather far-fetched and should not be pushed beyond where it ceases to be useful. The most cogent argument for maintaining capitalism is that it is the best known economic system for ensuring high productivity and efficiency which, in the end, translate into the satisfaction of human needs and desires. At the same time, hardly anyone would deny the need for

---

[2] From here on I will speak of "specialist science" rather than "normal science," so that my discussion would be acceptable even to those who reject Kuhn's particular ideas about normal science or paradigms.

philanthropy or a social welfare system that ameliorates the inevitable neglect of certain human needs and the unreasonable concentration of wealth in a capitalist economy. Likewise, we cannot do without specialist science because we do not know any other method of producing knowledge so effectively. At the same time, we also cannot deny the need to offset some of the noxious consequences of producing knowledge in that manner, including the neglect and suppression of certain questions and the unreasonable concentration of knowledge to a small intellectual elite. Forcing specialist science to be completely open would destroy it, and that would be analogous to anarchy. A better option would be to leave specialist science alone within reasonable limits, but to offset its undesirable effects by practicing complementary science alongside it. In that way HPS can maintain the spirit of open inquiry for general society while the specialist scientific disciplines pursue esoteric research undisturbed.

Philosophy, History, and Their Interaction in Complementary Science

Having explained my basic ideas about the complementary function of HPS, I would like to take a step back and consider more carefully what it means to do historical and philosophical studies of science. Consider philosophy first. It is often claimed that good science should be philosophical as well as technical, and indeed we are still less than two centuries away from the time when scientists routinely referred to themselves as "philosophers." On the other hand, it is also true that most scientists today would regard most discussions currently taking place in professional philosophy as utterly irrelevant to science. The relation between science and philosophy is certainly complex, and this complexity adds to the confusion in trying to see clearly what it is that we are trying to do in the philosophy of science.

I propose taking the philosophy of science as a field in which we investigate scientific questions that are not addressed in current specialist science—questions that could be addressed by scientists, but are excluded due to the necessities of specialization. In Kuhnian terms, science does not emerge from pre-science until the field of legitimate questions gets narrowed down with clearly recognized boundaries. For a long time it was common for one and the same treatise to contain tangled discussions of metaphysics, methodology, and what we would now identify as the proper "content" of science. Some may yearn for those good old days of natural philosophy, but it is not plausible to turn back the clock. Philosophy once aspired to encompass all knowledge, including what we now recognize as science. However, after various scientific disciplines (and other practices such as law and medicine) gradually carved themselves out, what is left under the rubric of philosophy is not the all-encompassing scholarship it once was. Our current academic discipline called "philosophy" became restricted and defined, as it were, against its own will. Philosophy as practiced now does not and cannot include science. But in my view that is just where its most important function now lies: to address what science and other specialisms neglect.

The last thought throws some interesting light on the general nature of philosophy. We tend to call something a question "philosophical" if it is something that we do not normally deal with in the course of routine action although, on reflection, it is relevant to the practice. Similarly, when we say "the philosophy of X," we often mean a discipline which deals with questions that are relevant to another discipline X but normally not addressed in X itself. There are various reasons why relevant questions may be excluded from a system of thought or practices. The questions may be too general; they may threaten some basic beliefs within the system; asking them may be pointless because every specialist knows and agrees on the correct answers; the answers may not make any significant practical difference; and so on. And in the end, questioning has to be selective because it is simply impossible to ask the infinity of all possible questions. But philosophy can function as the embodiment of the ideal of openness, or at least a reluctance to place restrictions on the range of valid questions.

Something very similar can also be said about the history of science. The similarity has two sources: in past science, there are some things that modern science regards as incorrect, and some other things that modern science regards as unnecessary. As scientific research moves on, much of science's past gets lost in a curious mix of neglect and suppression. Instrumental and mathematical techniques are often handed down to younger generations that happily disregard the arguments that had to be settled before those tools could be accepted. Awkward questions tend to be withdrawn after a period in which no clear answers are found, and defeated theories and worldviews are suppressed. Even when old facts and conclusions are retained, the assumptions, arguments, and methods that originally led to them may be rejected. The official "histories" that appear as mere garnishes in many science textbooks are more than happy to leave out all of these tedious or embarrassing elements of the past. They are left to the professional historians of science. Therefore, when the history of science asserts its independence from science itself, its domain is apt to be defined negatively, to encompass whatever elements of past science that current science cares not to retain in its institutional memory.

Given these considerations, it should not come as a surprise that philosophical questions about science and historical questions about science are co-extensive to a considerable degree. This area of overlap provides a strong rationale for practicing HPS as an integrated discipline, not as a mere juxtaposition of the history of science and the philosophy of science. What are regarded as philosophical questions nowadays are quite likely to have been asked in the past as scientific questions; if so, the philosophical questions are simultaneously topics for historical inquiry as well. Whether an investigation in HPS is initially stimulated by philosophical or historical considerations, the result may well be the same.

There are two obvious methods of initiating inquiry in the complementary mode of HPS, or, complementary science. They are obvious because they are rooted in very standard customs in philosophy and history of science. The first method, which has been my primary mode of questioning in this book, is to reconsider things that are taken for granted in current science. As anyone who has been exasperated by philosophers knows, skeptical scrutiny can raise doubts on just about anything. Some of these philosophical doubts can be fruitful starting points for historical inquiry, as it is quite possible that past scientists in fact addressed the same doubts in the process of the initial establishment of those taken-for-granted elements of modern science. This method is quite likely to focus attention on aspects of past science that may easily escape the notice of a historian who is not driven by the same problematic. After the historical record is established, philosophy can take its turn again to reassess the past arguments that have been unearthed. In that way philosophical analysis can initiate and guide interesting historical studies in the category of what I call "problem-centered narratives." This use of philosophy in history of science is very different from the use of historical episodes as empirical evidence in support of general philosophical theses about how science works.

The second method of initiating inquiry in complementary science is to look out for apparently unusual and puzzling elements in past science. This is something that historians of science have become very accustomed to doing in recent decades. History is probably one of the sharpest tools available to the philosopher wishing to explore the presuppositions and limitations of the forms of scientific knowledge that are almost universally accepted now. The historical record often shows us fresh facts, questions, and ways of thinking that may not occur to us even in the course of an open critical scrutiny of current science. In order to facilitate this possibility, we can actively seek elements of past science that have not survived into modern science. After those elements are identified, it is important to investigate the historical reasons for their rejection and assess the philosophical cogency of those reasons.

These processes of historical-philosophical inquiry are intertwined and self-perpetuating, since they will reveal further philosophical concerns and previously unknown bits of history that can stimulate other lines of inquiry. After some thinking about research in complementary science, and certainly while one is immersed in it, it becomes difficult to see where philosophy ends and history begins or vice versa. Philosophy and history work together in identifying and answering questions about the world that are excluded from current specialist science. Philosophy contributes its useful habits of organized skepticism and criticism, and history serves as the supplier of forgotten questions and answers. History of science and philosophy of science are inseparable partners in the extension and enrichment of scientific knowledge. I propose to call the discipline they form together *complementary science* because it should exist as a vital complement to specialist science.

The Character of Knowledge Generated by Complementary Science

Having explained the basic motivations for complementary science and the nature of the historical and philosophical studies that constitute it, I must now give a more detailed defense of the most controversial aspect of my vision. I have claimed that complementary science can *generate* scientific knowledge where science itself fails to do so. On the face of it, this sounds absurd. How could any knowledge about nature be generated by historical or philosophical studies? And if complementary science does generate scientific knowledge, shouldn't it just be counted as part of science, and isn't it foolhardy to suggest that such scientific activity could be undertaken by anyone but properly trained specialists? Such a sense of absurdity is understandable, but I believe it can be dispelled through a more careful consideration of what it means to generate knowledge. I will make such a consideration in this section, with illustrations from the material covered in previous chapters and occasional references to other works. There are three main ways in which complementary science can add to scientific knowledge, which I will address in turn.

Recovery

First of all, history can teach us about nature through the recovery of forgotten scientific knowledge. The potential for such recovery is shown amply in the material uncovered in chapter 1. Many investigators starting from De Luc in the late eighteenth century knew that pure water did not always boil at the "boiling point" even under standard pressure. They built up a growing and sophisticated body of knowledge about the "superheating" of water and other liquids that took place under various circumstances, and at least in one case observed that boiling could also take place slightly under the boiling point as well. But by the end of the nineteenth century we witness Aitken's complaint that authoritative texts were neglecting this body of knowledge, either through ignorance or through oversimplification. Personally, I can say that I have received a fair amount of higher education in physics at reputable institutions, but I do not recall ever learning about the superheating of water and the threat it might pose to the fixity of the boiling point. All I know about it has been learned from reading papers and textbooks from the eighteenth and nineteenth centuries. I predict that most readers of this book will have learned about it from here for the first time.

This is not to say that knowledge of superheating has been lost entirely to modern science. The relevant specialists do know that liquid water can reach temperatures beyond the normal boiling point without boiling, and standard textbooks of physical chemistry often mention that fact in passing.[3] Much less commonly noted is the old observation that water that is actually boiling can have various temperatures deviating from the standard boiling point. There are vast numbers of scientifically educated people today who do not know anything about these very basic and important phenomena. In fact, what they do claim to know is that superheating does not happen, when they unsuspectingly recite from their textbooks that pure water always boils at 100°C under standard atmospheric pressure. Most people are not taught about superheating because they do not need to know about it. As explained in "The Defense of Fixity" in chapter 1, the routine conditions under which thermometers are calibrated easily prevent superheating, so that people who use thermometers or even those who make thermometers need not

---

[3]See, for example, Oxtoby et al. 1999, 153; Atkins 1987, 154; Silbey and Alberty 2001, 190; Rowlinson 1969, 20. Interestingly, the explanations of superheating they offer are quite diverse, though not necessarily mutually contradictory. Silbey and Alberty attribute it to the collapse of nascent vapor bubbles due to surface tension (cf. De Luc's account of "hissing" before full boil). According to Atkins it occurs "because the vapor pressure inside a cavity is artificially low," which can happen for instance when the water is not stirred. But Oxtoby et al. imply that superheating can only occur when water is heated rapidly.

have any knowledge of superheating. Only those whose business it is to study changes of state under unusual circumstances need to be aware of superheating. This is a case of knowledge that is not widely remembered because knowing it does not help the pursuit of most of current specialist research.

There is another category of experimental knowledge that tends to get lost, namely facts that actively disturb our basic conceptual schemes. The best example of this category that I know is Pictet's experiment discussed in "Temperature, Heat, and Cold" in chapter 4, in which there is an apparent radiation and reflection of rays of cold, as well as rays of heat. This experiment received a good deal of attention at the time and it seems that most people who were knowledgeable about heat in the early nineteenth century knew about it, but gradually it became forgotten (see Chang 2002 and references therein). Nowadays only the most knowledgeable historians of that period of physics seem to know about this experiment at all. Unlike superheating, the radiation of cold is not a phenomenon recognized by most modern specialists on heat and radiation, to the best of my knowledge. It just does not fit into a scheme in which heat is a form of energy and cold can only be a relative deficit of energy, not something positive; remembering the existence of cold radiation will only create cognitive dissonance for the energy-based specialist.

When we make a recovery of forgotten empirical knowledge from the historical record, the claimed observation of the seemingly unlikely phenomenon is likely to arouse curiosity, if not suspicion. Can water really reach 200°C without boiling, as observed by Krebs?[4] Other people's observations can and should be subjected to doubt when there is good reason; otherwise we would have to take all testimony as equally valid, whether they be of N-rays, alien abductions, or spontaneous human combustion. Radical skepticism would lead us to conclude that there is no way to verify past observations, but more pragmatic doubts would lead to an attempt to re-create past experiments where possible.

In conducting the studies included in this book, I have not been in a position to make any laboratory experiments. However, historians of science have begun to re-create various past experiments.[5] Most of those works have not been carried out for complementary-scientific reasons, but the potential is obvious. One case that illustrates the potential amply is the replication of Pictet's experiment on the radiation and reflection of cold, published by James Evans and Brian Popp in the *American Journal of Physics* in 1985, in which they report (p. 738): "Most physicists, on seeing it demonstrated for the first time, find it surprising and even puzzling." Through this work, Evans and Popp brought back the apparent radiation and reflection of cold as a recognized real phenomenon (though they do not regard it as a manifestation of any positive reality of "cold"). However, all indications are that it

---

[4] Rowlinson (1969, 20) actually notes a 1924 experiment in which a temperature of 270°C was achieved.

[5] Salient examples include the replication of Coulomb's electrostatic torsion-balance experiment by Peter Heering (1992, 1994), and Joule's paddle-wheel experiment by H. Otto Sibum (1995). Currently William Newman is working on repeating Newton's alchemical experiment, and Jed Buchwald has been teaching laboratory courses at MIT and Caltech in which students replicate significant experiments from the history of science.

was quickly forgotten all over again, or not noticed very much. This is not only based on my own patchy impressions of what people do and do not seem to know. A search in the combined Science Citation Index (Expanded), the Social Sciences Citation Index and the Arts and Humanities Citation Index, conducted in March 2003, turned up only two citations. One was a one-paragraph query published in the Letters section of a subsequent number of the *American Journal of Physics* (Penn 1986), and the other was my own article on this subject (Chang 2002)!

The recovery of forgotten knowledge is not restricted to facts, but extends to ideas as well (and it is, after all, very difficult to separate facts and ideas cleanly). In fact, historians of science for many decades have made great efforts to remember all sorts of ideas that have been forgotten by modern science. This kind of recovery is the mainstay of the history of science, so much so that there is no point in picking out a few examples out of the great multitude. But in order for the recovered ideas to enter the realm of complementary science, we need to get beyond thinking that they are merely curious notions from the past that are either plainly incorrect or at least irrelevant to our own current knowledge of nature. I will be considering that point in more detail later.

The consideration of recovery raises a basic question about what it means for knowledge to exist. When we say we have knowledge, it must mean that *we* have knowledge; it is no use if the ultimate truth about the universe was known by a clan of people who died off 500 years ago without leaving any records or by some space aliens unknown to us. Conversely, in a very real sense, we create knowledge when we give it to more people. And the acquisition of the "same" piece of knowledge by every new person will have a distinct meaning and import within that individual's system of beliefs. When it comes to knowledge, dissemination is a genuine form of creation, and recovery from the historical record is a form of dissemination—from the past to the present across a gap created by institutional amnesia, bridged by the durability of paper, ink, and libraries.

Critical Awareness

Superficially, it might appear that much of the work in complementary science actually undermines scientific knowledge because it tends to generate various degrees of doubt about the accepted truths of science, as we have seen in each of the first three chapters of this book. Generating doubt may seem like the precise opposite of generating knowledge, but I would argue that constructive skepticism can enhance the quality of knowledge, if not its quantity. If something is actually uncertain, our knowledge is superior if it is accompanied by an appropriate degree of doubt rather than blind faith. If the reasons we have for a certain belief are inconclusive, being aware of the inconclusiveness prepares us better for the possibility that other reasons may emerge to overturn our belief. With a critical awareness of uncertainty and inconclusiveness, our knowledge reaches a higher level of flexibility and sophistication. Strictly speaking, complementary science is not necessary for such a critical awareness in each case; in principle, specialist scientists could take care not to forget the imperfection of existing knowledge. However, in practice it is going to be very difficult for specialists to maintain this kind of critical vigilance on the foundations of their own practice, except in isolated cases. The task is much more easily and naturally undertaken by philosophers and historians of science.

Even philosophers tend not to recognize critical awareness and its productive consequences as contributions to scientific knowledge. But there philosophy is underselling itself. There is a sense in which we do not truly know anything unless we know how we know it, and on reflection few people would doubt that our knowledge is superior when we are also aware of the arguments that support our beliefs, and those that undermine them. That is not incompatible with the fact that such superior knowledge can constitute a hindrance in the achievement of certain aims that require an effective non-questioning application of the knowledge. I am not able to give a full-fledged argument as to why critical awareness makes superior knowledge, but I will at least describe more fully what I believe in this regard, especially in relation to the fruits of complementary science.

For example, there is little that deserves the name of knowledge in being able to recite that the earth revolves around the sun. The belief carries more intellectual value if it is accompanied by the understanding of the evidence and the arguments that convinced Copernicus and his followers to reject the firmly established, highly developed, and eminently sensible system of geocentric astronomy established by Ptolemy, as detailed by Kuhn (1957) for instance. This is exactly the kind of scientific knowledge that is not available in current specialist science but can be given by HPS. There are many other examples in which work in HPS has raised and examined very legitimate questions about the way in which certain scientific controversies were settled. For example, many scholars have shown just how inconclusive Antoine Lavoisier's arguments against the phlogiston theory were.[6] Gerald Holton (1978) revealed that Robert Millikan was guided by an ineffable intuition to reject his own observations that seemed to show the existence of electric charges smaller than what he recognized as the elementary charge belonging to an individual electron. Allan Franklin (1981) has furthered this debate by challenging Holton's analysis (see also Fairbank and Franklin 1982). Klaus Hentschel (2002) has shown that there were sensible reasons for which John William Draper maintained longer than most physicists that there were three distinct types of rays in the sunbeam.[7] I once added a small contribution in this direction, by showing the legitimate reasons that prompted Herbert Dingle to argue that special relativity did not predict the effect known as the "twin paradox" (Chang 1993).

There is no space here to list all the examples of HPS works that have raised the level of critical awareness in our scientific knowledge. However, I cannot abandon the list without mentioning the thriving tradition in the philosophy of modern

---

[6]In my view, the most convenient and insightful overview of this matter is given by Alan Musgrave (1976). According to Musgrave, the superiority of Lavoisier's research program to the phlogiston program can only be understood in terms of Lakatos's criterion of progress. Morris (1972) gives a detailed presentation of Lavoisier's theory of combustion, including its many problems.

[7]See also Chang and Leonelli (forthcoming) for a further sympathetic discussion of Draper's reasons.

physics, in which a community of philosophers have been questioning and re-examining the orthodox formulations and interpretation of various theories, especially quantum mechanics. Works in this tradition are often criticized as being neither philosophy nor physics. I think that criticism is understandable, but misguided. Much of the work in the philosophy of modern physics should be regarded as valuable works of complementary science, not as poor pieces of philosophy that do not address general and abstract philosophical concerns sufficiently. An exemplary instance of what I have in mind is James Cushing's (1994) scrutiny of the rejection of the Bohmian formulation of quantum mechanics.

Coming back to the topics discussed in this book, the critical awareness achieved in complementary science is best illustrated in chapter 2. There it was revealed that scientists found it impossible to reach a conclusive positive solution to the problem of choosing the correct thermometric fluid, though Regnault's comparability criterion was effective in ruling out most alternatives except for a few simple gases. Similarly, in chapter 3 we saw that the extension of the thermometric scale to the realms of the very hot and the very cold suffered from similar problems, and that scientists forged ahead without being able to say conclusively which of the competing standards were correct. That is how matters stood at least until Kelvin's concept of absolute temperature was operationalized in the late nineteenth century, as discussed in chapter 4. But the discussion in that chapter showed the futility of the hope that a highly theoretical concept of temperature would eliminate the inconclusiveness in measurement, since the problem of judging the correctness of operationalization was never solved completely, though the iterative solution adopted by the end of the nineteenth century was admirable. And in chapter 1 it was shown that even the most basic task of finding fixed points for thermometric scales was fraught with difficulties that only had serendipitous solutions. I would submit that when we know everything discussed in the first four chapters of this book, our scientific knowledge of what temperature means and how it is measured is immeasurably improved.

New Developments

Recovery and critical awareness are valuable in themselves, but they can also stimulate the production of genuinely novel knowledge. Historians have generally shrunk from further developing the valid systems of knowledge that they uncover from the past record of science. The most emblematic example of such a historian is Kuhn. Having made such strenuous and persuasive arguments that certain discarded systems of knowledge were coherent and could not be pronounced to be simply incorrect, Kuhn gave no explicit indication that these theories deserved to be developed further. Why not? According to his own criterion of judgment, scientific revolutions constitute progress when the newer paradigm acquires a greater problem-solving ability than ever achieved by the older paradigm (Kuhn 1970c, ch. 13). But how do we know that the discrepancy in problem-solving ability is not merely a result of the fact that scientists abandoned the older paradigm and gave up the effort to improve its problem-solving ability? A similar question also arises at the conclusion of some other historians' works on scientific controversy. For example,

Steven Shapin and Simon Schaffer (1985) strongly challenged the received wisdom that Thomas Hobbes's ideas about pneumatics were rightly rejected, in favor of the superior knowledge advanced by Robert Boyle. But they gave no indication that it would be worthwhile to try developing Hobbes's ideas further.

The historian of science, of course, has an easy answer here: it is not the job of the historian to develop scientific ideas actively. But whose job is it? It is perfectly understandable that current specialist scientists would not want to be drawn into developing research programs that have been rejected long ago, because from their point of view those old research programs are, quite simply, wrong. This is where complementary science enters. Lacking the obligation to conform to the current orthodoxy, the complementary scientist is free to invest some time and energy in developing things that fall outside the orthodox domain. In this book, or elsewhere, I have not yet engaged very much in such new developments. That is partly because a great deal of confidence is required to warrant this aspect of complementary science, and I have only begun to gain such confidence in the course of writing this book. But some clues have already emerged for potential future work, which I think are worth noting here.

One clear step is to extend the experimental knowledge that has been recovered. We can go beyond simply reproducing curious past experiments. Historians of science have tended to put an emphasis on replicating the conditions of the historical experiments as closely as possible. That serves the purpose of historiography, but does not necessarily serve the purpose of complementary science. In complementary science, if a curious experiment has been recovered from the past, the natural next step is to build on it. This can be done by performing better versions of it using up-to-date technology and the best available materials, and by thinking up variations on the old experiments that would not only confirm but extend the old empirical knowledge. For example, various experiments on boiling, discussed in chapter 1, would be worth developing further. In another case, I have proposed some instructive variations of Count Rumford's ingenious experiments intended to demonstrate the positive reality of what he called "frigorific radiation," following Pictet's experiment on the apparent radiation of cold (Chang 2002, 163). I have not had the resources with which to perform those experiments, but I hope there will be opportunities to carry them out.

Less demanding of resources but mentally more daring would be new theoretical developments. For example, in "Theoretical Temperature without Thermodynamics?" in chapter 4, I made a brief suggestion on how a theoretical concept of temperature might be defined on the basis of the phenomenalistic physics of gases, without relying on thermodynamics or any other highly abstract theories. Less specifically, in my article on the apparent radiation of cold I registered a view that Rumford's theory of calorific-frigorific radiation would be worth developing further, just to see how far we could take it (Chang 2002, 164). Similarly, in a forthcoming article (Chang and Leonelli) on the debates on the nature of radiation, I make an allowance that there may be useful potential in reviving for further development the pluralistic theory postulating different sets of rays responsible for the illuminating, heating, and chemical effects of radiation. These are very tentative suggestions, and not necessarily very plausible lines of inquiry, but I mention them in order to illustrate the kind of developments that may be possible when complementary science reaches its maturity.

The realm of theoretical development is where the complementary scientist is likely to face the greatest degree of objection or incomprehension. If an idea proposed in complementary science does not conform to the currently orthodox view of the directions in which productive new developments are likely to come, specialists will dismiss it out of hand as wrong, implausible, or worthless in some unspecified way. But complementary science is inherently a pluralistic enterprise. Although there may be some past systems of knowledge that are quite beyond the horizon of meaningful revival because they have become so disconnected from even everyday beliefs of the modern world, there is no unthinking dismissal of theoretical possibilities in complementary science. If we look back at a decision made by past scientists and there seems to be room for reasonable doubt, that is a plausible indication that what was rejected in that decision may be worth reviving. When the complementary scientist picks up a rejected research program to explore its further potential, or suggests a novel research program, that is also not done with the crank's conviction that his particular heresy represents the only truth. And if specialists should ever choose to adopt an idea originating from complementary science, they may want to adopt it as the undisputed truth; however, that would still not change the fact that complementary science itself is not in the truth business.

Relations to Other Modes of Historical and Philosophical Study of Science

There are many modes of study that take place under the rubric of the history of science or the philosophy of science. My goal has been to articulate the complementary mode of HPS, not to deny the importance of other modes by any means. Conversely, the complementary mode must not be rejected simply because its aims are different from those adopted in other modes.

In this connection I have one immediate worry. To many historians of science, what I am proposing here will seem terribly retrograde. In recent decades many exciting works in the fields of history and sociology of science have given us valuable accounts of the sciences as social, economic, political, and cultural phenomena. HPS as I am proposing here may seem too internalistic, to the exclusion of the insights that can be gained from looking at the contexts in which science has developed and functioned. The important distinction to be stressed, however, is that HPS in its complementary mode is not *about* science. Its aims are continuous with the aims of science itself, although the specific questions that it addresses are precisely those not addressed by current science; that is why I call it complementary science. HPS in its complementary mode is not meant to be an incomplete sort of history that ignores the social dimension; it is ultimately a different kind of enterprise altogether from the social history of science. One might even say it is not history at all, because history does not in the first instance seek to further our understanding of nature, while complementary science does. I cannot emphasize too strongly that I do not intend to deny the essential importance of understanding science as a social phenomenon, but I also believe that the complementary function of HPS is a distinct and meaningful one.

If we grant that the complementary mode of HPS is legitimate and useful, it will be helpful to clarify its character further by comparing and contrasting it with some other modes of HPS that bear some similarity to it.

Sociology of scientific knowledge. Perhaps curiously, complementary science has one important aspect in common with the sociology of scientific knowledge (SSK): the questioning of accepted beliefs in science. The reinvestigation of familiar facts can be seen as a process of opening Bruno Latour's (1987) "black box" and revealing the character of "science in action." But there is a clear difference between the intended outcomes of such questioning in SSK and in complementary science. SSK deflates the special authority of science as a whole by reducing the justification of scientific beliefs to social causes. In contrast, the aim of skepticism and anti-dogmatism in complementary science is the further enhancement of particular aspects of scientific knowledge. In some cases work in complementary science may show some past scientific judgments to have been epistemically unfounded, but that is different from SSK's methodological refusal to recognize a distinction between epistemically well founded and unfounded beliefs.[8]

Internal history. From the concrete studies I have offered, it will be obvious that much of what I regard as the past achievement of HPS in its complementary mode comes from the tradition of the internal history of science. Is complementary science simply a continuation of that tradition, in which one tries to uncover and understand scientific knowledge faithfully as it existed in the past? There is one important reason why it is not. If we pursue internal history for its own sake, our ultimate aim must be the discovery of some objective historical truth, about what past scientists believed and how they thought. This is not the final aim of complementary science, which only makes use of the internal history of science in order to increase and refine our current knowledge. One significant difference stemming from this divergence of aims is that complementary science does not shrink from making normative epistemic evaluations of the quality of past science, which would be anathema to the "new" internal history of science.[9] Still, complementary science is by no means committed to Whiggism, since the judgments made by the historian-philosopher can very easily diverge from the judgments made by the current specialist scientists.

Methodology. Complementary science is also distinct from the search for "the scientific method," namely the most effective, reliable, or rational method of gaining

---

[8]David Bloor (1991, 7) states this point unequivocally, as one of the main tenets of the strong program in the sociology of scientific knowledge: "It would be impartial with respect to truth and falsity, rationality or irrationality, success or failure. Both sides of these dichotomies will require explanation." Moreover, "the same types of cause would explain, say, true and false beliefs."

[9]I am referring to the anti-Whiggish type of internal history that Kuhn once designated as the "new internal historiography" of science ([1968] 1977, 110).

knowledge about nature. This may sound puzzling, considering that a good deal of the discussion in my concrete studies was very much about scientific methodology, and all of chapter 5 was devoted to it. Studies in complementary science can and do involve questions about the methods of acquiring knowledge, but there is a significant difference of focus to be noted. The attitude toward methodology taken in complementary science is much like most practicing scientists' attitude toward it: methodology is not the primary or final goal of inquiry. What we call good methods are those methods that have produced useful or correct results; this judgment of goodness comes retrospectively, not prospectively. In other words, methodological insights are to be gained as by-products of answering substantive scientific questions; when we ask a question about nature, how we find an answer is part of the answer. In complementary science we do not set down general methodological rules for science to follow. We only recognize good rules by seeing them in action, as successful strategies perhaps worth trying elsewhere, too.

Naturalistic epistemology. Finally, complementary science must be distinguished from a strong trend in current philosophy of science, which is to give a characterization of science as a particular kind of epistemic activity, without a commitment to normative implications (see Kornblith 1985). This trend probably arises at least partly in reaction to the apparent futility of trying to dictate methodology to scientists. The "naturalistic" impulse is to an extent congenial to complementary science because it provides a strong motivation for an integrated HPS. But what naturalistic epistemology fosters is HPS in the *descriptive* mode, which takes science primarily as a naturally existing object of description. In contrast, for HPS in the complementary mode, the ultimate object of study is nature, not science.

A Continuation of Science by Other Means

In closing, I would like to return briefly to the relation between specialist science and complementary science. One big question that I have not discussed sufficiently so far is whether complementary science is an enterprise that is critical of orthodox specialist science, and more broadly, what normative dimensions there are to the complementary function of HPS. This is a difficult question to answer unequivocally, and I think the subtlety of the issue can be captured as follows: complementary science is *critical* but not *prescriptive* in relation to specialist science.

There are two different dimensions to the critical stance that complementary science can take toward specialist science. First, when complementary science identifies scientific questions that are excluded by specialist science, it is difficult to avoid the implication that we would like to have those questions answered. That is already a value judgment on science, namely that it does not address certain questions we consider important or interesting. However, at least in a large number of cases, this judgment also comes with the mitigating recognition that there are good reasons for specialist science to neglect those questions. That recognition prevents the step from judgment to prescription. The primary aim of complementary science is not to tell specialist science what to do, but to do what specialist science is presently unable to do. It is a shadow discipline, whose boundaries change exactly so as to encompass whatever gets excluded in specialist science.[10]

The second dimension of the critical stance is more controversial, as I have discussed in "The Character of Knowledge Generated by Complementary Science." On examining certain discarded elements of past science, we may reach a judgment that their rejection was either for *imperfect reasons* or for *reasons that are no longer valid*. Such a judgment would activate the most creative aspect of complementary science. If we decide that there are avenues of knowledge that were closed off for poor reasons, then we can try exploring them again. At that point complementary science would start creating parallel traditions of scientific research that diverge from the dominant traditions that have developed in specialist science. It is important to note that even such a step falls short of a repudiation of current specialist science. Since we do not know in advance whether and to what degree the complementary traditions might be successful, the act of creating them does not imply any presumption that it will lead to superior results to what the specialists have achieved since closing off the avenues that we seek to reopen. (All of this is not to deny that there are possible situations that would call for a prescriptive mode of HPS, in which we question whether science is being conducted properly, and propose external intervention if the answer is negative.)

Complementary science could trigger a decisive transformation in the nature of our scientific knowledge. Alongside the expanding and diversifying store of current specialist knowledge, we can create a growing complementary body of knowledge that combines a reclamation of past science, a renewed judgment on past and present science, and an exploration of alternatives. This knowledge would by its nature tend to be accessible to non-specialists. It would also be helpful or at least interesting to the current specialists, as it would show them the reasons behind the acceptance of fundamental items of scientific knowledge. It may interfere with their work insofar as it erodes blind faith in the fundamentals, but I believe that would actually be a beneficial effect overall. The most curious and exciting effect of all may be on education. Complementary science could become a mainstay of science education, serving the needs of general education as well as preparation for specialist training.[11] That would be a most far-reaching step, enabling the educated public to participate once again in building the knowledge of our universe.

---

[10]That is not to say that those boundaries are completely sharp. The boundaries of complementary science will be fuzzy, to the extent that the boundaries of science are fuzzy. But the existence of gray areas does not invalidate the distinction altogether. Also, someone who is primarily a specialist scientist may well engage in some complementary scientific work and vice versa; that is no stranger than a scientist exploring the artistic dimensions of scientific work.

[11]The importance of the history and philosophy of science to "liberal" science education has been argued by many authors, as documented thoroughly by Michael Matthews (1994). For me the chief inspiration comes from the vision behind James Bryant Conant's general education program at Harvard, and its extension by Gerald Holton and his associates (see the introduction to Conant 1957, and Holton 1952).

Glossary of Scientific, Historical, and Philosophical Terms

Items marked with an asterisk (*) are terms that I have coined myself or to which I have given nonstandard meanings; the rest are standard terms or terms introduced by other authors as specified.

Absolute temperature. When William Thomson (Lord Kelvin) crafted the concept of absolute temperature starting from 1848, his main intention was to make a temperature concept that did not refer to the properties of any particular material substances. In modern usage this meaning is conflated with the sense of counting up from the absolute zero, which indicates the complete absence of heat. The latter was in fact an earlier concept, advocated for example by Guillaume Amontons and William Irvine. Cf. *Amontons temperature*; *Irvinism*.

Absolute zero. See *absolute temperature*, *Amontons temperature*.

*Abstraction. The omission of certain properties in the description of an object. Abstraction is not to be confused with *idealization*.

Adiabatic gas law. The law describing the behavior of a gas expanding or contracting without any exchange of heat with the external environment. The standard expression is $pv^\gamma = $ constant, where γ is the ratio between the *specific heat* of the gas at constant pressure and the specific heat at constant volume.

Adiabatic phenomena. Phenomena occurring to a system that is thermally isolated from its environment. Best known were adiabatic heating and cooling, in which a gas is heated by compression without any heat being added to it and cooled by expansion without any heat being taken away from it. These phenomena are now commonly considered as striking demonstrations of the interconversion of heat and mechanical energy, but that interpretation was only proposed by James Joule around 1850. One of the most convincing explanations of adiabatic phenomena before Joule was that given by John Dalton, which was based on *Irvinist caloric theory*.

*Amontons temperature. Air-thermometer temperature counted from the absolute zero. The idea is due to Guillaume Amontons who, by extrapolation of the known trend of gases to lose pressure gradually by cooling, predicted that at a certain point the pressure value would hit zero, and understood such presumed disappearance of pressure as an indication of a complete absence of heat. Therefore, in the Amontons scale the zero-pressure point is recognized as the absolute zero of temperature.

Atomic heat. See *Dulong and Petit's law.*

Auxiliary hypothesis. An additional hypothesis that is used to enable the deduction of an observable consequence from the main hypothesis that one aims to test. When the main hypothesis is apparently falsified by empirical observations, one can always defend it by shifting the blame on to auxiliary hypotheses. Cf. *holism.*

Boiling point. The temperature at which a liquid boils. In the context of thermometry, an unspecified "boiling point" usually refers to the boiling point of water. Although the boiling point of a pure liquid under fixed pressure is now commonly assumed to be constant, that was widely known not to be the case during the eighteenth and nineteenth centuries.

Boyle's law, or *Mariotte's law.* A regularity attributed to Robert Boyle or Edme Mariotte, which states that the product of pressure and volume is constant for a body of gas at constant temperature ($pv = $ const.).

Bumping (*soubresaut* in French). A noisy and unstable type of boiling, in which large isolated bubbles of vapor rise occasionally, either one at a time or several in an irregular pattern. The temperature is unstable, dropping when the large bubbles are produced and rising again while no bubbles form.

Caloric. The material substance of heat. The term was coined by Antoine-Laurent Lavoisier as part of his new chemical nomenclature in the 1780s. Although the majority of chemists and physicists at the time agreed that heat was a material substance and many people used the term "caloric," there were considerable differences among the various theories of caloric. Cf. *chemical caloric theory; Irvinism.*

Calorimetry. The quantitative measurement of the amounts of heat. The most common methods in the eighteenth and nineteenth centuries were *ice calorimetry* and *water calorimetry.*

Cannon-boring experiment. The experiment in which Count Rumford (Benjamin Thompson) showed that a great deal of heat was generated in the process of grinding through a metal cylinder with a blunt borer. Rumford first noticed the heat generated by such friction while supervising the boring of cannons in the Munich arsenal, when he was charged with running the Bavarian army. Contrary to later legend, this experiment did not make a convincing refutation of the *caloric* theory, although it was well known and widely debated.

Cardinal scale. A scale of measurement that assigns numbers that are suitable for arithmetic operations. Cf. *ordinal scale.*

Carnot engine, Carnot cycle. The ideal and abstract heat engine first conceived by Sadi Carnot. See the section "William Thomson's Move to the Abstract" in chapter 4 for a full explanation.

Centigrade scale, Celsius scale. The centigrade scale is the common scale in which the freezing/melting point of water was defined as 0° and the boiling/steam point of water as 100°. This is commonly attributed to Anders Celsius, but his original scale, although centigrade, had the boiling point as 0° and the freezing point as 100°. There are disputes about who first conceived the centigrade scale with its modern direction of numbers.

Change of state. See *states of matter.*

Charles's law, or *Gay-Lussac's law.* A regularity attributed to J. A. C. Charles or Joseph-Louis Gay-Lussac, which states that the volume of a gas under fixed pressure varies linearly with temperature.

**Chemical caloric theory.* The tradition of caloric theory, most clearly originating from Antoine-Laurent Lavoisier but also Joseph Black, in which caloric was conceived as a substance capable of combining chemically with other substances (see *combined caloric*).

Coherentism. An epistemological standpoint according to which a belief is justified insofar as it belongs to a system of beliefs that are mutually supportive. Coherentists deny that any beliefs are self-justifying. Cf. *foundationalism.*

Combined caloric. In *chemical caloric theory*, caloric in the state of chemical combination with ordinary matter (cf. *free caloric*). In more phenomenological terms, combined caloric would be described as latent heat. Antoine-Laurent Lavoisier explained the heat in combustion as the release of combined caloric from oxygen gas.

Comparability. The requirement that a good measuring instrument should be self-consistent. When the instrument is taken as a type, comparability means that all instruments of the same type should agree with each other when applied to a given situation. This requirement is based on the *principle of single value*. Victor Regnault used comparability as his chief criterion for testing the correctness of thermometers.

Delisle (de l'Isle) scale. The temperature scale created by Joseph-Nicolas Delisle, French astronomer in Russia. Its zero-point was set at the boiling point of water, and the numbers increased with increasing cold, with the freezing point of water designated as 150°.

Duhem-Quine thesis. See *holism*.

Dulong and Petit's law, or *law of atomic heat*. The empirical observation made by Pierre Dulong and Alexis-Thérèse Petit that the product of atomic weight and *specific heat* (by weight) was roughly the same for all chemical elements. This was often taken to mean that an atom of any element had the same capacity for heat; hence the law was sometimes referred to as the law of "atomic heat."

Dynamic theory of heat, or *mechanical theory of heat*. Any theory of heat in which heat is conceived as a form of motion (cf. *material theory of heat*).

Ebullition. Generally, just another word for boiling. However, for Jean-André De Luc, "true ebullition" was a theoretical concept that designated the boiling that took place purely as an effect of heat, as opposed to ordinary boiling, which took place at a lower temperature because of the action of the air dissolved in water.

École Polytechnique. The elite technical university in Paris, founded in 1794 as part of the educational reforms of the French Revolution. A majority of leading French physical scientists in the first half of the nineteenth century taught or studied there.

*Epistemic iteration. A process in which successive stages of knowledge, each building on the preceding one, are created in order to enhance the achievement of certain epistemic goals. It differs crucially from mathematical iteration in that the latter is used to approach a correct answer that is known, or at least in principle knowable, by other means.

Expansive principle. James Watt's maxim that higher efficiency in a steam engine can be obtained if one allows steam to do work through expanding by its own power.

Fahrenheit scale. The popular temperature scale devised by Daniel Gabriel Fahrenheit, in which the freezing point of water is set at 32° and the boiling point of water at 212°. Fahrenheit's motivation in arriving at such a scale was complicated; see Middleton 1966 for a full account.

Fixed point. A reference point in a thermometric scale, which is given a fixed value by definition. The term is also used to refer to the natural phenomenon that defines such a point.

Foundationalism. The doctrine that epistemic justification has a hierarchical structure. According to foundationalism, some beliefs are self-justifying and as such constitute one's evidence base; others are justified only if they are appropriately supported by these basic beliefs. Cf. *coherentism*.

Free caloric. In *chemical caloric theory*, caloric that is not bound up in chemical combination with ordinary matter (cf. *combined caloric*). In more phenomenological terms, free caloric would be described as *sensible heat*.

Free caloric of space. Pierre-Simon Laplace's term to describe the caloric that was postulated to exist in intermolecular spaces. Such caloric was understood to be in the process of radiative transfer between molecules. Laplace defined temperature as the density of free caloric of space.

Free surface. A surface at which a *change of state* can occur. In John Aitken's view, a change of state did not necessarily take place when the temperature associated with it was reached; only the presence of a free surface enabled the change.

Freezing mixture. A mixture used to obtain very low temperatures. Generally it consisted of an acid mixed with ice (or snow). Freezing mixtures were used for a long time without any satisfactory theoretical understanding of their workings. With the advent of Joseph Black's concept of *latent heat*, it was understood that the addition of the acid melted the ice by lowering its freezing point, and the ice absorbed a great deal of heat from the surroundings in the process of melting.

Freezing point. The temperature at which a liquid freezes. In the context of thermometry, an unspecified "freezing point" usually refers to the freezing point of water. The question of the fixedness of the freezing point of a pure liquid is complicated by the phenomenon of *supercooling*. And although the freezing point of water is commonly considered to be the same as the melting point of ice, there have been doubts on that question.

Heat capacity. The capacity of a material object for containing heat. Operationally the heat capacity of a body is identified as the amount of heat required to raise its temperature by a unit amount, which is *specific heat*. In *Irvinist caloric theory* heat capacity and specific heat were explicitly identified. In the *chemical caloric theory*, however, the identification was denied, since it was considered possible that caloric could become combined (latent) as it entered a body, hence not affecting the thermometer at all. In modern thermal physics the concept of heat capacity is not considered cogent, although in common discourse the term is still often used to mean specific heat.

Hissing (sifflement in French). The pre-boiling phenomenon in which numerous bubbles of vapor rise partway through the body of water, but are condensed back into the liquid state before they reach the surface. This happens when the middle or upper layers of the water are cooler than the bottom layers. The familiar noise from a kettle before full boiling sets in is due to hissing.

Holism (in theory testing). The view, attributed most famously to Pierre Duhem and W. V. O. Quine, that it is impossible to subject an isolated hypothesis to an empirical test. Rather, only a sizeable group of hypotheses can have empirical significance and be subjected to empirical tests. Cf. *auxiliary hypotheses.*

Hypothetico-deductive (H-D) model of theory testing. According to the H-D model, the testing of a theory consists in deducing an observable consequence from it, which can then be compared with actual observations. All empirical theories are hypothetical and can be tested this way.

Ice calorimetry. The measurement of the quantity of heat through the melting of ice. A hot object is introduced into a chamber surrounded by ice at the melting point, and the amount of heat given from the hot object to the ice is calculated as the product of the weight of melted water and the *latent heat* of fusion by weight. The idea was first put into practice by Pierre-Simon Laplace and Antoine-Laurent Lavoisier.

Ideal gas, ideal gas law. The "ideal gas" has been conceived in various ways. The standard calorist conception was that gases came close to exhibiting the pure effect of caloric, since they contained a great deal of caloric and the molecules of ordinary matter were separated so far from each other that the attractions between them became negligible. It was easy to imagine an ideal gas in which that negligible intermolecular interaction vanished altogether; in such a case the thermal expansion of gases was imagined to be perfectly regular. In the *kinetic theory of gases*, the ideal gas was conceived as one in which the molecules were not only noninteracting (except by mutual collision) but pointlike (occupying no space) and perfectly elastic. It was seen that such an ideal gas would obey the ideal gas law, commonly written as $PV = RT$, where R is a constant. The ideal gas law could be considered a synthesis of *Boyle's law* and *Charles's law*.

Idealization. A description in which some properties of an object are set at certain values that are convenient or desirable, though not actually obtained. Idealization is not to be confused with *abstraction*.

Irvinist caloric theory (Irvinism). The influential caloric theory originally conceived by William Irvine. The heart of Irvinism was Irvine's doctrine of heat capacity, which stated that the total amount of caloric contained in a body was a simple product of its heat capacity and its absolute temperature (counted from absolute zero, signifying a complete absence of heat).

Isothermal phenomena. Phenomena that take place at a fixed temperature. For example, the production of steam from water under constant pressure is assumed to be an isothermal process; a gas that expands or contracts isothermally obeys *Boyle's law*.

Iteration. See *epistemic iteration*.

Joule's conjecture. As designated by William Thomson, the idea from James Joule that all of the mechanical work spent in the *adiabatic* compression of a gas is converted into heat. Julius Robert Mayer advanced a similar but broader idea, which came to be known as *Mayer's hypothesis*, of which there are many possible formulations (see Hutchison 1976a).

Joule-Thomson experiment. A series of experiments first performed by James Joule and William Thomson in the 1850s, also known as the porous plug experiment. Joule and Thomson pushed various gases through either a narrow opening or a porous plug and observed the resulting changes in their temperature. The amounts of temperature change were interpreted as indications of the extent to which the actual gases deviated from the *ideal gas*; hence, the results of the Joule–Thomson experiment were also used to calculate corrections of gas thermometer readings in order to obtain values of *absolute temperature*.

Kinetic theory of gases. The popular theory of gases fully developed in the second half of the nineteenth century, in which the gas molecules are understood to be in random motion interrupted by frequent collisions with the walls of containers and with each other. In the kinetic theory, temperature is interpreted to be proportional to the average kinetic energy of the molecules.

Laplacian physics. As characterized by the historian of science Robert Fox (1974), the tradition of physics spearheaded by Pierre-Simon Laplace. It was a "Newtonian" corpuscular tradition, in which various phenomena were understood as resulting from unmediated forces acting between pointlike particles of matter. Laplace reinterpreted the action of *caloric* in this manner, too.

Latent heat. Heat that is absorbed into a body without raising its temperature, for example, when a solid melts and a liquid boils. Latent heat phenomena were most notably observed by Joseph Black. In the *chemical caloric theory*, latent heat was understood as combined caloric. In *Irvinist caloric theory*, latent heat phenomena were attributed to changes in the heat capacities of bodies.

Material theory of heat. Any theory of heat, including any variety of the caloric theory, that regards heat as a material substance (cf. *dynamic theory of heat*).

Mayer's hypothesis. See *Joule's conjecture*.

Mechanical theory of heat. See *dynamic theory of heat*.

Melting point. The temperature at which a solid melts. In the context of thermometry, an unspecified "melting point" usually refers to the melting point of ice. Although the melting point of ice is commonly considered to be the same as the freezing point of water, there have been doubts on that question.

Method of mixtures. A calorimetric method of testing the correctness of thermometers, practiced most effectively by Jean-André De Luc. Measured-out portions of water at two previously known temperatures are mixed, and the calculated temperature of the mixture is compared with the reading given by a thermometer under test. By this method De Luc argued for the superiority of mercury to all other thermometric liquids.

Metrology. The science or art of making measurements.

*Metrological extension. A type of *operational extension* in which the measurement method for a concept is extended into a new domain.

*Metrological meaning. The meaning of a concept that is given by the methods of its measurement (cf. *operational meaning*).

Mixtures. See *method of mixtures.*

Nomic measurement. See *problem of nomic measurement.*

One-point method. Any method of graduating thermometers using only one *fixed point* (cf. *two-point method*). With this method, temperature is measured by noting the volume of the thermometric fluid in relation to its volume at the one fixed-point temperature.

*Ontological principle. Ontological principles are those assumptions that are commonly regarded as essential features of reality within an epistemic community, which form the basis of intelligibility. The justification of an ontological principle is neither by logic nor by experience.

*Operational extension. An aspect of the *semantic extension* of a concept, which consists of the creation of *operational meaning* in a domain where it was previously lacking.

*Operational meaning. The meaning of a concept that is embodied in the physical operations whose description involves the concept. It is broader than *metrological meaning*, although it has often been used to designate metrological meaning by many commentators including Percy Bridgman.

Operationalism, or *operationism.* The philosophical view that the meaning of a concept is to be found primarily or even solely in the methods of its measurement. It has often been attributed to Percy Bridgman, though Bridgman himself denied that he had meant to propose a systematic philosophy while he continued to advocate operational analysis as a method of clarifying one's thinking.

*Operationalization. The process of giving operational meaning to a concept where there was none before. Operationalization may or may not involve the specification of explicit measurement methods.

Ordinal scale. A scale of measurement that only gives an ordering, rather than assignments of full-fledged numbers. An ordinal scale may have numerals attached to it, but those numerals are really names, not true numbers. A *thermoscope* has an ordinal scale. Cf. *cardinal scale.*

Porous plug experiment. See *Joule-Thomson experiment.*

*Pressure-balance theory of boiling. The popular nineteenth-century theory that boiling took place when the vapor pressure of a liquid matched the external pressure. In other words, the liquid starts to boil when it reaches the temperature that allows it to produce vapor with sufficient pressure to overcome the external pressure that would confine it within the liquid.

*Principle of respect. The maxim that a previously established system of knowledge should not be discarded lightly. The acceptance of a previous system is necessary first of all because one needs some starting point in the process of inquiry. It is also based on the recognition that an established system of knowledge probably had some respectable merits that gave it a broad appeal in the first place.

*Principle of single value. The principle that a real physical property cannot take more than one definite value in a given situation. It is the principle behind the criterion of *comparability*. It is an example of an *ontological principle.*

*Problem of nomic measurement. The problem of circularity in attempting to justify a measurement method that relies on an empirical law that connects the quantity to be measured with another quantity that is (more) directly observable. The verification of the law would requires the knowledge of various values of the quantity to be measured, which one cannot reliably obtain without confidence in the method of measurement.

Quicksilver. Mercury, especially conceived as an essentially fluid metal.

Radiation, or *radiant heat.* An unmediated and instantaneous (or nearly instantaneous) transfer of heat across a distance. Radiant heat was studied systematically starting with Marc-Auguste Pictet's experiments and Pierre Prevost's theory of exchanges, around 1790. Well into the nineteenth century, it was commonly understood as *caloric* flying around at great speeds, despite Count Rumford's argument that it consisted in vibrations in the ether.

Réaumur scale. A thermometric scale that was popular in French-speaking Europe, commonly attributed to R. A. F. de Réaumur. The standardized form of this scale was actually due to Jean-André De Luc, who used mercury (Réaumur used alcohol) and graduated the instrument to read 0°R at the freezing/melting point and 80°R at the boiling point (Réaumur used a *one-point method*).

Reductive doctrine of meaning. The idea that the meaning of a concept does, or should, only consist in its method of measurement. This is often considered an essential part of Percy Bridgman's operationalism.

Respect. See *principle of respect.*

Reversibility, reversible engine. A reversible heat engine can be "run backwards," with an expenditure of the same amount of mechanical work that would have been produced by going through the operations in the normal way. In Sadi Carnot's theory, the ideal heat engine is postulated to be completely reversible. The idea behind that requirement was that a transfer of heat across a finite temperature gap (which would not be a reversible process) would involve a waste of potential mechanical effect that could be produced. An aspect of that intuition was preserved in later thermodynamics in the idea that a heat transfer across a temperature gap resulted in an increase in entropy.

*Royal Society committee on thermometry. A seven-person committee, chaired by Henry Cavendish and including Jean-André De Luc as a member, appointed by the Royal Society of London in 1776 to arrive at the definitive method of graduating thermometers, particularly setting the *fixed points.* Its report was published in the following year (Cavendish et al. 1777).

Royal Society scale. A thermometric scale used for a time (at least in the 1730s) in weather observations commissioned by the Royal Society of London. Its origin is unclear. It was marked 0° at "extreme heat" (around 90°F or 32°C), with numbers increasing with increasing cold. This scale has no relation to the later work of the *Royal Society committee on thermometry.*

Saturated vapor. If a body of water evaporates into an enclosed space as much as possible, then the space is said to be "saturated" with vapor. Similarly, if such a maximum evaporation would occur into an enclosed space containing air, the air is said to be saturated. Perhaps more confusingly, it is also said under those circumstances that the vapor itself is saturated.

*Semantic extension. The act of giving any sort of meaning to a concept in a domain where it did not have a clear meaning before. *Operational extension* is a type of semantic extension, and *metrological extension* is a type of operational extension.

Sensible heat. Heat that is perceivable by the senses or detectable by thermometers (cf. *latent heat*). In the *chemical caloric theory,* sensible heat was understood as *free caloric.*

Shooting. The sudden freezing of supercooled water (or another liquid), with ice crystals "shooting out" from a catalytic point. Shooting can be caused by many factors, such as mechanical agitation or the insertion of an ice crystal. The normal result of shooting is the production of just the right amount of ice to release enough *latent heat* to bring up the temperature of the whole to the normal freezing point.

Single value. See *principle of single value.*

Specific heat. The amount of heat required to raise the temperature of an object by a unit amount. The specific heat of a substance is expressed as the specific heat for a unit quantity of that substance. There were difficulties in determining whether the specific heat of a given object or substance was a function of its temperature. For gases there was an additional complication, which was noted when the experimental techniques attained sufficient precision: specific heat under constant pressure was greater than specific heat under constant volume. Cf. *heat capacity*.

Spirit (of wine). Ethyl alcohol, commonly obtained by distilling wine.

States of matter. Solid, liquid, and gas. It was common knowledge that changes of state were caused by the addition or abstraction of heat, but it was Joseph Black who first clearly conceptualized the *latent heat* involved in changes of state. Most calorists shared Lavoisier's idea that the repulsive force of caloric was responsible for the loosening of intermolecular attractions, which was required to turn solids into liquids and liquids into gases.

Steam point. The temperature of steam boiled off from water. There were debates on whether the steam point was the same as the *boiling point*. Henry Cavendish argued in the *Royal Society committee on thermometry* that the steam point (under fixed pressure) was more reliably fixed than the boiling point. Afterwards the steam point was more commonly used than the boiling point in making precision thermometers.

Subtle fluid. An all-pervasive fluid imperceptible to ordinary senses (although its effects may be perceivable). Examples of subtle fluids include caloric, phlogiston, electric and magnetic fluids, and ether. They were also often referred to as imponderable fluid, with a focus on their (apparent) lack of weight.

Supercooling. The cooling of a liquid below its normal freezing point. Supercooling was observed in the eighteenth century not only in water but also in molten metals. Supercooling in water, first recorded by Daniel Gabriel Fahrenheit in 1724, presented a challenge to the presumed fixity of the freezing point. Supercooling in mercury presented puzzling anomalies in the workings of the mercury thermometer.

Superheating. The heating of a liquid above its normal boiling point. The superheating of water presents a problem for the presumed fixity of the boiling point. In debates about superheating and its implications, it is important to keep in mind the difference between the temperature that water can withstand without boiling, and the temperature that water maintains while boiling. Superheating can happen in both of those senses, though the latter is more of a problem for the fixing of the boiling point.

*Thermoscope. A temperature-measuring instrument that indicates the relative changes or comparisons of temperatures, without giving numbers. A thermoscope has an *ordinal scale*, unlike a thermometer, which has a *cardinal scale*.

Two-point method. Any method of graduating thermometers using two *fixed points* (cf. *one-point method*). Most commonly, the scale was obtained by assuming that the thermometric fluid expanded linearly with temperature between and beyond the fixed points.

Underdetermination (of theory by evidence). The fact, or the fear, that there are always multiple theories all compatible with any given set of empirical evidence.

Vapor pressure. The pressure of a saturated vapor produced by evaporation from a liquid. From the late eighteenth century, vapor pressure was widely recognized to be a function of temperature, and of temperature only. That assumption played a key role not only in establishing the *pressure-balance theory of boiling*, but also in William Thomson's attempts to operationalize *absolute temperature*.

Water calorimetry. The measurement of the quantity of heat through the heating of water. A hot object is introduced into a measured amount of cold water at a previously known temperature, and the amount of heat given from the hot object to the cold water is calculated as the product of the temperature rise in the water and its *specific heat*. This simple scheme had to be modified when it was recognized that the specific heat of water was itself a function of temperature.

Bibliography

Aitken, John. 1878. "On Boiling, Condensing, Freezing, and Melting." *Transactions of the Royal Scottish Society of Arts* 9:240–287. Read on 26 July 1875.

———. 1880–81. "On Dust, Fogs, and Clouds." *Transactions of the Royal Society of Edinburgh* 30(1):337–368. Read on 20 December 1880 and 7 February 1881.

———. 1923. *Collected Scientific Papers of John Aitken*. Edited by C. G. Knott. Cambridge: Cambridge University Press.

Amontons, Guillaume. 1702. "Discours sur quelques propriétés de l'air, & le moyen d'en connoître la temperature dans tous les climats de la terre." *Histoire de l'Académie Royale des Sciences, avec les Mémoires de mathématique et de physique*, vol. for 1702, part for memoirs: 155–174.

———. 1703. "Le Thermomètre réduit à une mesure fixe & certaine, & le moyen d'y rapporter les observations faites avec les ancien Thermomètres." *Histoire de l'Académie Royale des Sciences, avec les Mémoires de mathématique et de physique*, vol. for 1703, part for memoirs: 50–56.

Atkins, P. W. 1987. *Physical Chemistry*. 3d ed. Oxford: Oxford University Press.

Bacon, Francis. [1620] 2000. *The New Organon*. Edited by Lisa Jardine and Michael Silverthorne. Cambridge: Cambridge University Press.

Ball, Philip. 1999. H_2O: *A Biography of Water*. London: Weidenfeld and Nicolson.

Barnett, Martin K. 1956. "The Development of Thermometry and the Temperature Concept." *Osiris* 12:269–341.

Beckman, Olof. 1998. "Celsius, Linné, and the Celsius Temperature Scale." *Bulletin of the Scientific Instrument Society*, no. 56:17–23.

Bentham, Jeremy. 1843. *The Works of Jeremy Bentham*. Edited by John Bowring. 11 vols. Edinburgh: William Tait.

Bergman, Torbern. 1783. *Outlines of Mineralogy*. Translated by William Withering. Birmingham: Piercy and Jones.

Biot, Jean-Baptiste. 1816. *Traité de physique expérimentale et mathématique*. 4 vols. Paris: Deterville.

Birch, Thomas. [1756–57] 1968. *History of the Royal Society of London*. 4 vols. New York: Johnson Reprint Co., 1968. Originally published in 1756 and 1757 in London by Millar.

[Black, Joseph.] 1770. *An Enquiry into the General Effects of Heat; with Observations on the Theories of Mixture.* London: Nourse.

———. 1775. "The Supposed Effect of Boiling upon Water, in Disposing It to Freeze More Readily, Ascertained by Experiments." *Philosophical Transactions of the Royal Society of London* 65:124–128.

———. 1803. *Lectures on the Elements of Chemistry.* Edited by John Robison. 2 vols. London: Longman and Rees; Edinburgh: William Creech.

Blagden, Charles. 1783. "History of the Congelation of Quicksilver." *Philosophical Transactions of the Royal Society of London* 73:329–397.

———. 1788. "Experiments on the Cooling of Water below Its Freezing Point." *Philosophical Transactions of the Royal Society of London* 78:125–146.

Bloor, David. 1991. *Knowledge and Social Imagery.* 2d ed. Chicago: University of Chicago Press.

Boerhaave, Herman. [1732] 1735. *Elements of Chemistry: Being the Annual Lectures of Hermann Boerhaave, M.D.* Translated by Timothy Dallowe, M.D. 2 vols. London: Pemberton. Originally published as *Elementa Chemiae* in 1732.

Bogen, James, and James Woodward. 1988. "Saving the Phenomena." *Philosophical Review* 97:302–352.

Bolton, Henry Carrington. 1900. *The Evolution of the Thermometer, 1592–1743.* Easton, Pa.: Chemical Publishing Company.

Boudouard, O.: see Le Chatelier and Boudouard.

Bouty, Edmond. 1915. "La Physique." In *La Science Française [à l'Exposition de San Francisco]*, vol. 1, 131–151. Paris: Larousse.

Boyd, Richard. 1990. "Realism, Approximate Truth, and Philosophical Method." In C. Wade Savage, ed., *Scientific Theories*, 355–391. Minneapolis: University of Minnesota Press.

Boyer, Carl B. 1942. "Early Principles in the Calibration of Thermometers." *American Journal of Physics* 10:176–180.

Brande, William Thomas. 1819. *A Manual of Chemistry; Containing the Principal Facts of the Science, Arranged in the Order in which they are Discussed and Illustrated in the Lectures at the Royal Institution of Great Britain.* London: John Murray.

Braun: see Watson.

Bridgman, Percy Williams. 1927. *The Logic of Modern Physics.* New York: Macmillan.

———. 1929. "The New Vision of Science." *Harper's* 158:443–454. Also reprinted in Bridgman 1955.

———. 1955. *Reflections of a Physicist.* New York: Philosophical Library.

———. 1959. *The Way Things Are.* Cambridge, Mass.: Harvard University Press.

Brown, Harold I. 1993. "A Theory-Laden Observation Can Test the Theory." *British Journal for the Philosophy of Science* 44:555–559.

Brown, Sanborn C.: see also Rumford 1968.

Brown, Sanborn C. 1979. *Benjamin Thompson, Count Rumford.* Cambridge, Mass.: MIT Press.

Brush, Stephen G. 1976. *The Kind of Motion We Call Heat: A History of the Kinetic Theory of Gases in the 19th Century.* 2 vols. Amsterdam: North-Holland.

———, ed. 1965. *Kinetic Theory.* Vol. 1, *The Nature of Gases and of Heat.* Oxford: Pergamon Press.

Burton, Anthony. 1976. *Josiah Wedgwood: A Biography.* London: André Deutsch.

Callendar, Hugh Longbourne. 1887. "On the Practical Measurement of Temperature: Experiments Made at the Cavendish Laboratory, Cambridge." *Philosophical Transactions of the Royal Society of London* A178:161–230.

Cardwell, Donald S. L. 1971. *From Watt to Clausius.* Ithaca, N.Y.: Cornell University Press.

———. 1989. *James Joule: A Biography.* Manchester: Manchester University Press.

———, ed. 1968. *John Dalton and the Progress of Science*. Manchester: Manchester University Press.

Carnot, Nicolas-Léonard-Sadi. [1824] 1986. *Reflections on the Motive Power of Fire*. Translated and edited by Robert Fox, along with other manuscripts. Manchester: Manchester University Press. Originally published in 1824 as *Réflexions sur la puissance motrice du feu et sur les machines propres à développer cette puissance* in Paris by Bachelier.

Cartwright, Nancy. 1983. *How the Laws of Physics Lie*. Oxford: Clarendon Press.

———. 1999. *The Dappled World: A Study of the Boundaries of Science*. Cambridge: Cambridge University Press.

Cartwright, Nancy, Jordi Cat, Lola Fleck, and Thomas E. Uebel. 1996. *Otto Neurath: Philosophy between Science and Politics*. Cambridge: Cambridge University Press.

Cat, Jordi: see Cartwright et al.

Cavendish, Henry. [1766] 1921. "Boiling Point of Water." In Cavendish 1921, 351–354.

———. 1776. "An Account of the Meteorological Instruments Used at the Royal Society's House." *Philosophical Transactions of the Royal Society of London* 66:375–401. Also reprinted in Cavendish 1921, 112–126.

———. 1783. "Observations on Mr. Hutchins's Experiments for Determining the Degree of Cold at Which Quicksilver Freezes." *Philosophical Transactions of the Royal Society of London* 73:303–328.

———. 1786. "An Account of Experiments Made by John McNab, at Henley House, Hudson's Bay, Relating to Freezing Mixtures." *Philosophical Transactions of the Royal Society of London* 76:241–272.

———. [n.d.] 1921. "Theory of Boiling." In Cavendish 1921, 354–362. Originally unpublished manuscript, probably dating from c. 1780.

———. 1921. *The Scientific Papers of the Honourable Henry Cavendish, F. R. S.* Vol. 2, *Chemical and Dynamical*. Edited by Edward Thorpe. Cambridge: Cambridge University Press.

Cavendish, Henry, William Heberden, Alexander Aubert, Jean-André De Luc, Nevil Maskelyne, Samuel Horsley, and Joseph Planta. 1777. "The Report of the Committee Appointed by the Royal Society to Consider of the Best Method of Adjusting the Fixed Points of Thermometers; and of the Precautions Necessary to Be Used in Making Experiments with Those Instruments." *Philosophical Transactions of the Royal Society of London* 67:816–857.

Chaldecott, John A. 1955. *Handbook of the Collection Illustrating Temperature Measurement and Control*. Part II, *Catalogue of Exhibits with Descriptive Notes*. London: Science Museum.

———. 1975. "Josiah Wedgwood (1730–95)—Scientist." *British Journal for the History of Science* 8:1–16.

———. 1979. "Science as Josiah Wedgwood's Handmaiden." *Proceedings of the Wedgwood Society*, no. 10:73–86.

Chang, Hasok. 1993. "A Misunderstood Rebellion: The Twin-Paradox Controversy and Herbert Dingle's Vision of Science." *Studies in History and Philosophy of Science* 24:741–790.

———. 1995a. "Circularity and Reliability in Measurement." *Perspectives on Science* 3:153–172.

———. 1995b. "The Quantum Counter-Revolution: Internal Conflicts in Scientific Change." *Studies in History and Philosophy of Modern Physics* 26:121–136.

———. 1997. "[Review of] Deborah Mayo, *Error and the Growth of Experimental Knowledge*." *British Journal for the Philosophy of Science* 48:455–459.

———. 1999. "History and Philosophy of Science as a Continuation of Science by Other Means." *Science and Education* 8:413–425.

———. 2001a. "How to Take Realism beyond Foot-Stamping." *Philosophy* 76:5–30.

———. 2001b. "Spirit, Air and Quicksilver: The Search for the 'Real' Scale of Temperature." *Historical Studies in the Physical and the Biological Sciences* 31(2):249–284.

———. 2002. "Rumford and the Reflection of Radiant Cold: Historical Reflections and Metaphysical Reflexes." *Physics in Perspective* 4:127–169.

Chang, Hasok, and Sabina Leonelli. Forthcoming. "Infrared Metaphysics."

Chang, Hasok, and Sang Wook Yi. Forthcoming. "Measuring the Absolute: William Thomson and Temperature." *Annals of Science*.

Cho, Adrian. 2003. "A Thermometer beyond Compare?" *Science* 299:1641.

Clapeyron, Benoit-Pierre-Émile. [1834] 1837. "Memoir on the Motive Power of Heat." *Scientific Memoirs, Selected from the Transactions of Foreign Academies of Science and Learned Societies and from Foreign Journals* 1:347–376. Translated by Richard Taylor. Originally published as "Mémoire sur la puissance motrice de la chaleur," *Journal de l'École Polytechnique* 14 (1834): 153–190.

Conant, James Bryant, ed. 1957. *Harvard Case Histories in Experimental Science*. 2 vols. Cambridge, Mass.: Harvard University Press.

Crawford, Adair. 1779. *Experiments and Observations on Animal Heat, and the Inflammation of Combustible Bodies*. London: J. Murray.

———. 1788. *Experiments and Observations on Animal Heat, and the Inflammation of Combustible Bodies*. 2d ed. London: J. Johnson.

Crosland, Maurice. 1967. *The Society of Arcueil: A View of French Science at the Time of Napoleon I*. London: Heinemann.

———. 1978. *Gay-Lussac: Scientist and Bourgeois*. Cambridge: Cambridge University Press.

Cushing, James T. 1994. *Quantum Mechanics: Historical Contingency and the Copenhagen Hegemony*. Chicago: University of Chicago Press.

Dalton, John. 1802a. "Experimental Essays on the Constitution of Mixed Gases; on the Force of Steam or Vapour from Water and Other Liquids in Different Temperatures, Both in a Torricellian Vacuum and in Air; on Evaporation; and on the Expansion of Gases by Heat." *Memoirs and Proceedings of the Manchester Literary and Philosophical Society* 5(2):535–602.

———. 1802b. "Experiments and Observations on the Heat and Cold Produced by the Mechanical Condensation and Rarefaction of Air." *[Nicholson's] Journal of Natural Philosophy, Chemistry, and the Arts* 3:160–166. Originally published in the *Memoirs and Proceedings of the Manchester Literary and Philosophical Society* 5(2):515–526.

———. 1808. *A New System of Chemical Philosophy*. Vol. 1, part 1. Manchester: R. Bickerstaff.

Daniell, John Frederick. 1821. "On a New Pyrometer." *Quarterly Journal of Science, Literature, and the Arts* 11:309–320.

———. 1830. "On a New Register-Pyrometer, for Measuring the Expansions of Solids, and Determining the Higher Degrees of Temperature upon the Common Thermometric Scale." *Philosophical Transactions of the Royal Society of London* 120:257–286.

Day, A. L., and R. B. Sosman. 1922. "Temperature, Realisation of Absolute Scale of." In Richard Glazebrook, ed., *A Dictionary of Applied Physics*, Vol. 1, *Mechanics, Engineering, Heat*, 836–871. London: Macmillan.

Delisle [or, De l'Isle], Joseph-Nicolas. 1738. *Memoires pour servir a l'histoire & au progrès de l'astronomie, de la géographie, & de la physique*. St. Petersburg: l'Academie des Sciences.

De Luc, Jean-André. 1772. *Recherches sur les modifications de l'atmosphère*. 2 vols. Geneva: n.p. Also published in Paris by La Veuve Duchesne, in 4 vols., in 1784 and 1778.

———. 1779. *An Essay on Pyrometry and Areometry, and on Physical Measures in General*. London: J. Nichols. Also published in the *Philosophical Transactions of the Royal Society of London* in 1778.

De Montet, Albert. 1877–78. *Dictionnaire biographique des Genevois et des Vaudois*. 2 vols. Lausanne: Georges Bridel.

De Morveau: see Guyton de Morveau.
Donny, François. 1846. "Mémoire sur la cohésion des liquides, et sur leur adhérence aux corps solides." *Annales de chimie et de physique*, 3d ser., 16:167–190.
Dörries, Matthias. 1997. *Visions of the Future of Science in Nineteenth-Century France (1830–1871)*. Habilitation thesis. Munich.
———. 1998a. "Easy Transit: Crossing Boundaries between Physics and Chemistry in mid-Nineteenth Century France." In Jon Agar and Crosbie Smith, eds., *Making Space for Science: Territorial Themes in the Shaping of Knowledge*, 246–262. Basingstoke: Macmillan.
———. 1998b. "Vicious Circles, or, The Pitfalls of Experimental Virtuosity." In Michael Heidelberger and Friedrich Steinle, eds., *Experimental Essays—Versuche zum Experiment*, 123–140. Baden-Baden: NOMOS.
Draper, John William. 1847. "On the Production of Light by Heat." *Philosophical Magazine*, 3d ser., 30:345–360.
[Du Crest, Jacques-Barthélemi Micheli.] 1741. *Description de la methode d'un thermomètre universel*. Paris: Gabriel Valleyre.
Dufour, Louis. 1861. "Recherches sur l'ébullition des liquides." *Archives des sciences physiques et naturelles*, new ser., 12:210–266.
———. 1863. "Recherches sur la solidification et sur l'ébullition." *Annales de chimie et de physique*, 3d ser., 68:370–393.
Duhem, Pierre. 1899. "Usines et Laboratoires." *Revue Philomathique de Bordeaux et du Sud-Ouest* 2:385–400.
———. [1906] 1962. *The Aim and Structure of Physical Theory*. Translated by Philip P. Wiener. New York: Atheneum.
Dulong, Pierre-Louis, and Alexis-Thérèse Petit. 1816. "Recherches sur les lois de dilatation des solids, des liquides et des fluides élastiques, et sur la mesure exacte des températures." *Annales de chimie et de physique*, 2d ser., 2:240–263.
———. 1817. "Recherches sur la mesure des températures et sur les lois de la communication de la chaleur." *Annales de chimie et de physique*, 2d ser., 7:113–154, 225–264, 337–367.
Dumas, Jean-Baptiste. 1885. "Victor Regnault." In *Discours et éloges académiques*, vol. 2, 153–200. Paris: Gauthier-Villars.
Dyment, S. A. 1937. "Some Eighteenth Century Ideas Concerning Aqueous Vapour and Evaporation." *Annals of Science* 2:465–473.
Ellis, Brian. 1968. *Basic Concepts of Measurement*. Cambridge: Cambridge University Press.
Ellis, George E. 1871. *Memoir of Sir Benjamin Thompson, Count Rumford*. Boston: American Academy of Arts and Sciences.
Evans, James, and Brian Popp. 1985. "Pictet's Experiment: The Apparent Radiation and Reflection of Cold." *American Journal of Physics* 53:737–753.
Fahrenheit, Daniel Gabriel. 1724. "Experiments and Observations on the Freezing of Water in Vacuo." *Philosophical Transactions of the Royal Society of London, Abridged*, vol. 7 (1724–34), 22–24. Originally published in Latin in the *Philosophical Transactions of the Royal Society of London* 33:78–84.
Fairbank, William M., Jr., and Allan Franklin. 1982. "Did Millikan Observe Fractional Charges on Oil Drops?" *American Journal of Physics* 50:394–397.
Farrer, Katherine Eufemia, ed. 1903–06. *Correspondence of Josiah Wedgwood*. 3 vols. Manchester: E. J. Morten.
Feigl, Herbert. 1970. "The 'Orthodox' View of Theories: Remarks in Defense as well as Critique." In Michael Radner and Stephen Winokur, eds., *Analyses of Theories and Methods of Physics and Psychology* (Minnesota Studies in the Philosophy of Science, vol. 4), 3–16. Minneapolis: University of Minnesota Press.

———. 1974. "Empiricism at Bay?" In R. S. Cohen and M. Wartofsky, eds., *Methodological and Historical Essays in the Natural and Social Sciences*, 1–20. Dordrecht: Reidel.

Feyerabend, Paul. 1975. *Against Method*. London: New Left Books.

Fitzgerald, Keane. 1760. "A Description of a Metalline Thermometer." *Philosophical Transactions of the Royal Society of London* 51:823–833.

Fleck, Lola: see Cartwright et al.

Foley, Richard. 1998. "Justification, Epistemic." In Edward Craig, ed., *Routledge Encyclopedia of Philosophy*, vol. 5, 157–165. London: Routledge.

Forbes, James David. 1860. "Dissertation Sixth: Exhibiting a General View of the Progress of Mathematical and Physical Science, Principally from 1775 to 1850." In *Encyclopaedia Britannica*, 8th ed., vol. 1, 794–996.

Fourier, Joseph. [1822] 1955. *The Analytic Theory of Heat*. Translated by Alexander Freeman. New York: Dover. Originally published in 1822 as *Théorie analytique de la chaleur* in Paris by Didot.

Fox, Robert. 1968. "Dalton's Caloric Theory." In Cardwell, ed., 1968, 187–202.

———. 1971. *The Caloric Theory of Gases from Lavoisier to Regnault*. Oxford: Clarendon Press.

———. 1974. "The Rise and Fall of Laplacian Physics." *Historical Studies in the Physical Sciences* 4:89–136.

Frängsmyr, Tore, J. L. Heilbron, and Robin E. Rider, eds. 1990. *The Quantifying Spirit in the 18th Century*. Berkeley: University of California Press.

Frank, Philipp G., ed. 1954. *The Validation of Scientific Theories*. Boston: Beacon Press.

Franklin, Allan: see also Fairbank and Franklin.

Franklin, Allan. 1981. "Millikan's Published and Unpublished Data on Oil Drops." *Historical Studies in the Physical Sciences* 11(2):185–201.

Franklin, Allan, et al. 1989. "Can a Theory-Laden Observation Test the Theory?" *British Journal for the Philosophy of Science* 40:229–231.

Galison, Peter. 1997. *Image and Logic: A Material Culture of Microphysics*. Chicago: University of Chicago Press.

Gay-Lussac, Joseph-Louis. 1802. "Enquiries Concerning the Dilatation of the Gases and Vapors." [Nicholson's] *Journal of Natural Philosophy, Chemistry, and the Arts* 3:207–16, 257–67. Originally published as "Sur la dilation des gaz et des vapeurs," *Annales de chimie* 43:137–175.

———. 1807. "Premier essai pour déterminer les variations des température qu'éprouvent les gaz en changeant de densité, et considérations sur leur capacité pour calorique." *Mémoires de physique et de chimie, de la Société d'Arcueil* 1:180–203.

———. 1812. "Sur la déliquescence des corps." *Annales de chimie* 82:171–177.

———. 1818. "Notice Respecting the Fixedness of the Boiling Point of Fluids." *Annals of Philosophy* 12:129–131. Abridged translation of "Sur la fixité du degré d'ébullition des liquides," *Annales de chimie et de physique*, 2d ser., 7:307–313.

Gernez, Désiré. 1875. "Recherches sur l'ébullition." *Annales de chimie et de physique*, 5th ser., 4:335–401.

———. 1876. "Sur la détermination de la température de solidification des liquides, et en particulier du soufre." *Journal de physique théorique et appliquée* 5:212–215.

Gillispie, Charles Coulston. 1997. *Pierre-Simon Laplace 1749–1827: A Life in Exact Science*. With Robert Fox and Ivor Grattan-Guinness. Princeton: Princeton University Press.

Glymour, Clark. 1980. *Theory and Evidence*. Princeton: Princeton University Press.

Gray, Andrew. 1908. *Lord Kelvin: An Account of His Scientific Life and Work*. London: J. M. Dent & Company.

Grünbaum, Adolf. 1960. "The Duhemian Argument." *Philosophy of Science* 27:75–87.
Guerlac, Henry. 1976. "Chemistry as a Branch of Physics: Laplace's Collaboration with Lavoisier." *Historical Studies in the Physical Sciences* 7:193–276.
Guthrie, Matthieu [Matthew]. 1785. *Nouvelles expériences pour servir à déterminer le vrai point de congélation du mercure & la difference que le degré de pureté de ce metal pourroit y apporter*. St. Petersburg: n.p.
Guyton [de Morveau], Louis-Bernard. 1798. "Pyrometrical Essays to Determine the Point to which Charcoal is a Non-Conductor of Heat." *[Nicholson's] Journal of Natural Philosophy, Chemistry, and the Arts* 2:499–500. Originally published as "Essais pyrométriques pour déterminer à quel point le charbon est non-conducteur de chaleur," *Annales de chimie et de physique* 26:225ff.
———. 1799. "Account of Certain Experiments and Inferences Respecting the Combustion of the Diamond, and the Nature of Its Composition." *[Nicholson's] Journal of Natural Philosophy, Chemistry, and the Arts* 3:298–305. Originally published as "Sur la combustion du diamant," *Annales de chimie et de physique* 31:72–112; also reprinted in the *Philosophical Magazine* 5:56–61, 174–188.
———. 1803. "Account of the Pyrometer of Platina." *[Nicholson's] Journal of Natural Philosophy, Chemistry, and the Arts*, new ser., 6:89–90. Originally published as "Pyromètre de platine," *Annales de chimie et de physique* 46:276–278.
———. 1811a. "De l'effet d'une chaleur égale, longtemps continuée sur les pièces pyrométriques d'argile." *Annales de chimie et de physique* 78:73–85.
———. 1811b. "Suite de l'essai de pyrométrie." *Mémoires de la Classe des Sciences Mathématiques et Physiques de l'Institut de France* 12(2):89–120.
Hacking, Ian. 1983. *Representing and Intervening*. Cambridge: Cambridge University Press.
Hallett, Garth. 1967. *Wittgenstein's Definition of Meaning as Use*. New York: Fordham University Press.
Halley, Edmond. 1693. "An Account of Several Experiments Made to Examine the Nature of the Expansion and Contraction of Fluids by Heat and Cold, in order to ascertain the Divisions of the Thermometer, and to Make that Instrument, in all Places, without Adjusting by a Standard." *Philosophical Transactions of the Royal Society of London* 17:650–656.
Haüy, (l'Abbé) René Just. [1803] 1807. *An Elementary Treatise on Natural Philosophy*. 2 vols. Translated by Olinthus Gregory. London: George Kearsley. Originally published in 1803 as *Traité élémentaire de physique* in Paris.
———. 1806. *Traité élémentaire de physique*. 2d ed. 2 vols. Paris: Courcier.
Heering, Peter. 1992. "On Coulomb's Inverse Square Law." *American Journal of Physics* 60:988–994.
———. 1994. "The Replication of the Torsion Balance Experiment: The Inverse Square Law and its Refutation by Early 19th-Century German Physicists." In Christine Blondel and Matthias Dörries, eds., *Restaging Coulomb: Usages, controverses et réplications autour de la balance de torsion*. Florence: Olschki.
Heilbron, John L.: see also Frängsmyr et al.
Heilbron, John L. 1993. *Weighing Imponderables and Other Quantitative Science around 1800*, supplement to *Historical Studies in the Physical and Biological Sciences*, vol. 24, no. 1.
Heisenberg, Werner. 1971. *Physics and Beyond*. Translated by A. J. Pomerans. London: George Allen & Unwin.
Hempel, Carl G. 1965. *Aspects of Scientific Explanation, and Other Essays in the Philosophy of Science*. New York: Free Press.
———. 1966. *Philosophy of Natural Science*. Englewood Cliffs, N.J.: Prentice-Hall.

Henry, William. 1802. "A Review of Some Experiments, which Have Been Supposed to Disprove the Materiality of Heat." *Memoirs of the Literary and Philosophical Society of Manchester* 5(2):603–621.

Hentschel, Klaus. 2002. "Why Not One More Imponderable? John William Draper's Tithonic Rays." *Foundations of Chemistry* 4:5–59.

Herschel, William. 1800a. "Experiments on the Refrangibility of the Invisible Rays of the Sun." *Philosophical Transactions of the Royal Society of London* 90:284–292.

———. 1800b. "Investigation of the Powers of the Prismatic Colours to Heat and Illuminate Objects; with Remarks, that Prove the Different Refrangibility of Radiant Heat. To which is Added, an Inquiry into the Method of Viewing the Sun Advantageously, with Telescopes of Large Apertures and High Magnifying Powers." *Philosophical Transactions of the Royal Society of London* 90:255–283.

Holton, Gerald. 1952. *Introduction to Concepts and Theories in Physical Science*. Cambridge: Addison-Wesley.

———. 1978. "Subelectrons, Presuppositions, and the Millikan-Ehrenhaft Dispute." *Historical Studies in the Physical Sciences* 9:161–224.

Holton, Gerald, and Stephen G. Brush. 2001. *Physics, the Human Adventure*. New Brunswick, N.J.: Rutgers University Press. This is the third edition of Holton 1952.

Humphreys, Paul. 2004. *Extending Ourselves: Computational Science, Empiricism, and Scientific Method*. Oxford: Oxford University Press.

Husserl, Edmund. 1970. *The Crisis of European Sciences and Transcendental Phenomenology*. Translated by David Carr. Evanston, Ill.: Northwestern University Press.

Hutchins, Thomas. 1783. "Experiments for Ascertaining the Point of Mercurial Congelation." *Philosophical Transactions of the Royal Society of London* 73:*303–*370.

Hutchison, Keith. 1976a. "Mayer's Hypothesis: A Study of the Early Years of Thermodynamics." *Centaurus* 20:279–304.

———. 1976b. "W. J. M. Rankine and the Rise of Thermodynamics." D. Phil. diss., Oxford University.

Jaffe, Bernard. 1976. *Crucibles: The Story of Chemistry from Ancient Alchemy to Nuclear Fission*. 4th ed. New York: Dover.

Joule, James Prescott. 1845. "On the Changes of Temperature Produced by the Rarefaction and Condensation of Air." *Philosophical Magazine*, 3d ser., 26:369–383. Also reprinted in *Scientific Papers of James Prescott Joule* (London, 1884), 172–189.

Joule, James Prescott, and William Thomson. [1852] 1882. "On the Thermal Effects Experienced by Air in Rushing through Small Apertures." In Thomson 1882, 333–345 (presented as "preliminary" of a composite article, Art. 49: "On the Thermal Effects of Fluids in Motion"). Originally published in the *Philosophical Magazine* in 1852.

———. [1853] 1882. "On the Thermal Effects of Fluids in Motion." In Thomson 1882, 346–356. Originally published in the *Philosophical Transactions of the Royal Society of London* 143:357–365.

———. [1854] 1882. "On the Thermal Effects of Fluids in Motion, Part 2." In Thomson 1882, 357–400. Originally published in the *Philosophical Transactions of the Royal Society of London* 144:321–364.

———. [1860] 1882. "On the Thermal Effects of Fluids in Motion, Part 3. On the Changes of Temperature Experienced by Bodies Moving through Air." In Thomson 1882, 400–414. Originally published in the *Philosophical Transactions of the Royal Society of London* 150:325–336.

———. [1862] 1882. "On the Thermal Effects of Fluids in Motion, Part 4." In Thomson 1882, 415–431. Originally published in the *Philosophical Transactions of the Royal Society of London* 152:579–589.

Jungnickel, Christa, and Russell McCormmach. 1999. *Cavendish: The Experimental Life.* Rev. ed. Lewisburg, Pa.: Bucknell University Press.

Klein, Herbert Arthur. 1988. *The Science of Measurement: A Historical Survey.* New York: Dover.

Klickstein, Herbert S. 1948. "Thomas Thomson: Pioneer Historian of Chemistry." *Chymia* 1:37–53.

Knight, David M. 1967. *Atoms and Elements.* London: Hutchinson.

Knott, Cargill G. 1923. "Sketch of John Aitken's Life and Scientific Work." In Aitken 1923, vii–xiii.

Kornblith, Hilary, ed. 1985. *Naturalizing Epistemology.* Cambridge, Mass.: MIT Press.

Kosso, Peter. 1988. "Dimensions of Observability." *British Journal for the Philosophy of Science* 39:449–467.

———. 1989. *Observability and Observation in Physical Science.* Dordrecht: Kluwer.

Kuhn, Thomas S. 1957. *The Copernican Revolution.* Cambridge, Mass.: Harvard University Press.

———. [1968] 1977. "The History of Science." In *The Essential Tension: Selected Studies in Scientific Tradition and Change,* 105–126. Chicago: University of Chicago Press. Originally published in 1968 in the *International Encyclopedia of the Social Sciences,* vol. 14.

———. 1970a. "Logic of Discovery or Psychology of Research?" In Lakatos and Musgrave 1970, 1–23.

———. 1970b. "Reflections on My Critics." In Lakatos and Musgrave 1970, 231–278.

———. 1970c. *The Structure of Scientific Revolutions.* 2d ed. Chicago: University of Chicago Press.

———. 1977. "Objectivity, Value Judgment, and Theory Choice." In *The Essential Tension: Selected Studies in Scientific Tradition and Change,* 320–339. Chicago: University of Chicago Press.

Lafferty, Peter, and Julian Rowe, eds. 1994. *The Hutchinson Dictionary of Science.* Rev. ed. Oxford: Helicon Publishing Inc.

Lakatos, Imre. 1968–69. "Criticism and the Methodology of Scientific Research Programmes." *Proceedings of the Aristotelian Society* 69:149–186.

———. 1970. "Falsification and the Methodology of Scientific Research Programmes." In Lakatos and Musgrave 1970, 91–196.

———. [1973] 1977. "Science and Pseudoscience." In *Philosophical Papers,* vol. 1, 1–7. Cambridge: Cambridge University Press. Also reprinted in Martin Curd and J. A. Cover, eds., *Philosophy of Science: The Central Issues* (New York: Norton, 1998), 20–26.

———. 1976. "History of Science and its Rational Reconstructions." In Colin Howson, ed., *Method and Appraisal in the Physical Sciences,* 1–39. Cambridge: Cambridge University Press.

Lakatos, Imre, and Alan Musgrave, eds. 1970. *Criticism and the Growth of Knowledge.* Cambridge: Cambridge University Press.

Lambert, Johann Heinrich. 1779. *Pyrometrie, oder vom Maaße des Feuers und der Wärme.* Berlin: Haude und Spener.

Lamé, Gabriel. 1836. *Cours de physique de l'École Polytechnique.* 3 vols. Paris: Bachelier.

Langevin, Paul. 1911. "Centennaire de M. Victor Regnault." *Annuaire de Collège de France* 11:42–56.

Laplace, Pierre-Simon: see also Lavoisier and Laplace.

Laplace, Pierre-Simon. 1796. *Exposition du systême du monde.* 2 vols. Paris: Imprimerie du Cercle-Social.

———. 1805. *Traité de mécanique céleste.* Vol. 4. Paris: Courcier.

———. 1821. "Sur l'attraction des Sphères, et sur la répulsion des fluides élastiques." *Connaissance des Tems pour l'an 1824* (1821), 328–343.

———. [1821] 1826. "Sur l'attraction des corps spheriques et sur la répulsion des fluides élastiques." *Mémoires de l'Académie Royale des Sciences de l'Institut de France* 5:1–9. Although Laplace's paper was presented in 1821, the volume of the *Mémoires* containing it was not published until 1826.

———. [1823] 1825. *Traité de mécanique céleste.* Vol. 5, book 12. Paris: Bachelier. The volume was published in 1825, but book 12 is dated 1823.

Latour, Bruno. 1987. *Science in Action.* Cambridge, Mass.: Harvard University Press.

Laudan, Laurens. 1973. "Peirce and the Trivialization of the Self-Correcting Thesis." In Ronald N. Giere and Richard S. Westfall, eds., *Foundations of Scientific Method: The Nineteenth Century*, 275–306. Bloomington: Indiana University Press.

Lavoisier, Antoine-Laurent. [1789] 1965. *Elements of Chemistry.* Translated in 1790 by Robert Kerr, with a new introduction by Douglas McKie. New York: Dover. Originally published in 1789 as *Traité élémentaire de chimie* in Paris by Cuchet.

Lavoisier, Antoine-Laurent, and Pierre-Simon Laplace. [1783] 1920. *Mémoire sur la chaleur.* Paris: Gauthier-Villars.

Le Chatelier, Henri, and O. Boudouard. 1901. *High-Temperature Measurements.* Translated by George K. Burgess. New York: Wiley.

Lide, David R., and Henry V. Kehiaian. 1994. *CRC Handbook of Thermophysical and Thermochemical Data.* Boca Raton, Fla.: CRC Press.

Lilley, S. 1948. "Attitudes to the Nature of Heat about the Beginning of the Nineteenth Century." *Archives internationales d'histoire des sciences* 1(4):630–639.

Lodwig, T. H., and W. A. Smeaton. 1974. "The Ice Calorimeter of Lavoisier and Laplace and Some of its Critics." *Annals of Science* 31:1–18.

Lycan, William G. 1988. *Judgement and Justification.* Cambridge: Cambridge University Press.

———. 1998. "Theoretical (Epistemic) Virtues." In Edward Craig, ed., *Routledge Encyclopedia of Philosophy* 9:340–343. London: Routledge.

Mach, Ernst. [1900] 1986. *Principles of the Theory of Heat (Historically and Critically Elucidated).* Edited by Brian McGuinness; translated by Thomas J. McCormack, P. E. B. Jourdain, and A. E. Heath; with an introduction by Martin J. Klein. Dordrecht: Reidel. Initially published in 1896 as *Die Prinzipien der Wärmelehre* in Leipzig by Barth; this translation is from the 2d German edition of 1900.

McCormmach, Russell: see Jungnickel and McCormmach.

McKendrick, Neil. 1973. "The Rôle of Science in the Industrial Revolution: A Study of Josiah Wedgwood as a Scientist and Industrial Chemist." In Mikulás Teich and Robert Young, eds., *Changing Perspectives in the History of Science: Essays in Honour of Joseph Needham*, 274–319. London: Heinemann.

Magie, William Francis, ed. 1935. *A Source Book in Physics.* New York: McGraw-Hill.

Marcet, François. 1842. "Recherches sur certaines circonstances qui influent sur la température du point d'ébullition des liquides." *Bibliothèque universelle*, new ser., 38:388–411.

Margenau, Henry. 1963. "Measurements and Quantum States: Part I." *Philosophy of Science* 30:1–16.

Martine, George. [1738] 1772. "Some Observations and Reflections Concerning the Construction and Graduation of Thermometers." In *Essays and Observations on the Construction and Graduation of Thermometers*, 2d ed., 3–34. Edinburgh: Alexander Donaldson.

———. [1740] 1772. "An Essay towards Comparing Different Thermometers with One Another." In *Essays and Observations on the Construction and Graduation of Thermometers*, 2d ed., 37–48. Edinburgh: Alexander Donaldson.

Matousek, J. W. 1990. "Temperature Measurements in Olden Tymes." *CIM Bulletin* 83(940): 110–115.

Matthews, Michael. 1994. *Science Teaching: The Role of History and Philosophy of Science*. New York: Routledge.
Maxwell, Grover. 1962. "The Ontological Status of Theoretical Entities." In H. Feigl and G. Maxwell, eds., *Scientific Explanation, Space and Time* (Minnesota Studies in the Philosophy of Science, vol. 3), 3–27. Minneapolis: University of Minnesota Press.
Mayo, Deborah. 1996. *Error and the Growth of Experimental Knowledge*. Chicago: University of Chicago Press.
Meikle, Henry. 1826. "On the Theory of the Air-Thermometer." *Edinburgh New Philosophical Journal* 1:332–341.
———. 1842. "Thermometer." *Encyclopaedia Britannica*, 7th ed., 21:236–242.
"Memoirs of Sir Benjamin Thompson, Count of Rumford." 1814. *Gentleman's Magazine, and Historical Chronicle* 84(2):394–398.
Mendoza, Eric. 1962–63. "The Surprising History of the Kinetic Theory of Gases." *Memoirs and Proceedings of the Manchester Literary and Philosophical Society* 105:15–27.
Micheli du Crest: see Du Crest.
Middleton, W. E. Knowles. 1964a. "Chemistry and Meteorology, 1700–1825." *Annals of Science* 20:125–141.
———. 1964b. *A History of the Barometer*. Baltimore: Johns Hopkins Press.
———. 1965. *A History of the Theories of Rain and Other Forms of Precipitation*. London: Oldbourne.
———. 1966. *A History of the Thermometer and Its Use in Meteorology*. Baltimore: Johns Hopkins Press.
Morgan, Mary: see Morrison and Morgan.
Morrell, J. B. 1972. "The Chemist Breeders: The Research Schools of Liebig and Thomas Thomson." *Ambix* 19:1–46.
Morris, Robert J. 1972. "Lavoisier and the Caloric Theory." *British Journal for the History of Science* 6:1–38.
Morrison, Margaret. 1998. "Modelling Nature: Between Physics and the Physical World." *Philosophia Naturalis* 35(1):65–85.
Morrison, Margaret, and Mary S. Morgan. 1999. "Models as Mediating Instruments." In Mary S. Morgan and Margaret Morrison, eds., *Models as Mediators*, 10–37. Cambridge: Cambridge University Press.
Mortimer, Cromwell. [1735] 1746–47."A Discourse Concerning the Usefulness of Thermometers in Chemical Experiments. . . . " *Philosophical Transactions of the Royal Society of London* 44:672–695.
Morton, Alan Q., and Jane A. Wess. 1993. *Public and Private Science: The King George III Collection*. Oxford: Oxford University Press, in association with the Science Museum.
Morveau: see Guyton de Morveau.
Murray, John. 1819. *A System of Chemistry*. 4th ed. 4 vols. Edinburgh: Francis Pillans; London: Longman, Hurst, Rees, Orme & Brown.
Musgrave, Alan: see also Lakatos and Musgrave.
Musgrave, Alan. 1976. "Why Did Oxygen Supplant Phlogiston? Research Programmes in the Chemical Revolution." In C. Howson, ed., *Method and Appraisal in the Physical Sciences*, 181–209. Cambridge: Cambridge University Press.
Neurath, Otto. [1932/33] 1983. "Protocol Statements [1932/33]." In *Philosophical Papers 1913–1946*, ed. and trans. by Robert S. Cohen and Marie Neurath, 91–99. Dordrecht: Reidel.
[Neurath, Otto, et al.] [1929] 1973. "Wissenschaftliche Weltauffassung: Der Wiener Kreis [The Scientific Conception of the World: The Vienna Circle]." In Otto Neurath,

Empiricism and Sociology, 299–318. Dordrecht: Reidel. This article was originally published in 1929 as a pamphlet.

Newton, Isaac. [1701] 1935. "Scala Graduum Caloris. Calorum Descriptiones & Signa." In Magie 1935, 125–128. Originally published anonymously in *Philosophical Transactions of the Royal Society of London* 22: 824–829.

Oxtoby, David W., H. P. Gillis, and Norman H. Nachtrieb. 1999. *Principles of Modern Chemistry*. 4th ed. Fort Worth: Saunders College Publishing.

Pallas, Peter Simon: see Urness.

Péclet, E. 1843. *Traité de la chaleur considérée dans ses applications*. 2d ed. 3 vols. Paris: Hachette.

Peirce, Charles Sanders. [1898] 1934. "The First Rule of Logic." In *Collected Papers of Charles Sanders Peirce*, vol. 5, edited by C. Hartshorne and P. Weiss, 399ff. Cambridge, Mass.: Harvard University Press.

Penn, S. 1986. "Comment on Pictet Experiment." *American Journal of Physics* 54(2):106. This is a comment on Evans and Popp 1985, followed by a response by James Evans.

Petit, Alexis-Thérèse: see Dulong and Petit.

Pictet, Mark Augustus [Marc-Auguste]. 1791. *An Essay on Fire*. Translated by W. B[elcome]. London: E. Jeffery. Originally published in 1790 as *Essai sur le feu* in Geneva.

Poisson, Siméon-Denis. 1835. *Théorie mathématique de la chaleur*. Paris: Bachelier.

Polanyi, Michael. 1958. *Personal Knowledge: Towards a Post-Critical Philosophy*. Chicago: University of Chicago Press.

Popper, Karl R. 1969. *Conjectures and Refutations*. 3d ed. London: Routledge and Kegan Paul.

———. 1970. "Normal Science and its Dangers." In Lakatos and Musgrave 1970, 51–58.

———. 1974. "Replies to My Critics." In Paul A. Schilpp, ed., *The Philosophy of Karl Popper*, vol. 2, 961–1200. La Salle, Ill.: Open Court.

Pouillet, Claude S. M. M. R. 1827–29. *Élémens de physique expérimentale et de météorologie*. 2 vols. Paris: Béchet Jeune.

———. 1836. "Recherches sur les hautes températures et sur plusieurs phénomènes qui en dépendent." *Comptes rendus hebdomadaines* 3:782–790.

———. 1837. "Déterminations des basses températures au moyen du pyromètre à air, du pyromètre magnétique et du thermomètre à alcool." *Comptes rendus hebdomadaines* 4:513–519.

———. 1856. *Élémens de physique expérimentale et de météorologie*. 7th ed. 2 vols. Paris: Hachette.

Press, William H., Brian P. Flannery, Saul A. Teukolsky, and William T. Vetterling. 1988. *Numerical Recipes in C: The Art of Scientific Computing*. Cambridge: Cambridge University Press.

Preston, Thomas. 1904. *The Theory of Heat*. 2d ed. Revised by J. Rogerson Cotter. London: Macmillan.

Prevost, Pierre. 1791. "Sur l'équilibre du feu." *Journal de physique* 38:314–323.

Prinsep, James. 1828. "On the Measurement of High Temperatures." *Philosophical Transactions of the Royal Society of London* 118:79–95.

Psillos, Stathis. 1999. *Scientific Realism: How Science Tracks Truth*. London: Routledge.

Pyle, Andrew J. 1995. *Atomism and its Critics: Problem Areas Associated with the Development of the Atomic Theory of Matter from Democritus to Newton*. Bristol: Thoemmes Press.

Quine, Willard van Orman. [1953] 1961. "Two Dogmas of Empiricism." In *From a Logical Point of View*, 2d ed., 20–46. Cambridge, Mass.: Harvard University Press. The first edition of this book was published in 1953.

Réaumur, René Antoine Ferchault de. 1739. "Observations du thermomètre pendant l'année M.DCCXXXIX, faites à Paris et en différent pays." *Histoire de l'Académie Royale des Sciences, avec les Mémoires de mathématique et de physique*, vol. for 1739, 447–466.

Regnault, (Henri) Victor. 1840. "Recherches sur la chaleur spécifique des corps simples et composés, premier mémoire." *Annales de chimie et de physique*, 2d ser., 73:5–72.

———. 1842a. "Recherches sur la dilatation des gaz." *Annales de chimie et de physique*, 3d ser., 4:5–67.

———. 1842b. "Recherches sur la dilatation des gaz, 2e mémoire." *Annales de chimie et de physique*, 3d ser., 5:52–83.

———. 1842c. "Sur la comparaison du thermomètre à air avec le thermomètre à mercure." *Annales de chimie et de physique*, 3d ser., 5:83–104.

———. 1847. "Relations des expériences entreprises par ordre de Monsieur le Ministre des Travaux Publics, et sur la proposition de la Commission Centrale des Machines à Vapeur, pour déterminer les principales lois et les données numériques qui entrent dans le calcul des machines à vapeur." *Mémoires de l'Académie Royale des Sciences de l'Institut de France* 21:1–748.

Reiser, Stanley Joel. 1993. "The Science of Diagnosis: Diagnostic Technology." In W. F. Bynum and Roy Porter, eds., *Companion Encyclopedia of the History of Medicine* 2:824–851. London: Routledge.

Rider, Robin E.: see Frängsmyr et al.

Rostoker, William, and David Rostoker. 1989. "The Wedgwood Temperature Scale." *Archeomaterials* 3:169–172.

Rottschaefer, W. A. 1976. "Observation: Theory-Laden, Theory-Neutral or Theory-Free?" *Southern Journal of Philosophy* 14:499–509.

Rowe, Julian: see Lafferty and Rowe.

Rowlinson, J. S. 1969. *Liquids and Liquid Mixtures*. 2d ed. London: Butterworth.

Rumford, Count (Benjamin Thompson). [1798] 1968. "An Experimental Inquiry Concerning the Source of the Heat which is Excited by Friction." In Rumford 1968, 3–26. Originally published in *Philosophical Transactions of the Royal Society of London* 88:80–102.

———. [1804a] 1968. "Historical Review of the Various Experiments of the Author on the Subject of Heat." In Rumford 1968, 443–496. Originally published in *Mémoires sur la chaleur* (Paris: Firmin Didot), vii–lxviii.

———. [1804b] 1968. "An Inquiry Concerning the Nature of Heat, and the Mode of its Communication." In Rumford 1968, 323–433. Originally published in *Philosophical Transactions of the Royal Society of London* 94:77–182.

———. 1968. *The Collected Works of Count Rumford*. Vol. 1. Edited by Sanborn C. Brown. Cambridge, Mass.: Harvard University Press.

Salvétat, Alphonse. 1857. *Leçons de céramique profesées à l'École centrale des arts et manufactures, ou technologie céramique* . . . Paris: Mallet-Bachelier.

Sandfort, John F. 1964. *Heat Engines: Thermodynamics in Theory and Practice*. London: Heinemann.

Sargent, Rose-Mary. 1995. *The Diffident Naturalist: Robert Boyle and the Philosophy of Experiment*. Chicago: University of Chicago Press.

Saunders, Simon. 1998. "Hertz's Principles." In Davis Baird, R. I. G. Hughes, and Alfred Nordmann, eds., *Heinrich Hertz: Classical Physicist, Modern Philosopher*, 123–154. Dordrecht: Kluwer.

Schaffer, Simon: see Shapin and Schaffer.

Scherer, A. N. 1799. "Sur le pyromètre de Wedgwood et de nouveaux appareils d'expériences." *Annales de chimie* 31:171–172.

Schlick, Moritz. [1930] 1979. "On the Foundations of Knowledge." In *Philosophical Papers*, vol. 2 (1925–1936), edited by H. L. Mulder and B. F. B. van de Velde-Schlick, 370–387. Dordrecht: Reidel.

Schmidt, J. G. 1805. "Account of a New Pyrometer, which is Capable of Indicating Degrees of Heat of a Furnace." *[Nicholson's] Journal of Natural Philosophy, Chemistry, and the Arts* 11:141–142.

Schofield, Robert E. 1963. *The Lunar Society of Birmingham*. Oxford: Clarendon Press.

Shapere, Dudley. 1982. "The Concept of Observation in Science and Philosophy." *Philosophy of Science* 49:485–525.

Shapin, Steven, and Simon Schaffer. 1985. *Leviathan and the Air-Pump: Hobbes, Boyle, and the Experimental Life*. Princeton: Princeton University Press.

Sharlin, Harold I. 1979. *Lord Kelvin: The Dynamic Victorian*. In collaboration with Tiby Sharlin. University Park: Pennsylvania State University Press.

Sibum, Heinz Otto. 1995. "Reworking the Mechanical Value of Heat: Instruments of Precision and Gestures of Accuracy in Early Victorian England." *Studies in History and Philosophy of Science* 26:73–106.

Silbey, Robert J., and Robert A. Alberty. 2001. *Physical Chemistry*. 3d ed. New York: Wiley.

Sklar, Lawrence. 1975. "Methodological Conservatism." *Philosophical Review* 84:374–400.

Smeaton, W. A.: see Lodwig and Smeaton.

Smith, Crosbie. 1998. *The Science of Energy: A Cultural History of Energy Physics in Victorian Britain*. London: Athlone Press and the University of Chicago Press.

Smith, Crosbie, and M. Norton Wise. 1989. *Energy and Empire: A Biographical Study of Lord Kelvin*. Cambridge: Cambridge University Press.

Smith, George E. 2002. "From the Phenomenon of the Ellipse to an Inverse-Square Force: Why Not?" In David B. Malament, ed., *Reading Natural Philosophy: Essays in the History and Philosophy of Science and Mathematics*, 31–70. Chicago: Open Court.

Sosman, R. B.: see Day and Sosman.

Taylor, Brook. 1723. "An Account of an Experiment, Made to Ascertain the Proportion of the Expansion of the Liquor in the Thermometer, with Regard to the Degrees of Heat." *Philosophical Transactions of the Royal Society of London* 32:291.

Thenard, Louis-Jacques. 1813. *Traité de chimie élémentaire, théoretique et pratique*. 4 vols. Paris: Crochard.

Thompson, Silvanus P. 1910. *The Life of William Thomson, Baron Kelvin of Largs*. 2 vols. London: Macmillan.

Thomson, Thomas. 1802. *A System of Chemistry*. 4 vols. Edinburgh: Bell & Bradfute, and E. Balfour.

———. 1830. *An Outline of the Sciences of Heat and Electricity*. London: Baldwin & Cradock; Edinburgh: William Blackwood.

Thomson, William: see also Joule and Thomson.

Thomson, William (Lord Kelvin). [1848] 1882. "On an Absolute Thermometric Scale Founded on Carnot's Theory of the Motive Power of Heat, and Calculated from Regnault's Observations." In Thomson 1882, 100–106. Originally published in the *Proceedings of the Cambridge Philosophical Society* 1:66–71; also in the *Philosophical Magazine*, 3d ser., 33:313–317.

———. [1849] 1882. "An Account of Carnot's Theory of the Motive Power of Heat; with Numerical Results Deduced from Regnault's Experiments on Steam." In Thomson 1882, 113–155. Originally published in the *Transactions of the Royal Society of Edinburgh* 16:541–574.

———. [1851a] 1882. "On the Dynamical Theory of Heat, with Numerical Results Deduced from Mr Joule's Equivalent of a Thermal Unit, and M. Regnault's Observations on

Steam." In Thomson 1882, 174–210 (presented as Parts 1–3 of a composite paper, Art. 48 under this title). Originally published in the *Transactions of the Royal Society of Edinburgh* in March 1851; also in the *Philosophical Magazine*, 4th ser., 4 (1852).

———. [1851b] 1882. "On a Method of Discovering Experimentally the Relation between the Mechanical Work Spent, and the Heat Produced by the Compression of a Gaseous Fluid." In Thomson 1882, 210–222 (presented as Part 4 of Art. 48). Originally published in the *Transactions of the Royal Society of Edinburgh* 20 (1851).

———. [1851c] 1882. "On the Quantities of Mechanical Energy Contained in a Fluid in Different States, as to Temperature and Density." In Thomson 1882, 222–232 (presented as Part 5 of Art. 48). Originally published in the *Transactions of the Royal Society of Edinburgh* 20 (1851).

———. [1854] 1882. "Thermo-Electric Currents." In Thomson 1882, 232–291 (presented as Part 6 of Art. 48). Originally published in the *Transactions of the Royal Society of Edinburgh* 21 (1854).

———. [1879–80a] 1911. "On Steam-Pressure Thermometers of Sulphurous Acid, Water, and Mercury." In Thomson 1911, 77–87. Originally published in the *Proceedings of the Royal Society of Edinburgh* 10:432–441.

———. [1879–80b] 1911. "On a Realised Sulphurous Acid Steam-Pressure Differential Thermometer; Also a Note on Steam-Pressure Thermometers." In Thomson 1911, 90–95. Originally published in the *Proceedings of the Royal Society of Edinburgh* 10:523–536.

———. 1880. *Elasticity and Heat* (Being articles contributed to the Encyclopaedia Britannica). Edinburgh: Adam and Charles Black.

———. 1882. *Mathematical and Physical Papers*. Vol. 1. Cambridge: Cambridge University Press.

———. 1911. *Mathematical and Physical Papers*. Vol. 5. Cambridge: Cambridge University Press.

Tomlinson, Charles. 1868–69. "On the Action of Solid Nuclei in Liberating Vapour from Boiling Liquids." *Proceedings of the Royal Society of London* 17:240–252.

Truesdell, Clifford. 1979. *The Tragicomical History of Thermodynamics, 1822–1854*. New York: Springer-Verlag.

Tunbridge, Paul A. 1971. "Jean André De Luc, F.R.S." *Notes and Records of the Royal Society of London* 26:15–33.

Tyndall, John. 1880. *Heat Considered as a Mode of Motion*. 6th ed. London: Longmans, Green and Company.

Uebel, Thomas: see Cartwright et al.

Urness, Carol, ed. 1967. *A Naturalist in Russia: Letters from Peter Simon Pallas to Thomas Pennant*. Minneapolis: University of Minnesota Press.

Van der Star, Pieter, ed. 1983. *Fahrenheit's Letters to Leibniz and Boerhaave*. Leiden: Museum Boerhaave; Amsterdam: Rodopi.

Van Fraassen, Bas C. 1980. *The Scientific Image*. Oxford: Clarendon Press.

Van Swinden, J. H. 1788. *Dissertations sur la comparaison des thermomètres*. Amsterdam: Marc-Michel Rey.

Voyages en Sibérie, extraits des journaux de divers savans voyageurs. 1791. Berne: La Société Typographique.

Walter, Maila L. 1990. *Science and Cultural Crisis: An Intellectual Biography of Percy Williams Bridgman (1882–1961)*. Stanford, Calif.: Stanford University Press.

Watson, William. 1753. "A Comparison of Different Thermometrical Observations in Sibiria [sic]." *Philosophical Transactions of the Royal Society of London* 48:108–109.

———. 1761. "An Account of a Treatise in Latin, Presented to the Royal Society, intituled, De admirando frigore artificiali, quo mercurius est congelatus, dissertatio,

&c. a J. A. Braunio, Academiae Scientiarum Membro, &c." *Philosophical Transactions of the Royal Society of London* 52:156–172.

Wedgwood, Josiah. 1782. "An Attempt to Make a Thermometer for Measuring the Higher Degrees of Heat, from a Red Heat up to the Strongest that Vessels Made of Clay can Support." *Philosophical Transactions of the Royal Society of London* 72:305–326.

———. 1784. "An Attempt to Compare and Connect the Thermometer for Strong Fire, Described in Vol. LXXII of the Philosophical Transactions, with the Common Mercurial Ones." *Philosophical Transactions of the Royal Society of London* 74:358–384.

———. 1786. "Additional Observations on Making a Thermometer for Measuring the Higher Degrees of Heat." *Philosophical Transactions of the Royal Society of London* 76:390–408.

Wess, Jane A.: see Morton and Wess.

Wise, M. Norton: see also Smith and Wise.

Wise, M. Norton, ed. 1995. *The Values of Precision*. Princeton: Princeton University Press.

Wittgenstein, Ludwig. 1969. *On Certainty (Über Gewissheit)*. Edited by G. E. M. Anscombe and G. H. von Wright. Dual language edition with English translation by Denis Paul and G. E. M. Anscombe. New York: Harper.

Woodward, James: see also Bogen and Woodward.

Woodward, James. 1989. "Data and Phenomena." *Synthese* 79:393–472.

Young, Thomas. 1807. *Lectures on Natural Philosophy and the Mechanical Arts*. 2 vols. London: Joseph Johnson. The first volume contains the main text, and the second volume contains a very extensive bibliography.

Zemansky, Mark W., and Richard H. Dittman. 1981. *Heat and Thermodynamics*. 6th ed. New York: McGraw-Hill.

Index

The letter "n" after a page number indicates reference to a footnote on that page; "f" indicates a reference to a figure; "t" indicates a reference to a table.

Absolute temperature
  Amontons's conception of, 182
  Callendar and Le Chatelier on, 213–215
  Irvinist conception of, 65, 168–170, 198
  Thomson's conception of, 174–175, 181–186, 215
  Thomson's first definition of, 182: operationalization of, 187–192, 208
  Thomson's second definition of, 185–186, 185n: operationalization of, 192–197, 201, 208–215; relation with the first definition, 183n
  Thomson's work of 1880 on, 204–205, 209–210
Absolute zero, 182
  Amontons's conception of, 108–109
  considered meaningless by Rumford and Davy, 172
  Irvinist conception of, 168, 198
  Irvinist estimates of, 170
  non-existent in Thomson's first absolute temperature, 190, 191t
  see also absolute temperature
Abstraction, in philosophical ideas, 6–7, 233–234, 245
Abstraction, in scientific theories, 202, 217
  in Carnot's theory, 178, 187, 203
  operationalization of, 207–208, 213, 216
  Thomson's preference for, 174, 186, 202–203
Académie des Sciences, 58, 71n, 75, 76, 98n
Accademia del Cimento, 10
Accuracy, 45, 217, 226–227, 231
Adams, George (?–1773), 12, 13f, 16, 50
Adhesion, 22, 30–32
Adiabatic gas law, 181, 214, 216

Adiabatic phenomena, 66, 180, 181, 194
  Irvinist explanation of, 180n
Affirmation (of existing system of knowledge), 213, 220, 224–227, 231–233
Affirmations (as defined by Schlick), 222
Air thermometer, 60, 120
  comparability of, 80–83, 96
  compared with mercury thermometer, 71, 79, 89, 100–101
  compared with spirit thermometer, 117–118, 154
  compared with Thomson's first absolute temperature, 184, 188, 190, 191t, 204
  compared with Thomson's second absolute temperature, 196–197, 214–215
  constant-pressure, 81, 137n, 214
  constant-volume, 80, 87, 88f, 196t, 214
  De Luc and Haüy critical of, 62n, 67n
  for high temperatures, 137–139
  Lamé on, 75
  Laplace's arguments for, 70–74
  for low temperatures, 116–118
  Regnault on, 80–81, 83, 87–89, 137
  as whole-range standard, 154
  see also under expansion
Aitken, John (1839–1919), 18n, 19n, 35–39, 50–51, 55, 241
Alcohol. See spirit (of wine)
Alcohol thermometer. See spirit thermometer
Amontons, Guillaume (1663–1738)
  double-scaled spirit thermometer, 162–163
  on expansion of air, 61, 62n, 99
  fixed points used by, 10t
Amontons temperature, 108, 182, 194, 218. See also absolute temperature; absolute zero
Anac, 108

275

Animal heat, 169
Aniseed oil, 9
Anna Ivanovna, empress of Russia, 104
Applicability (of abstract ideas), 234
Approximate truth, 52, 185, 190, 228
Approximation, 203, 206, 208
  successive, 45, 214–216, 225–226 (*see also* epistemic iteration)
Aqua fortis, 105
Arago, François (1786–1853), 96
Archimedes's law, 52
Areometry, 78n, 79
Aristotle, 162
Atkins, P. W., 241n
Atomic heat. *See* Dulong and Petit's law
Aubert, Alexander, 11n
Auxiliary hypotheses, 30, 94. *See also* holism

Bacon, Francis (1561–1626), 162
Banks, Joseph, 121
Barometers (and manometers), 12n, 14, 16, 76, 78, 87, 105, 200
Beckman, Olof, 160
Bentham, Jeremy (1748–1832), 106
Bentley, Thomas, 119
Bergman, Torbern, 113, 123
Bering, Captain Vitus, 104
Bernoulli, Daniel (1700–1782), 172
Berthelot, Marcelin, 75, 99
Berthollet, Claude-Louis (1748–1822), 70, 135, 171n
Bertrand, Louis (1731–1812), 166
Bichat, Xavier, 86
Biot, Jean-Baptiste (1774–1862), 12n, 21, 51, 76, 116, 174
Black, Joseph (1728–1799), 56, 62, 65, 123, 176, 177
  apparatus for freezing mercury, 110, 114
  on latent heat, 26–27, 54
Blagden, Charles (1748–1820)
  on freezing of mercury, 104–105, 111–115
  on supercooling, 53–56
Blainville, H. M. D. de, 97n
Blood heat, 10–11, 40–42
Bloor, David, 236n, 248n
Boat metaphor. *See* Neurath, Otto
Boerhaave, Herman (1668–1738), 57–58, 60, 225
Bogen, James, 52
Bohm, David, 245
Bohr, Niels, 220
Boiling, 8–39, 246
  degree of, 12, 16, 20, 49–50
  De Luc's phenomenology of, 20–21
  and evaporation, 38
  facilitated by: air, 18–19, 22, 32–34, 37, 51; dust, 34, 51; solid surfaces, 22
  pressure-balance theory of, 29–34, 38, 48
  theories of, by various authors: Cavendish, 24–27, 32, 35–36, 51; De Luc, 17, 24; Donny, 33; Gernez, 32–34; Tomlinson, 34; Verdet, 32

  superheated, 23–28
  see also boiling point of water; ebullition; evaporation; steam; superheating
Boiling point of mercury, 119, 126t, 133
Boiling point of spirit, 9
Boiling point of water, 11–17, 241–242
  affected by: adhesion, 22, 30–32; cohesion, 30–31; depth of water, 17, 49; material of vessel, 21–23, 31, 49; pressure, 12, 17, 29; sulphur coating, 31
  Aitken's definition of, 38
  considered indefinite, 12–14, 16, 25
  defined without boiling, 35
  fixity of, 48–51
  as two points, 12, 13f
  see also boiling; steam point; superheating
Bolton, Henry Carrington, 50
Bootstrapping, 45
Boulton, Matthew, 176
Bouty, Edmond, 98
Boyd, Richard, 228n
Boyer, Carl B., 11n, 50
Boyle, Robert (1627–1691), 9, 162, 162n, 246
Boyle's law, 71, 99, 181, 190
Brass, 121, 126t, 130t
Braun, Joseph Adam (1712?–1768), 105–108, 142
Bridge St. Bridge, 229
Bridgman, Percy Williams (1882–1961), 58, 141–153, 200, 203n, 222–223. *See also* meaning; operationalism
British Association for the Advancement of Science, 182
Brown, Harold, 95
Brush, Stephen G., 72n, 172–173
Bubbling, 20
Buchwald, Jed, 242
Building, as metaphor for foundation of knowledge, 223–224
Bumping, 20–21. *See also* superheating
Butter, 10

Cagniard de la Tour, Charles (1777?–1859), 22
Callendar, Hugh Longbourne (1863–1930), 213–215, 230
Caloric. *See* caloric theory; combined caloric; free caloric; free caloric of space; radiant caloric
Caloric theory, 51, 65–67, 168–171
  Carnot's acceptance of, 178
  chemical, 65–67, 170–171, 202
  Irvinist, 65, 168–170, 171n, 202
  Laplace's, 71–74
  Lavoisier's, 65–66
Calorimetry, 77n, 139, 154, 175, 192, 202
  by ice, 134–136, 154, 156, 192
  for pyrometry, 134–137
  by water, 134, 136, 139, 154, 192, 201
Cannon-boring experiment, 171
Carbon, 129
Carbonic acid gas (carbon dioxide), 81, 215
Cardinal scale, 229

Cardwell, Donald S. L., 66, 178, 179, 181
Carnelly, 55n
Carnot, Sadi (1796–1832), 97–98, 175, 232
Carnot engine, 97, 175–176, 178–181, 191
   as an abstraction, 203
   as a cycle, 181f, 187, 189f, 218
   as an idealization, 188, 191, 208–210
   Thomson's concrete models of, 187–190, 206, 208–210: gas-only system, 188n, 190n; steam-water system, 188–191, 209–210
   Thomson's reformulation of, 183–184
Carnot's coefficient (or multiplier), 183
Carnot's function, 183–185
Cartwright, Nancy, 52, 156n, 202n, 206
Catherine the Great, 106
Cavendish, Henry (1731–1810), 8, 11–12, 16, 54n, 162n
   advocacy of steam point, 24–27
   on boiling, 24–27, 32, 35–36, 51
   on freezing of mercury, 110–115, 154,
Cavendish, Lord Charles (1704–1783), 29
Cellars. See deep places, temperatures of
Celsius, Anders (1701–1744), 10t, 11, 60, 160
Centigrade scale, 59, 60, 160
Chaldecott, John, 120n, 140n
Changes of state, 29n, 72, 137, 242
   Aitken's view on, 37–38
   in chemical caloric theory, 65
   Irvinist explanation of, 168–170
Charles's law. See Gay-Lussac's law
Charlotte, Queen, 14, 119
Chatelier. See Le Chatelier, Henri-Louis
Chemical caloric theory, 65–67, 170–171, 202
Chirikov, Captain Alexei, 104
Circularity, 62n, 95, 146, 156, 229
   inherent in empiricist justification, 99, 220–221
   in operationalization, 192, 201, 203, 205, 208
   in problem of nomic measurement, 59–60, 71n, 80, 83, 89–90, 136
Clapeyron, Benoit-Pierre-Émile (1799–1864), 97, 98, 180–181
Clausius, Rudolf (1822–1888), 173, 194
Clay pyrometer. See under Wedgwood, Josiah
Cleghorn, William, 106
Clément, Nicolas (1778/9–1841), 135–137, 140, 157
Cloud chamber, 35–36
Cogency (of abstract ideas), 234
Cohen, Leonard, 39
Coherence, 47, 157–158, 221
Coherentism, 6, 156, 220–221, 231, 233
   round-earth, 223–224
Cohesion, 30–31, 194
Cold
   artificial, 105, 116 (see also freezing mixtures)
   positive reality of, 160–168, 242, 246
   radiation of, 166–167, 242–243, 246
   Rumford on, 167, 172
   in Siberia, 104
   see also low temperatures, production and measurement of

Collège de France, 75, 76
Combined caloric, 66–67, 71–72, 170–171. See also latent heat
Comparability, 77–83, 89–92, 139–140, 152, 193, 221, 223, 230, 245
   of air thermometer, 80–83, 96, 137
   De Luc's use of, 78
   of generalized gas thermometer, 81–82
   of mercury thermometer, 79–80
   Regnault's use of, 77, 79–83, 175
   of spirit thermometer, 78–79
   of Wedgwood's clay pyrometer, 120–121, 125–126, 139–140, 155–156
   see also single value, principle of
Complementary science, 3, 5–6, 235–250
   character of knowledge generated by, 240–247
   new knowledge generated by, 245–247
   relation to other modes of science studies, 247–249
   relation to specialist science, 247, 249–250
Comte, Auguste (1798–1851), 74, 97–98
Conant, James Bryant, 250n
Concepts. See entries for abstraction; extension; meaning; operationalism; theoretical concepts
Conceptual extension. See extension
Concrete image (of an abstract system). See imaging
Concretization. See imaging
Condensation. See under steam
Condenser, separate. See under Watt, James
Confirmation, criteria of, 227. See also epistemic virtues; justification; testing
Committee of Public Safety, 129
Conformity condition, 152–155
Conservation of energy. See under energy
Conservation of heat. See under heat
Conservatism, 224–225, 231
Consistency, logical, 44, 227, 231, 233
Consistency, physical, 90, 92–93, 118, 155. See also single value, principle of
Constructs, mental, 151
Contraction of clay by heat. See under linearity; Wedgwood
Contraction of fluids by cold. See expansion
Convention, 147, 152
Conventionalism, 58–59, 211, 222
Convergence, 145, 148, 156, 157, 212, 215, 217, 221. See also overlap condition
Cooling, Newton's law of, 120, 137
Cooling by expansion. See adiabatic phenomena; Joule-Thomson experiment
Copernicanism, 244
Copper, 126t, 131t, 136, 138n
Correction, 16, 43, 204, 212–213, 215–216, 221, 230. See also self-correction
Correctness, 6, 204, 206, 207, 211, 216, 232,
   of measurement methods, 221, 222, 224n
   of operationalization, 208, 212–213, 245
Correspondence, 208, 216
Coulomb, [Charles de], 242n
Crawford, Adair (1748–1795), 66, 68, 169
Credulity, principle of, 225

278 Index

Crell, Lorenz, 123
Critical awareness, 243–245
Critical point, 22
Criticism, 235, 237, 240, 249–250. See also critical awareness
Cushing, James T., 245

Dalencé, Joachim (1640–1707?), 10
Dalton, John (1766–1844)
  on atoms, 71n
  criticised by Dulong and Petit, 101
  on expansion of gases, 69, 70n, 99
  on freezing point of mercury, 116
  and Irvinist caloric theory, 169–170, 180n
  on method of mixtures, 66, 69, 94
  temperature scale, 66, 116
  on vapor pressure, 29, 30t
Daniell, John Frederic (1790–1845)
  platinum pyrometry, 129–134, 136, 138, 139, 156
  temperature scale, 133
  on Wedgwood pyrometer, 122, 140n, 149, 153
Davy, Humphry (1778–1829), 164n, 172
Day, A. L., 214, 215
Deep places, temperatures of, 9, 61n
Definition. See meaning; operationalism
Degree of boiling. See under boiling
Delahire. See La Hire, Phillipe de
Delisle (De l'Isle), Joseph-Nicolas (1688–1768), 10t, 13f, 105, 160–162
De l'Isle de la Croyere, Louis, 105n
De Luc, François, 14
De Luc, Guillaume-Antoine, 14
De Luc (Deluc), Jean-André (1727–1817), 11, 14–16, 72n, 99, 176
  on boiling, 16–27, 32–35, 38, 50, 51, 241
  on comparability, 78–79
  on freezing and supercooling, 53–56
  on freezing of mercury, 107–109
  on mercury thermometer, 62–64, 114–115
  method of mixtures, 62–64, 93–94: critics of, 67–69, 96, 101, 136, 171
  on spirit thermometer, 64t, 78
De Mairan, Jean-Jacques d'Ortous (1678–1771), 164
De Morveau. See Guyton de Morveau, Louis-Bernard
Desormes, Charles-Bernard (1777–1862), 135–137, 140, 157
Diamond, 129
Dingle, Herbert, 244
Dishwasing, as metaphor for iterative progress, 220
Distance. See length, Bridgman's discussion of
Dogmatism, 237, 248
Donny, François Marie Louis (1822–?), 22, 33
Dörries, Matthias, 75, 100
Draper, John William, 244
Drebbel, Cornelius, 105n
Dry ice, 116
Du Crest (Ducrest), Jacques Barthélemi Micheli (1690–1766), 10t, 61

Dufour, Louis (1832–1892), 22, 31, 32–35, 37n, 38, 54
Duhem, Pierre, 92, 95, 98, 100, 221, 224n, 232
Duhem-Quine thesis. See holism.
Dulong, Pierre (1785–1838), 74, 116, 134
  in post-Laplacian empiricism, 96–98, 232
  on thermometry, 79, 100–102, 136, 137
Dulong and Petit's law, 97, 98
Dumas, Jean-Baptiste, 75, 76, 99
Dust, 34, 36–38, 50–51
Dynamic theory of heat, 84, 162, 167, 171–173, 183

Earth. See round-earth coherentism
Ebullition
  normal (Gernez), 33–34
  true (De Luc), 17–21, 25
  see also boiling
École Polytechnique, 74, 75, 129, 135, 180
Efficiency (of engines), 178, 180, 183–184, 187, 191, 201, 203n
Einstein, Albert, 144, 147
Electric-resistance thermometry, 213
Ellis, Brian, 41n, 90n
Empirical adequacy, 228
Empirical laws, 49, 52, 59, 66n, 74–75, 98–99, 190, 221. See also middle-level regularities; phenomenological laws
Empiricism, 84–85, 92, 97–98, 221. See also Bridgman, Percy Williams; phenomenalism; positivism; post-Laplacian empiricism; Regnault, Henri Victor
Encyclopaedia Britannica, 10t, 11, 23, 40, 60, 115, 118, 164, 204, 210
Energy, 84, 172–173, 242
  conservation of, 149, 182, 183, 186, 209
  internal, 194, 210, 211
  in quantum mechanics, 52n, 157n
Enrichment (by iteration), 228–230
Epistemic iteration, 6, 44–48, 158, 212–217, 220–221, 226–231, 233, 234, 245
Epistemic values, 46, 227–228, 233–234. See also epistemic virtues
Epistemic virtues, 90, 156, 158, 227–230, 232–233. See also epistemic values
Equilibrium
  in changes of state, 32, 54–56
  in mixtures, 67n, 134
  of radian heat, 71
  thermodynamic, 178, 185n
Eschinardi, Francesco, 10t, 47t
Ether, 172, 173
Evans, James, 167n, 172n, 242
Evaporation, 18, 20–21, 26n, 27–28, 32–33, 38
Expansion (thermal)
  Dalton on, 116
  Dulong and Petit on, 101–102
  Fourier unconcerned about, 98
  Haüy's theory of, 67–68
  of reservoirs of thermometers, 79–80, 137n, 138

of various substances: air, 74–75, 83, 137, 190; gases (in general), 69–75, 83, 98–99, 101–102, 138, 192, 211–212, 216, 218; glass, 58, 79–80, 83, 101n; gold, 138; iron, 130t, 134; mercury, 59, 62–64, 67–68, 114, 133, 153; metals (in general), 101, 120, 154; platinum, 128–129, 133–134; silver, 123–125, 127, 153; solids (in general), 61, 134; spirit, 61, 64t, 78–79, 108–109, 114–115, 117, 227
Expansive principle. *See under* Watt, James
Explanatory power, 227
Explosion, 20
Extension (of concepts), 141–142, 146–147, 148–155
  metrological, 150, 152, 157, 229–230, 245: of temperature, 117, 152–155, 221, 223
  operational, 150
  semantic, 149–151
Extrapolation, 108–109, 114–115, 134

Fabri, Honoré, 10t
Fahrenheit, Daniel Gabriel (1686–1736), 12n, 58, 60, 77, 79, 105n
  on mixing experiments, 225
  on supercooling, 53–55
  temperature scale, 10t, 13f, 40–41
Fairbank, William M. Jr., 244
Falsification, 94, 227, 230
Feigl, Herbert, 52, 198
Ferdinand, Grand Duke (Medici), 10t
Feyerabend, Paul, 52, 233
Fitzgerald, Keane, 120n
Fixed points
  for pyrometry, 126, 129, 133
  various, 8–11, 10t
  see also boiling point of water; fixity; freezing point
Fixity, 9, 23, 39–40, 52, 55, 221
  defence of, 48–52
Fog, 36
Foley, Richard, 221, 223
Forbes, James David, 76
Fordyce, George (1736–1802), 119–120
Foundationalism, 6, 42, 156, 220–224, 227
Fourier, Jean-Baptiste-Joseph (1768–1830), 74, 96–97, 98, 174, 232
Fourmi, 140
Fowler, John, 10t
Fox, Robert, 65, 70, 73n, 96–98, 100
Franklin, Allan, 95, 244
Free caloric, 66–67, 71–73, 170–171
Free caloric of space, 72
Free surface, 19n, 37–38, 55. *See also* Aitken, John; changes of state
Freezing mixtures, 105–106, 109, 113, 115, 166
Freezing of mercury. *See under* mercury
Freezing point
  of aniseed oil, 9
  considered indefinite, 9, 106
  of mercury, 107–118
  of water, 9, 11, 53–56, 135, 162n

French Revolution, 75, 128–129
Fresnel, Augustin (1788–1827), 96
Frigorific particles, 162–163
Frigorific radiation, 167–168, 246
Fruitfulness, 227

Galileo, 8, 47t, 226
Galison, Peter, 36n, 52
Gases
  expansion of, 69–75, 83, 98–99, 101–102, 138, 192, 211–212, 216, 218; Dalton and Gay-Lussac on, 69; Regnault on, 98–99
  Regnault's technique for weighing, 76n
  special status of (in caloric theory), 69–70, 101
  specific heat of, 181, 204, 205, 210
  see also gas thermometer; ideal gas; kinetic theory of gases; Laplace, Pierre-Simon
Gassendi, Pierre (1592–1655), 162
Gas thermometer (generalized), 81–82, 116, 171, 217–218
  as approximate indicator of absolute temperature, 192–197, 210–216
  see also Joule-Thomson experiment
Gay-Lussac, Joseph-Louis (1778–1850)
  on boiling, 21, 23, 30–31
  on expansion of gases, 69, 99, 101n
  on thermometry, 57, 70n, 71, 100
Gay-Lussac's law, 190
George III, 12, 14, 122f, 123
Gernez, Désiré-Jean-Baptiste (1834–1910), 32–35, 54–55,
Glass, 58, 113
  expansion of, 58, 79–80, 83, 101n
  vessels for boiling water, 18, 21–23, 31, 49
Glasses, as an illustration of self-correction, 230
Gmelin, Johann Georg (1709–1755), 104–106
Gold, 126t, 129, 131t, 138, 141n
Graham, Andrew, 110
Gray, Andrew, 193n
Greenwich. *See* National Maritime Museum
Grünbaum, Adolf, 95
Guericke, Otto von, 10t
Guthrie, Matthew (1732–1807), 107, 114, 115n
Guyton de Morveau, Louis-Bernard (1737–1816), 128–129
  on calorimetric pyrometry, 135
  platinum pyrometry, 128–129, 133, 156–157
  and Wedgwood's clay pyrometer, 120n, 123, 140

Hacking, Ian, 52, 85–86, 166
Halley, Edmond (1656–1742), 8–9, 10t, 41
Hanson, Norwood Russell, 52
Haüy, René-Just (1743–1794), 67–69, 70, 71, 72n, 78, 171
Heat
  in chemical reactions, 168
  conservation of, 94, 134, 183, 184, 187
  mechanical equivalent of, 185, 204
  operationalization of, 201–202

Heat (Continued)
  relation with work, 175, 180, 183, 186, 194, 209 (see also efficiency)
  see also caloric theory; dynamic theory of heat
Heat capacity, 41, 65, 66, 68, 98, 168–170, 198, 227
  and specific heat, 170, 171, 202
Heat engine. See Carnot engine; steam engine
Heberden, William, 11n
Heering, Peter, 242n
Heights, measurement by pressure, 14
Heilbron, John L., 73n
Heisenberg, Werner, 147n, 220
Hempel, Carl G., 114, 224n, 227
Henry, William (1774–1836), 171n
Hentschel, Klaus, 244
Herapath, John (1790–1868), 172–173
Herschel, William (1738–1822), 166
Hertz, Heinrich, 231
Hesse, Mary, 52
Hierarchy. See under justification
High-level theories, 51–52
High temperatures, measurement of. See pyrometry
Hissing (sifflement), 16, 20
History and philosophy of science, 5–7
  complementary function of, 236–238, 248, 249
  descriptive mode of, 249
  as an integrated discipline, 239, 249
  prescriptive mode of, 249–250
History of science
  internal, 248
  relation to philosophy, 233–234, 236, 239–240
  relation to science, 239
  social, 247–248
Hobbes, Thomas, 246
Holism (in theory testing), 92–94. See also auxiliary hypotheses; Duhem, Pierre; Quine, Willard van Orman
Holton, Gerald, 142, 244, 250n
Hooke, Robert (1635–1703), 9, 10t, 162n
Hope, Thomas, 123
Horsley, Samuel, 11n
Hudson's Bay, 110–111, 115
Human body temperature. See blood heat
Humphreys, Paul, 217n
Huntoon, R. D., 186
Husserl, Edmund, 47, 224
Hutchins, Thomas (?–1790), 110–115
Hutchison, Keith, 194n, 195n
Huygens, Christiaan, 10t, 226
Hydrogen, 81, 205
Hypothetico-deductive (H-D) model of theory testing, 93, 227

Ice
  calorimeter, 134–136, 154, 156, 192
  latent heat in melting of, 54: Irvinist account of, 65, 168–170
  melting point of, 53, 55, 135 (see also freezing point of water)

superheating of, 37, 55
see also supercooling
Ideal gas, 211–218
  as an abstraction, 203–204
  as indicator of absolute temperature, 192–195
Ideal gas law, 193, 203–204, 231
Ideal-gas temperature, 218
Idealization, 202–204, 206
  Carnot engine as an, 188, 191, 208–210
Imaging, 206, 208–210, 216. See also operationalization
Independence, 95
Indicator diagram. See under Watt, James
Industrial Revolution, 176
Infinite regress, 40, 43, 224
Infrared radiation, 166
Institut de France, 129
Intelligibility, 91–92
Internal history of science, 248
International Practical Scale of Temperature, 154, 187
Inverted temperature scales, 105, 160–161
Iron
  for air thermometers, 138
  cast, 119, 126t, 129, 141n, 149
  expansion of, 130t, 134
  melting point(s) of, 119, 126t, 129, 131t–132t, 136, 139, 141n, 149
  soft, 136
  welding heat of, 126, 131t, 149
Irvine, William (1743–1787), 65, 168. See also Irvinist caloric theory
Irvinist caloric theory, 65, 168–170, 171n, 202
  adiabatic phenomena explained by, 180n
  Dalton's advocacy of, 66, 180n
Isothermal expansion, 209–210
  in Carnot cycle, 181
  of an ideal gas, 193–194
Iteration. See epistemic iteration; mathematical iteration

Joule, James Prescott (1818–1889), 173, 182, 218, 242n
  collaboration with Thomson, 182–186, 192 (see also Joule-Thomson experiment)
Joule's conjecture, 184–185
  see also Mayer's hypothesis
Joule-Thomson experiment (and Joule-Thomson effect), 185, 194–197, 195f, 201–202, 203–205, 211–214, 216, 218
Justification (epistemic), 6, 31, 220–225, 248
  avoidance of, 222–223
  coherentist, 220, 221, 223–224
  foundationalist, 42, 220, 221–222: not required in epistemic iteration, 213, 215–216, 224, 227
  hierarchical, 221, 223–224
  of measurement methods/standards, 40–44, 61, 86, 155, 175, 208, 221–222
  of metrological extension, 150
  of ontological principles, 91

of operationalization of temperature, 205, 206, 208
  see also testing; validity

Kant, Immanuel, 91
Kelvin, Lord. See Thomson, William
Kinetic theory of gases, 172–173, 201, 211, 218
Kinetic theory of heat. See dynamic theory of heat
Kirwan, Richard, 115
Klein, Herbert Arthur, 186–187
Knott, Cargill, 35
Kosso, Peter, 86n, 95
Krebs, Georg (1833–1907), 22, 242
Kuhn, Thomas S.
  on affirmation 224
  on epistemic values, 227, 228
  on normal science, 29, 231, 236–237
  on paradigms, 231–232
  on theory-ladenness of observation, 52
  on validity of past science, 244, 245

La Hire, Philippe de, 10t
Lakatos, Imre, 84, 228, 233, 244n
Lambert, Johann Heinrich (1728–1777), 61
Lamé, Gabriel (1795–1870), 58t, 74–75, 78n, 98n
Langevin, Paul, 75
Laplace, Pierre-Simon (1749–1827), 67, 96, 134–135, 170–171
  on caloric theory, 70–74
  on gases, 73
  ice calorimeter, 134–136, 154, 156, 192
Laplacian physics, 70. See also post-Laplacian empiricism
Latent caloric. See combined caloric
Latent heat
  in chemical caloric theory, 67, 170–171
  of fusion/freezing, 54, 105n, 110, 114, 134
  Irvinist conception of, 65, 168–170, 198
  of steam, 26, 177, 189–190
  see also combined caloric; supercooling
Latour, Bruno, 248
La Tour, Cagniard de. See Cagniard de la Tour, Charles
Lavoisier, Antoine-Laurent (1743–1794), 14, 29n, 33n, 123, 244n
  collaborators of, 67, 128, 134
  caloric theory, 65–66, 67, 170–171
  ice calorimeter, 134–136, 154, 156, 192
Law of cooling. See cooling, Newton's law of
Lead, 52, 130t
Leapfrogging, 154–155
Le Chatelier, Henri-Louis (1850–1936), 127, 139, 140n, 141, 213–216, 230
Length, Bridgman's discussion of, 144–146, 222–223
Leonelli, Sabina, 244n, 246
Le Sage (the Younger), George-Louis (1724–1803), 62

Leslie, John (1766–1832), 169
Light, as a chemical element, 65
Linearity (with respect to temperature)
  of contraction of clay, 126–128, 133–134, 153
  of thermometers, 59, 60–61, 89–90, 93, 101–102, 117–118
  see also expansion
Linseed oil, 61
Locke, John, 106
Lodwig, T. H., 135
Low-level laws. See middle-level regularities
Low temperatures, production and measurement of, 104–118
Lunar Society of Birmingham, 14, 113, 176
Lycan, William G., 225, 227, 228

Mach, Ernst, 43n, 97, 232
Manometers. See barometers
Marcet, Alexandre, 22
Marcet, François (1803–1883), 22–23, 27–28, 31, 51
Marcet, Jane, 22
Margenau, Henry, 198, 199f
Mariotte, Edme, 9
Mariotte's law. See Boyle's law
Martine, George (1702–1741), 61, 161t
Maskelyne, Nevil, 11n
Matching, 206, 215. See also operationalization
Material realization of a definition, 203
Material theory of heat. See caloric theory
Mathematical iteration, 45–46
Matousek, J. W., 103, 140n
Matthews, Michael, 250n
Maxwell, Grover, 85
Maxwell, James Clerk, 39
Mayer, Julius Robert (1814–1878), 194
Mayer's hypothesis, 194, 211. See also Joule's conjecture
Mayo, Deborah, 226
McKendrick, Neil, 127
McNab, John, 115
Meaning, 51, 157, 208n
  Bridgman on, 142–152
  and definition, 149–151
  operational, 149–152, 197, 199, 203
  reductive doctrine of, 148–152
  as use, 150–151
  verification theory of, 148
  see also extension
Measurement methods. See under justification; testing; validity. See also operationalization; standards
Mechanical effect. See work
Mechanical philosophy, 162, 171
Mechanical theory of heat. See dynamic theory of heat
Mechanical work. See work
Medical thermometry, 40n, 60
Meikle, Henry, 73n
Melloni, Macedonio, 140n

Melting point, 126t, 130t-132t, 139
  brass, 121, 126t, 130t
  butter, 10
  copper, 126t, 131t, 136, 138n
  gold, 126t, 129, 131t, 141n
  ice, 53, 55, 135 (see also freezing point of water)
  iron, 119, 126t, 129, 131t-132t, 136, 139, 141n, 149
  lead, 52, 130t
  silver, 126t, 131t, 138, 138n, 141n, 149, 153
  zinc, 130t, 137
Mendeléeff, Dmitri, 75
Mercury
  boiling point of, 119, 126t, 133
  considered essentially fluid, 105, 227
  expansion of, 59, 62–64, 67–68, 114, 133, 153
  freezing of, 104–107
  freezing point of, 107–118
  needed for air thermometers, 120
  supercooling of, 53
Mercury thermometer
  comparison with air thermometer, 71, 79, 89, 100–101
  comparison with platinum pyrometer, 133
  comparison with spirit thermometer, 58–59, 64t, 78t, 107–109, 115, 154
  comparison with Wedgwood pyrometer, 123–125, 133, 149
  criticized, 67–68, 71, 79–80, 94
  De Luc's arguments for, 62–64, 107–109
  failing at high temperatures, 103, 119
  failing at low temperatures, 103, 105–106, 142–143
  favored over others, 60, 61, 62–64, 78n, 152
  used in air thermometers, 87–89
  used in Joule-Thomson experiment, 197, 202, 204–205, 216
  used to measure freezing point of mercury, 110–115
  see also under expansion
Mersenne, Marin (1588–1648), 162
Metallic thermometers. See under pyrometry
Metallic vessels for boiling water, 16, 21, 25f, 49
Meteorology, 14, 35–36, 50, 160
Method of mixtures. See mixtures, method of
Methodology, 248–249
Metrological extension, 150, 152, 157
Micheli du Crest. See du Crest
Microscopes, 85–86, 166
Middle-level regularities (or low-level laws), 52, 218. See also empirical laws, phenomenological laws
Middleton, W. E. Knowles, 11n, 14n, 19n, 41, 61, 200n
Millikan, Robert, 166, 244
Minimalism, 77, 94–96
Mixtures, Fahrenheit's experiments on, 225
Mixtures, method of, 60–68, 93–94, 108, 136
  Black's use of, 62
  Crawford's use of, 66, 68
  Dalton's critique of, 66
  De Luc's use of, 62–64, 68, 93–94, 136
  Haüy's critique of, 67–68
  Le Sage's suggestion of, 62
  Taylor's use of, 61
Models, 186–190, 206
  semi-concrete, 188
  see also imaging
Molecules, 56, 137, 212
  action of: in boiling, 31–32, 37n, 38; in thermal expansion, 63, 67
  in dynamic theories of heat, 172–173, 201
  in Laplace's caloric theory, 71–73
Moore, G. E., 43
Morgan, Mary, 206n
Morris, Robert J., 244n
Morrison, Margaret, 206n
Mortimer, Cromwell, 120n, 132n
Morveau. See Guyton de Morveau, Louis-Bernard
Murray, John (1778?–1820), 119n, 122, 167, 169
Muschenbroek, Petrus van (1692–1761), 164
Musgrave, Alan, 244n
Mutual grounding, 117–118, 156–158, 223

Napoleon, 67, 75
Narratives, 233
National Maritime Museum (Greenwich), 160
Naturalistic epistemology, 249
Neurath, Otto, 156–157, 222n, 223, 236n
Newcomen engine, 176
Newman, William, 242
Newton, Isaac (1642–1727)
  on blood heat, 10–11, 40–41
  on boiling point of water, 12, 13f, 50
  law of cooling, 120, 137
  temperature scale, 11–13
Nomic measurement, problem of, 59, 89–90, 120
Normal science, 29, 231, 236–237. See also specialist science
Normative judgments (on science), 248, 249
Northfield Mount Hermon School, 229
Novel predictions, 228
Numerical thermometers. See thermometers, numerical

Observability (and unobservability), 59, 84–86, 96–97, 134, 147n, 166, 171, 197–198, 232
Observation. See observability; sensation; theory-ladenness of observation
Oils as thermometric fluids, 60, 61, 64
One-point method, 11n
Ontological principles, 91–92
Operational extension, 150
Operational meaning, 149–152, 197, 199, 203
Operationalism (operationism), 58, 142–147, 148, 151, 208n. See also Bridgman, Percy Williams
Operationalization, 173, 197–219, 222
  of heat, 201–202
  reductive view of, 197–202
  of Thomson's first absolute temperature, 187–192, 201

of Thomson's second absolute temperature, 192–197, 212f
as a two-step process, 206–208 (*see also* imaging; matching)
validity of, 205–212
Ordinal scale, 41, 87, 221, 229
Osborn, H. C., 141n
Overdetermination, 93–94
Overlap condition, 152–155

Pallas, Peter Simon (1741–1811), 106–107
Paradigm, 231–232, 245
Paris Observatory, 9
Pasteur, Louis, 32
Péclet, E., 141
Peirce, Charles Sanders, 45, 46f, 226
Petit, Alexis-Thérèse (1791–1820), 74, 134
in post-Laplacian empiricism, 96–98, 232
on thermometry, 79, 100–102, 136, 137
Phase changes. *See* changes of state
Phenomenalism, 96–98, 203, 232
Phenomenological laws, 154–155, 157, 218. *See also* empirical laws, middle-level regularities
Philosophy
relation to history of science, 233–234, 236, 239–240
relation to science, 238–239
Philosophy of modern physics, 244–245
Phlogiston, 244
Pictet, Marc-Auguste (1752–1825), 67n, 123, 164–167, 242, 246
Platinum, 128
in air thermometer, 138–139, 154
expansion of, 128–129, 133–134
melting point of, 132t
pyrometry, 128–134, 156–157
specific heat of, 139
Plausible denial, 49, 55
Pluralism, 228, 231–233, 247
Pluralistic traditionalism, 232–233
Poincaré, Henri, 59, 91, 222
Poisson, Siméon-Denis (1781–1840), 73, 98n
Polanyi, Michael, 224
Popp, Brian, 167n, 172n, 242
Popper, Karl R., 52, 82, 95, 169, 224, 235, 237
Porous-plug experiment. *See* Joule-Thomson experiment
Positivism, 74, 97, 144, 232
Post-Laplacian empiricism, 75, 96–98, 232
Pouillet, Claude-Servais-Mathias (1790–1868), 116–118, 119n, 137n, 139, 140, 154
Pragmatic virtues, 227
Precision, 21, 44, 80, 90, 139, 158
assumed in theoretical concepts, 145, 201
as an epistemic virtue, 227, 229, 231
in post-Laplacian empiricism, 97–98
Regnault's achievement of, 75–76, 98–100, 174
Pressure
as an abstract concept, 200–201
in Bridgman's work, 142–143
see also under boiling; boiling point of water; steam; vapor
Pressure-balance theory of boiling. *See under* boiling
Preston, Thomas, 34, 198–201
Prevost, Pierre (1751–1839), 71
Principle of respect. *See* respect, principle of
Principle of single value. *See* single value, principle of
Prinsep, James (1799–1840), 138–139
Problem-centered narrative, 5, 240
Problem of nomic measurement. *See* nomic measurement, problem of
Problem-solving ability, 228, 245
Progress, scientific
by epistemic iteration, 6, 44–46, 220, 223–234
imperative of, 44
by mutual grounding, 157
and observations, 14, 85–87, 99–100
through revolutions, 245
Psillos, Stathis, 228n
Ptolemy, 244
Pyle, Andrew J., 162n
Pyramid, 11
Pyrometry, 118–141, 149
air, 137–139
clay, 60, 118–128, 153–157
convergence of results in, 130t–132t, 140–141
by cooling, 120, 136–137
electric-resistance, 140n
by ice calorimetry, 134–136
metallic, 120, 154
optical, 140n
platinum, 128–134, 156
radiation, 140n
thermoelectric, 140n
by water calorimetry, 136

Quantum mechanics, 52n, 157n, 245
Quicksilver. *See* mercury
Quine, Willard van Orman, 156–157, 223

Radiant caloric, 71–72
Radiant cold, 166–168
Radiant heat, 140n, 164–168
Radiation, nature of, 246
Rain, 14, 36
Realism, 46, 59, 85, 90, 94, 151, 166, 217
Réaumur, R. A. F. de (1683–1757), 58–59, 60
temperature scale, 10t, 13f, 16n, 63, 107n
Recovery of forgotten knowledge, 5, 237, 240, 241–243
Red heat, 125, 126, 126t, 130t, 136
Reduction. *See* operationalization: reductive view of
Reductive doctrine of meaning. *See under* meaning
Regnault, Henri Victor (1810–1878), 75–102, 174–175, 203
on air thermometer, 80–83, 87–89, 137
career, 75
on comparability, 77, 79–83, 89–92, 155, 223, 245

Regnault (*Continued*)
  data on steam, 29, 30t, 75, 190, 210
  on expansion of gases, 81–82, 211, 216
  experimental style, 76
  on mercury thermometer, 79–80
  minimalism, 94–96
  and post-Laplacian empiricism, 98–102
  precision, 75–76, 98–100, 174
  against theories 76, 82–83, 86, 98–100, 232
  on thermometric standards, 86–89
Regularity of expansion. *See* expansion; linearity
Relativism, 233
Renaldini, Carlo, 10t, 47t
Replication of past experiments, 242–243, 246
Respect, principle of, 43–46, 155, 225–226, 231
Reversibility, 185, 187, 192, 206, 208, 209–210, 218
Robustness, 28, 51–52, 55, 218
Rømer, Ole, 10t
Rottschaefer, W. A., 95n
Round-earth coherentism, 223–224
Rousseau, Jean-Jacques, 14
Rowlinson, J. S., 241n, 242n
Royal Society, 9n, 24, 109–110, 121, 160n, 162n, 173
  committee on thermometry, 11–12, 16–17, 21, 24–27, 35, 38, 49–51, 111
  thermometer (temperature scale), 160–162
  various fellows and officers of, 8, 14, 53, 54, 61, 121
Rumford, Count (1753–1814), 123, 164n, 167–168, 171–172, 246

Sage. *See* Le Sage
Salvétat, Alphonse, 141n
Sanctorius, 10t, 41n
Saturated steam. *See* steam
Saturated vapor. *See* vapor
Saunders, Simon, 231
Savery engine, 176n
Scales. *See* cardinal scale; ordinal scale
Scepticism. *See* skepticism
Schaffer, Simon, 246
Scherer, A. N., 129n
Schlick, Moritz, 222
Schmidt, J. G., 138
Science education, 250
Scientific method. *See* methodology
Scientific progress. *See* progress, scientific
Scientific realism. *See* realism
Scientific revolutions, 245
Scope (as an epistemic virtue), 227, 230
Seebeck, Thomas, 140n
Séguin, Armand, 123
Self-correction, 6, 44–45, 226, 228, 230–231
Self-destruction (or self-contradiction), 226–227.
  See also self-correction
Self-evidencing explanation, 114
Self-improvement, 44, 221
Self-justifying beliefs, 221–222, 223, 227
Self-repulsion (of caloric), 65

Semantic extension, 149–151
Sensation, 42–44, 47, 84–86, 120, 157, 221, 229–230
Sense-data, 52
Sensible caloric. *See* free caloric
Sensible heat, 67
Separate condenser. *See under* Watt, James
Serendipity, 24, 37, 50–51, 55–56
Shapere, Dudley, 86n
Shapin, Steven, 246
Sharlin, Harold I., 174n, 225
Shooting. *See under* supercooling
Siberia, expeditions to, 104, 107
Sibum, Heinz Otto, 242n
Siemens, William, 140n
Sifflement. *See* hissing
Silver
  expansion of, 123–125, 127, 153
  melting point of, 126, 129, 131t, 138, 138n, 141n, 149, 153
  in Wedgwood's thermometry, 123–125, 126, 127
Simplicity, 222, 227
Simultaneity, 144, 147
Singing (of the kettle), 20
Single value, principle of, 77, 90–91, 93, 211, 215. *See also* comparability; ontological principles
Skepticism, 239–240, 242, 243, 248
Sklar, Lawrence, 224
Smeaton, John, 11n
Smeaton, William, 129n, 135
Smith, Charles Piazzi (1819–1900), 11
Smith, George E., 226
Snell's law, 52
Social history of science, 247–248
Sociology of scientific knowledge, 236n, 248
Sosman, R. B., 214, 215
Soubresaut. *See* bumping
Southern, John, 181
Specialist science, 237–238. *See also* normal science
Specific heat
  in chemical caloric theory, 66–68, 171
  and Dulong and Petit's law, 97, 98
  of gases, 181, 204, 205, 210
  Irvinist conception of, 168–170, 202
  of platinum, 139
  temperature-dependence of, 64, 68, 94, 136
Spietz, Lafe, 154n
Spirit (of wine)
  boiling point of, 9, 19n
  expansion of, 61, 64t, 78–79, 108–109, 114–115, 117, 227
Spirit thermometer
  Amontons's, 162
  Braun's, 107, 108
  comparison with air thermometer, 117–118, 154
  comparison with mercury thermometer, 58–59, 64t, 78t, 107–109, 115, 154

Index 285

criticized, 61, 64t, 78–79, 94, 114–115,
De Luc's, 19n
Du Crest's advocacy of, 61
Fahrenheit's, made for Wolff, 77
see also under expansion
Standards
choice of, 90
extension of, 107, 152
of fixity, 39
improvement of, 44–48, 84, 228–231 (*see also* self-improvement)
justification of, 40–44, 155, 175, 208, 221–222
mutual grounding of, 156–158
of temperature, 9, 109, 115, 128, 159
validity of, 40–44, 47–48, 55
whole-range, 154, 154n
States of matter. *See* changes of state
Steam
Cavendish on, 25–26
condensation of, 176
De Luc on, 26–27
density of, 190, 210
latent heat of, 26, 177, 189–190
pressure-temperature relation, 28–30, 35–36, 188–191
Regnault's data on, 29, 30t, 75, 190, 210
saturated, 28–29, 188, 190
supersaturated, 27, 35–37, 51
see also steam engine; steam point; steam-water system; vapor
Steam engine, 75–76, 175–179, 180–181
Steam point, 35, 48, 51
Cavendish's advocacy of, 24–28
De Luc's skepticism about, 26–27
dust needed for fixity of, 36, 50
Marcet on fixity of, 27–28
see also boiling; boiling point of water; supersaturation
Steam-water system, 178–179, 188–191, 209–210
Steam-water thermometer, 210
Stevens, S. S., 90n
Strong program. *See* sociology of scientific knowledge
Subtle fluids, 65
Sulphur, 31
Sulphuric acid, 60, 81–82, 115
Supercooling, 37, 53–56, 113–114
broken by shooting, 54–56
Fahrenheit's discovery of, 53–55
of mercury, 53
Superheating, 17–24, 27–28, 31, 241–242
of ice, 37, 55
modern explanations of, 241n
prevention of, 21, 23–24, 33–34, 241
while boiling vs. without boiling, 23
see also boiling point of water
Supersaturation
of solutions, 34–35
of steam, 27, 35–37, 51
Surchauffer, 18n
Surfusion, 53n

Taylor, Brook (1685–1731), 61–62
Temperature, theoretical concepts of, 159–160, 167–173, 217–219, 245, 246. *See also* absolute temperature
Temperature scales. *See* absolute temperature; Amontons temperature; Celsius, Anders; centigrade scale; Dalencé, Joachim; Dalton, John; Daniell, John Frederic; Delisle; Fahrenheit, Daniel Gabriel; fixed points; International Practical Scale of Temperature; inverted temperature scales; Newton, Isaac; one-point method; Réaumur, R. A. F. de; Royal Society; two-point method; Wedgwood, Josiah
Testability, 146, 169–171, 197, 213, 227
Testing (empirical), 52, 59, 89–96, 221, 231
circularity in, 99, 127–128, 192, 221
by comparability, 89–96, 155
hypothetico-deductive (H-D) model of, 93, 227
Regnault's insistence on, 76–77, 82
see also justification; testability; validity
Thenard, Louis-Jacques (1777–1857), 69
Theoretical concepts, 202–203, 206, 208n
Theoretical unity, 155, 157n, 158
Theory-ladenness of observation, 52, 60, 95, 222
Thermal expansion. *See* expansion (thermal)
Thermocouple, 117
Thermodynamics. *See* absolute temperature; Carnot engine; Thomson, William
Thermometers, numerical, 41, 47–48, 55, 57, 87–89, 102, 221, 229
Thermometric fluids
choice of, 62, 83, 89–90, 100–101, 155, 245
comparison between various, 58t, 61, 64t, 161t
comparison between air and mercury, 71, 79, 89, 100–101
comparison between air and spirit, 117–118, 154
comparison between mercury and spirit, 58–59, 64t, 78t, 107–109, 115, 154
variety of, 60
see also air thermometer; gas thermometer; mercury thermometer; oils as thermometric fluids; pyrometry; spirit thermometer; water
Thermoscopes, 41–44, 47, 60, 87, 89, 102, 221, 229, 230
Thilorier, A., 116–117
Thompson, Benjamin. *See* Rumford, Count
Thomson, Thomas (1773–1852), 71, 164
Thomson, William (Lord Kelvin) (1824–1907), 173–174
absolute temeprature: theoretical concept of, 159, 173–186, 215, 217–218; operationalization of, 186–197, 201–202, 204–205, 208–212, 216
on abstraction in scientific theories, 174, 186, 202–203, 215

Thomson (*Continued*)
  collaboration with Joule, 182–186, 192 (*see also* Joule-Thomson experiment)
  and Regnault, 75, 83, 174–175
  and thermodynamics, 97
  see also absolute temperature; Joule-Thomson experiment
Time, measurement of, 231
Tomlinson, Charles (1808–1897), 34–35, 51
Tradition, 225–226, 232–233, 237, 250
Truesdell, Clifford, 73n
Truth, 46, 217, 227, 228, 232, 247, 248
Twin paradox, 244
Two-point method, 59
Tyndall, John, 39, 164–165

Underdetermination, 155, 157–158
Uniformity of expansion. *See* expansion; linearity
Unity. *See* theoretical unity
Universality, 234
Unobservable entities and properties. *See* observability
Upside-down scales. *See* inverted temperature scales
Ure, Andrew, 139

Validation. *See* justification; validity
Validity
  of measurement methods/standards, 40–44, 47–48, 55, 151–152, 213, 226, 233, 242
  of metrological extension, 152–155
  of operationalization, 205–206, 208
Values, epistemic. *See* epistemic values
Van Fraassen, Bas, 85–86, 93n, 227, 228
Vapor
  pressure, 28–29, 30t, 32, 38
  saturated, 28–29
  see also boiling; evaporation; steam
Verdet, Marcel Émile (1824–1866), 22, 24, 32–34
Vienna Circle manifesto, 236n
Virtues, epistemic. *See* epistemic virtues
Volta, Alessandro, 67n

Walpole, Horace, 50
Walter, Maila L., 142n
Water
  calorimetry, 134, 136, 139, 154, 192, 201
  incredulity about freezing of, 106
  as a thermometric fluid, 42, 58t, 60, 64t
  see also boiling; boiling point of water; freezing point; ice; latent heat; specific heat; steam; supercooling; superheating; vapor
Water-steam system. *See* steam-water system
Waterston, John James (1811–1883), 173
Waterwheel, 178
Watson, William (1715–1787), 104–106
Watt, James (1736–1819), 29, 30t, 176–178
  expansive principle, 178, 179
  indicator diagram, 180–181
  separate condenser, 177–178
Wax, 11
Weather. *See* meteorology
Wedgwood, Josiah (1730–1795), 119, 176
  clay pyrometer, 60, 118–128, 153–157: alum-clay mixture used in, 126; compared with mercury thermometer, 123–125, 133, 149; criticized, 129–141; difficulties in standardizing, 125–126; fixed points used in, 126
  colored-clay pyrometer, 120
  on ice calorimeter, 134
  and operationalism, 148–149
  patch, 153
  temperature scale, 121–122: connection with Fahrenheit scale, 123–125; modified by Guyton, 129
Welfare state, analogous to complementary science, 237–238
Whiggism, 248
Whole-range standards, 154, 154n
Wilson, C. T. R., 35
Withering, William, 113
Wittgenstein, Ludwig, 42–43, 50, 150–151, 224
Wolff, Christian Freiherr von, 77
Wollaston, William Hyde (1766–1828), 123, 128, 232
Woodward, James, 52
Work, 175, 178, 209–210. *See also* heat: relation with work
Wunderlich, Carl (1815–1877), 40n

Yi, Sang Wook, 159n, 194n, 195n, 205n, 209n, 214n, 216n
Young, Thomas (1773–1829), 61, 167

Zeeman effect, 198
Zinc, 130t, 137

Printed in the United Kingdom
by Lightning Source UK Ltd.
122815UK00001B/51/A